THE

UNITED STATES

AND THE

PACIFIC

History of a Frontier

JEAN HEFFER

translated by W. Donald Wilson

UNIVERSITY OF NOTRE DAME PRESS

Notre Dame, Indiana

English Language Edition Copyright © 2002
University of Notre Dame
Notre Dame, Indiana 46556
All Rights Reserved
http://www.undpress.nd.edu
Manufactured in the United States of America

Translated by W. Donald Wilson from *Les États-Unis et le Pacifique: Histoire d'une frontière*, published by Éditions Albin Michel, Paris, France, 1995.

The publisher is grateful to The French Ministry of Culture—
Centre National de Livre for support of the costs of translation.

© Éditions Albin Michel S.A., 1995

Library of Congress Cataloging-in-Publication Data
Heffer, Jean.
[Etats-Unis et la pacifique, English]
The United States and the Pacific : history of a frontier / Jean Heffer;
translated by W. Donald Wilson.
p. cm.
ISBN 0-268-04308-6 (cloth : alk. paper)
1. Pacific Area—Foreign relations—United States. 2. United States—
Foreign relations—Pacific Area. I. Wilson, William Donald, 1938– II. Title.
DU30. H4413 2002
303.48'27301823'09—dc21   2001006777

∞ *This book is printed on acid-free paper.*

CONTENTS

ACKNOWLEDGMENTS

— There is a long list of people to thank in connection with a work which has taken me more than eight years to complete. I must first express my gratitude to my home institution, the Ecole des Hautes Etudes en Sciences Sociales (EHESS), which each year allowed me to visit the United States to conduct my research. Special thanks go to the Chair of Cultural Area Studies, the late Denys Lombard. In my seminars on this topic, I have benefited from the comments and opinions of my listeners, and particularly of Annick Foucrier and François Weil, now teaching at the Université de Paris-Nord and EHESS, respectively. In 1988–1989 I taught a course on a part of this topic at the University of Geneva, where I was given a warm welcome, particularly by the Dean, Jean-Claude Favez. I have also benefited greatly from the seminars given by Christian Sautter and Claude Meyer at EHESS.

In writing a synthesis, the historian must depend on libraries and archives. I found a great deal of information in the principal libraries in Paris: the Maison des Sciences de l'Homme, the National Library, the Sorbonne Library, and the Cujas Library. But it was also necessary to spend considerable amounts of time in America: at the Library of Congress, at Harvard (working in the Widener, Houghton, and Baker libraries), and at the University of Chicago. Everywhere I found marvelous working conditions of a kind we would do well to emulate in France.

Paradoxically, my research trips never brought me to the Pacific region. They made it possible for me to savor the charms of New England, particularly in New Bedford, the former headquarters of the whale fishery, where the Old Dartmouth Historical Society and the Free Public Library contain many scholarly treasures, as does the Whaling Museum at Sharon, on the outskirts of Boston.

I have more personal debts toward my family, which has tolerated my absences, and toward Caroline Béraud-Kaufmann, the secretary of the Center for North American Studies at the EHESS, who prepared the manuscript with her usual competence. Finally, I dedicate this study to Professor Akira Iriye, of Harvard University, and formerly of the University of Chicago. It was a great pleasure to make his acquaintance and sit in on his seminar when he came to the EHESS as Frendi-American Foundation Visiting Professor in 1986–1987. His teaching was an inspiration to me, and I feel deeply honored by his friendship. He and his wife Matsuko have always given me a warm welcome in both Cambridge and Chicago. I owe them an immense debt of gratitude.

One final comment, to clarify my intentions. It would have been unrealistic to have thought of undertaking a work of original research based on new archival sources. Ten lifetimes would have been insufficient for such an endeavour! The directors of the collection *L'Évolution de l'humanité* asked me to provide a synthesis. That is what I have attempted to do, while viewing the subject from a different perspective—that of the frontier, and by trying to include the entire Pacific region, whereas most studies have—indeed not unreasonably—concentrated on the Far East. The choice of such an approach risks arousing the ire of specialists, displeased that such and such an essential aspect has been left out. Choices are sometimes difficult, for the available space is limited. Sometimes the omissions must be ascribed to an inevitable ignorance, or to time constraints. Nevertheless, I wish to thank these specialists in advance, for without them this study would have been without substance. When I undertook it in 1985, there was no synthesis of this nature in English, and ten years later it still does not seem to me that any comparable ones exist, in spite of the recent books by Arthur P. Dudden (1992), Frank Gibney (1992), and Walter A. McDougall (1993). In French, this is even more the case. In the circumstances, I do not at all feel that my efforts have been in vain.

Jean Heffer

THE
UNITED STATES
AND THE
PACIFIC

Introduction:
The Pacific as a Frontier

— In Europe, the notion of "frontier" corresponds to a precise line
along which two sovereign states meet. In American history, particularly
since Frederick Turner at the end of the last century, "frontier" signifies
a zone of varying width separating "civilization" from "savagery."
Turner's frontier did not correspond to a line drawn on a map, but occupied a vast, ill-defined space where the manifestations of civilized and
uncivilized life merged to differing degrees. In some cases there were
only subtle degrees of shading between the two, while elsewhere a few
oases offered a startling contrast with the surrounding milieu. It was
here, Turner suggests, in this constantly shifting region between the Atlantic and Pacific and not in some Germanic forest that American
democracy was born. This "frontier" on land was not the only one,
however, for in the course of their history Americans have explored
other frontiers, maritime in nature, of which the most extensive was the
Pacific Ocean.[1]

The Dynamics of the Frontier

Applied to the Pacific, the notion of frontier in the "European" sense
would signify only the coastline between California and Alaska, or possibly the outer edge of the territorial waters. Using Turner's definition,

however, the term can also be applied to the vast ocean which washes the shores of the American continent, Asia, and Australia. Its some 70 million square miles provided the geographical setting where, throughout two centuries of history, American "civilization" and "savagery" were to meet. But this frontier was far from fixed. It was constantly evolving, and, like the land frontier, it was destined to disappear. Its primary characteristic was its situation at the perimeter far from the main hub of activity. This hub, whose location can be determined by calculating the mean center of the population, did indeed advance westward along the 38th and 39th parallels. However, for a long time it remained east of the Appalachians, which it did not cross until 1860, and only reached the other side of the Mississippi in 1980. Until recent times, the nerve center of the United States was basically located in the Northeast, far from the Pacific.

The characteristic traits of the frontier, economic, political, and cultural, were determined by its situation on the margin. Economically, since the frontier was not served by a close-knit transport network, its distance from the center resulted in high transportation costs. This being so, trade tended to favor types of production in which one of the partners enjoyed an absolute advantage, whereas, according to the Ricardian model, trade can develop on condition that there appears to be a relative advantage of some kind. Furthermore, where production was concerned, a loose-knit trade network tended to favor even low-density exploitation of natural resources. A low return relative to surface area was acceptable, and in some cases little care was taken to maintain ecological equilibrium. Trade focused on a limited range of goods. These characteristics of the frontier economy do not imply that the margin was subordinate to the center, or that the center exerted control over distant territories. In fact, each party was able to maximize its own utility.

Nevertheless, however important they may have been, the economic interests of the frontier were not vital to the center, which could always find alternatives that would allow it to dispense with such relatively marginal markets. Some interests would, of course, be affected by such a withdrawal from economic activity on the frontier, but little damage would result to the nation as a whole. In the circumstances, political and strategic concerns for the frontier could not be expected to enjoy a high priority. It was not on the frontier that the principal enemy was to be found, in opposition to whom major elements of the fleet would have to be diverted. So it was natural that the Pacific maritime frontier be granted only limited importance, since it had little to contribute to national survival in case of war. However, strategic decisions were not motivated solely by economic advantage or by what appeared to be the most rational defensive options. They were also influenced by a sense of national honor, or by a certain ideological conception of the future of the country and its civilization. In the eyes of politicians and soldiers a combination of such factors may well endow a frontier with greater importance than it would merit from a strictly economic point of view.

Distance also made cultural contact more tenuous and increased the likelihood of encountering other civilized nations with alien value systems, with the result that any interchanges were likely to be superficial. At the center, the exotic and picturesque qualities of the cultures encountered on the frontier would merely inspire reactions of indifference or mild curiosity rather than of antagonism, for a threat always seems less dangerous from afar. On the frontier itself, the conflict of cultures intensified as relations developed—although since American occupation of the area was, by definition, relatively sparse, a genuine mingling of cultures was unlikely.

So, in its initial phase the American frontier was economically, politically, and culturally marginal. It is possible, conversely, to outline the state of affairs prevailing at a time when the frontier is vanishing. Because of the transport revolution, distance has ceased to be an obstacle. Economically, the trade network is more closely intermeshed, and goods are produced wherever the comparative advantage of each country allows. The exploitation of natural resources has become more concentrated, and an ever greater range of goods is traded. The initial barter system is replaced by an increasingly complex commercial geography. Henceforth, political and military decision makers must pay close attention to the region, which has become a strategic priority. Cultural exchanges also increase, with varying degrees of reciprocity, thanks to the constant intercourse between individuals and groups. But the disappearance of the frontier does not necessarily bring with it a state of peace. An intensification of relations can lead to an increased likelihood of disputes and conflicts no less than to a better understanding of others. When, in 1893, shortly after a census had been taken, Turner declared that there was no longer a land frontier in the United States, some felt that this would mark the beginning of a period of social tension and class conflict which could no longer be avoided by dispatching troublemakers to occupy empty lands in the West.[2] But history, as we know, has never unfolded quite so predictably, and there is no inevitable correlation between closer contacts and an increase in hostility.

In the nineteenth and twentieh centuries, all frontiers have lacked stability. Technical progress and demographic growth have propelled them from an initially marginal state, through phases of expansion and of assimilation, toward ultimate disappearance. Any analysis of the Pacific as a frontier of the United States must focus on the complex mechanisms underlying this inexorable dynamic, while not forgetting the part played by individuals, who always provide an element of unpredictability.

The conceptual framework outlined above calls for two preliminary remarks. First, what the reader will find here is not a history of the U.S. Pacific coast states. This will certainly come as a disappointment to Californians, who tend to think of their state as the hub of the universe. In fact, if we take a long-term view, encompassing more than two centuries, it can be seen that Americans developed an interest in the Pacific long before the coastal areas were fully occupied. Furthermore, early in the twentieth century the Pacific seaboard was home to only 3.5 percent of the population, while at present, despite strong growth since World War II, it is so to only 16 percent (though it does contribute a slightly larger percentage of the national revenue). Without denying the role the Pacific states have occasionally played in relations between the United States and other Pacific Rim countries, it is easy to show that the principal decision makers were usually located in the East, in the country's real center of gravity (cf. Appendices, Table 1). The second comment is a warning to the reader not to expect any attempt at historical synthesis. The three levels of analysis adopted—economic, political, and cultural—inevitably overlap, but it would be presumptuous to assume any close relationship between them, in some certain sequence of cause and effect. Each is sufficiently independent of the others to deserve separate treatment.

THE PACIFIC AS AN ENTITY

It is often said that the Pacific will become the center of the world in the twenty-first century. Many economic and geostrategic studies speak of an Asia-Pacific Zone. However, the Pacific they usually speak of is rarely considered in its entirety, but as though it were amputated of its Latin American coastline and the archipelagoes scattered across it.

Numerous studies deal with relations between the United States and the countries of the Far East, while a much smaller number examine United States relations with Australia, New Zealand, and the South Sea Islands, or with the Latin American nations of the continent's western seaboard. But the whole of this vast ocean is rarely considered as a single unit of study.

Perhaps this is simply because it is impossible to speak of a "Pacific world" in the way it is of an Atlantic one. Whereas European civilization appears to be firmly rooted on both sides of the Atlantic, the Pacific seems to fall into two distinct cultural universes. It is true that the Atlantic is much smaller, but its area of 41 million square miles is still more than half (59 percent) that of the Pacific. The shape of its coastlines plays a more important role. In particular, the American continent is out of true by 100° of longitude between Cape Horn and the Bering Strait.[3] If we take a globe and look at the Atlantic from above the meridian at 30° W, we can see all of South America, Africa, Western Europe, and the eastern half of North America. The narrowest part, between Brazil and Guinea, is only about 1,800 miles wide, and we can take in all the 7,800 miles separating the mouths of the Congo and the Rio Grande. Now let us turn the globe 140° to the east, so as to view the Pacific with the meridian passing through the Bering Strait in front of us. South America is out of sight, as is the continent of Asia. Only Japan, Australasia, and the coast between Alaska and Mexico emerge from the vast expanse of water. The distance from Singapore to Panama is almost 12,500 miles. So, our examination of the globe gives the impression that the Atlantic is an ocean on a more human scale. Its dimensions and topography seem to favor commerce between the continents, while the Pacific's are such as to discourage it. It is hardly surprising, then, that historians have rarely studied the latter as a complete unit, taking all of its coasts into account. In the present study, we speak of the Pacific region as it is usually defined by oceanologists, i.e., ending at the Bering Strait in the north and including the seas along the Asian continent in the west, from the Sea of Okhotsk to the Java Sea, and washing the eastern coast of Australia. Only where it meets the Antarctic Ocean are its boundaries ill-defined. (However, among the countries which are part of the Pacific region today, I have left out Canada and Mexico since they form part of the same landmass as the United States.)

But in analyzing the Pacific as a whole we cannot forget that it is made up of several regions. The numerous clusters of archipelagoes flung out across the western half contrast with the emptiness of its eastern counterpart. However, given favorable winds, sailing ships were able to cover great distances in a few weeks. They could make the entire ocean their own. This frontier, as I have defined it, is of gargantuan proportions, but it is not too large to cross. Its unity, which we have described in such pedestrian terms, is rendered much more vividly in all its mystic aura by Melville in the chapter of *Moby Dick* where Captain Ahab's *Pequod*, in pursuit of the white whale, sails out of the China Sea between Taiwan and the Philippines, toward its inevitable fate in the Pacific:

> When gliding by the Bashee isles we emerged at last upon the great South Sea; were it not for other things, I could have greeted my dear Pacific with uncounted thanks, for now the long supplication of my youth was answered; that serene ocean rolled eastwards from me a thousand leagues of blue.
>
> There is, one knows not what sweet mystery about this sea, whose gently awful stirrings seem to speak of some hidden soul beneath; like those fabled undulations of the Ephesian sod over the buried Evangelist St. John. And meet it is, that over these sea-pastures, wide-rolling watery prairies and Potters' Fields of all four continents, the waves should rise and fall, and ebb and flow unceasingly; for here, millions of mixed shades and shadows, drowned dreams, somnambulisms, reveries; all that we call lives and souls, lie dreaming, dreaming, still; tossing like slumberers in their beds; the ever-rolling waves but made so by their restlessness.
>
> To any meditative Magian rover, this serene Pacific, once beheld, must ever after be the sea of his adoption. It rolls the midmost waters of the world, the Indian ocean and the Atlantic being but its arms. The same waves wash the moles of the new-built Californian towns, but yesterday planted by the recentest race of men and lave the faded but still gorgeous skirts of Asiatic lands, older than Abraham; while all between float milkyways of coral isles, and low-lying, endless, unknown Archipelagoes, and impenetrable Japans. Thus this mysterious, divine Pacific zones the world's whole bulk about; makes all coasts one bay to it; seems the tide-beating heart of earth. Lifted by those eternal swells, you needs must own the seductive god, bowing your head to Pan.[4]

However, our analysis cannot be restricted to the ocean expanse itself. The sea provides resources, but more than that it is an expanse traversed by myriad shipping lanes, leading from port to port. The countries of the rim, each with its own history, are focal points for economic, political, and cultural activity. So, the history of the ocean is necessarily closely linked with that of the lands ringing its shores. Even sailors who never went ashore could not remain unaware of events on the coasts they sailed along. United States activity in the Pacific Ocean involved not only the exploitation of the sea's natural resources but also trade in commodities with all the peoples of the region, diplomatic and military relations with coastal countries, and cultural contact with civilizations having extraordinarily varied and disparate levels of technical development. In tracing two centuries of the American presence, from China to Chile, from New Zealand to Siberia, from Central America to Malaysia, not forgetting the Hawaiian islands, Fiji, and the Marquesas, we must make sense of a patchwork of events which, initially, it could hardly have been predicted to lead to the situation which exists today.

When we link the two themes outlined above—the frontier and the unity of the ocean—we can identify three major chronological periods since 1784, the year when the first ship flying the American flag reached China. The first period, lasting until the eve of the Civil War, is that of what we shall call the "Great Frontier," during which the American presence was expanding throughout the entire ocean and evolving in the way we have outlined above. The second, from the 1860s to Pearl Harbor (1941), witnessed a simultaneous contraction of the area within which various American interests were active, and its gradual integration of the frontier region. Finally, World War II marked the beginning of the third period, still going on, during which the entire Pacific becomes an "American lake" and the former frontier tends to disappear. Thus, if Turner's frontier evolved along a relatively straight path, the same cannot be said for the American Pacific, which underwent alternate phases of expansion and contraction. The interweaving of the two gave rise to a complex dynamic, whose various conformations we shall attempt to trace in the following chapters.

PART I

THE GREAT FRONTIER
(1784–1867)

During the initial phase of its history, the Pacific frontier was characterized by the low density of economic relations within it, by its political marginality, and by the superficiality of cultural contacts across it. These three elements are interconnected. Since only a limited range of goods and services was traded, there was no reason for the center to devote significant strategic resources to a region it considered marginal, nor did the different cultures have much opportunity for dialogue, or even a permanent confrontation. Nevertheless, it would be a mistake to conclude that during this phase the economic space lacked any pattern, or that it was structured by intermittent relations, with no real connection among them. The fact is that if all the routes followed by American ships were traced on a map, certain preferences would emerge—a certain number of directions in which traffic was more intense. What was unusual about the period between 1784 (the year when the first ship flying the American flag entered the Pacific) and the end of the Civil War, and the Alaska Purchase, is the fact that there was simply not a single region of the ocean in which ships from New England or the mid-Atlantic states were not to be found at any given moment. It is in this sense that we can speak of a "great" frontier, for the Yankees were active throughout the entire expanse of the ocean. They could just as easily be found off New Zealand or Chile as whaling in the Bering Sea, or in China—a magnet for traders eager for profit and for Protestant missionaries in search of souls to save—as in California or in Oregon, which came under United States control in 1846–1848. The frontier was "great," like Turner's West, because of its vast extent, the consequences of which were a dispersal of effort and the absence of any vital interests.

ONE

The Chinese Magnet

— By the time Americans first ventured across the Pacific, as soon as their nation became independent, the major features of the ocean were already known. In the eighteenth century numerous voyages of discovery had been authorized by various European governments. The results, published without delay, filled in the vast uncharted regions extended across the South Seas.[1] A large part of its unknown character disappeared as a result of the British and French expeditions of Anson, Byron, Wallis, and Carteret, Bougainville, and, above all, Cook. The expeditions carried out by *La Boudeuse* and *L'Étoile* in 1766–1769, Cook's three voyages on board the *Endeavour* in 1768–1771, on the *Adventure* and the *Resolution* in 1772–1775, and on the *Resolution* again, together with the *Discovery*, in 1776–1779, contributed a detailed knowledge of coastlines, winds and currents, natural resources, and native life. Nevertheless, many details remained to be filled in—and details could often make the difference between life and death. Coastlines were often drawn inaccurately, too many islands were misplaced or still had not been discovered, and reefs, which represented such a danger to shipping, were often not indicated. No way had yet been found to calculate a ship's longitude with complete accuracy. The tables of angular distances, published by the *Nautical Almanac* from 1767 on, made it possible, but only by carrying out long and tiresome calculations. Consequently, many navigators still preferred dead reckoning (i.e., using the log) to the lunar distance method, despite the margin of error involved. In a word, sailing the Pacific was still a risky business.[2]

This still little-known ocean lay between two continents which, to all intents and purposes, knew nothing about each other. To the east lay the coast of America, ruled by the Spanish, while to the west lay the shores of Asia, where Chinese power was at the height of its splendor, despite the increasing encroachments of European colonization. Perhaps the most striking contrast was in the relative size of the populations. The immense Spanish Empire, lying between the Chilean island of Chiloé and San Francisco Bay, ruled over 15 to 16 million subjects, while the Chinese Empire had approximately 300 million subjects, making it a potential market twenty times greater. It is hardly surprising that it was Canton rather than Valparaiso, Lima, or Acapulco which featured in the dreams of Yankee traders. San Francisco had recently been founded to counter a possible Russian threat from the north. The *presidio* was established there on September 17, 1776, and the Franciscan mission of Dolores was founded in October, two months after the United States Declaration of Independence, just as American troops were engaging the British on the Harlem heights and at Manhattan. This was not the full extent of Spanish ambitions, however, for the Crown in Madrid laid claim to any land which one of its mariners sighted. This meant that the Spanish Empire could have extended from Tierra del Fuego to the Queen Charlotte Islands, now part of Canada. The official mercantilism underlying its trade policy meant that—with the exception of the Manila galleon—the viceroyalties into which the Pacific coast was divided were really oriented toward the Atlantic and the ports of the Iberian peninsula (mainly Seville and Cadiz), which, in spite of this, lost their monopoly in 1778.

Far to the north, the Russians were beginning to advance along the coast of Alaska. The two expeditions undertaken on behalf of the czar by the Dane Vitus Bering, in 1725–1728 and in 1741, discovered the population of sea otters, whose fur was highly prized in China. Trappers and traders made annual expeditions to the Aleutian Islands, massacring many of the natives and destroying the natural resources without any concern whatsoever for conservation. By 1784 this made a change of policy necessary if the fur trade was to survive. Permanent bases were established to serve as trading stations and military forts, rational management of stocks of marine animals was introduced, and regular communications with Siberia were established. Despite the failure of their application to Czarina Catherine II for a monopoly, a project developed by two Irkutsk merchants resulted that same year in the establishment of the first permanent Russian settlement in North America, on Kodiak Island. To the south, as far as San Francisco, not a single settlement had been established by any of the European powers, so the coastline, with its long, winding fjords, was left to the numerous Indian tribes.

On the western side of the Pacific, European penetration was still restricted to the coasts. However, colonial preserves had been established in Indonesia and in the Philippines. In the former, the Dutch East India Company had taken over almost the entire island of Java. Elsewhere, it contented itself with setting up trading posts where it could buy the spices (pepper, nutmeg, and cloves) which were in such great demand among European consumers. The company experienced some temporary difficulties, for its monopoly was being undermined by smuggling, as well as by the provisions of the recent Treaty of Paris, under which the Dutch possessions were opened to British trade, and its profitability declined due to higher operating costs and rampant corruption.

The Spaniards had been established in the Philippines since the sixteenth century and had turned them into a bastion of Christianity surrounded by a sea of heathens. For a long time the archipelago was to remain a dependency of New Spain (later Mexico). Starting in 1602, a regular shipping link between Manila and Acapulco was set up in the form of the famous Manila galleon, which made the only regular sailings across the Pacific. It was, however, of limited importance. The galleon made one or two round trips each year conducting a trade in precious objects over which the traders of Seville and Cadiz—a most effective lobby group—attempted to gain control at the expense of their counterparts in the New World and in Manila. However, the long-established mercantilism seems to have been weakened under the influence of the new governor general, Don Jose Basco y Vargas (1778–1786). Direct communications with Spain were improved, thanks to the ships of the Royal Philippines Company founded in 1785, while at the same time the port of Manila—unlike ports of the Dutch and British colonies—was opened to all Asian nations. This victory for "free trade" was the first sign of the decline of the Manila galleon, which would be abolished by the Cortes of Cadiz in 1813.[3]

In Asia proper, the tiny peninsula of Macao, near the entrance to Canton, was a Portuguese possession, while the Siberian shoreline north of the Amur basin from the Sea of Okhotsk to the Bering Strait along the Kamchatka Peninsula had been occupied off and on for the past century by the Russians. The rest of Asia was ruled by empires too powerful to allow European nations any hope of establishing colonial-style rule in their territory. Indeed, the Europeans failed to persuade them to even open their frontiers to foreign trade, so that commercial relations, when any existed, were restricted to a single port. Since 1641, Japan had been completely closed to foreigners, except for the island of Deshima, opposite Nagasaki, where the regime controlled by the Tokugawa Shogunate allowed a Dutch ship from Batavia and a few Chinese junks to dock each year—an average of thirteen vessels a year between 1769 and 1789, which was a very small number for a nation with a population of between 26 and 27 million.[4] On the opposite shore of the Sea of Japan, Korea, under the Yi dynasty, was just as closed, for the official neoconfucian dogma had no difficulty containing the influence of the reformist Sihak school. To the south, contacts with Westerners were also minimal. In Vietnam, the civil peace established around 1680 had made it possible to get along without such contacts, and it was only following the disturbances in the south, after the overthrow of the Nguyen regime by the Tay Son brothers, at Hué (1777), that relations were resumed in 1784 between the heir to the fallen dynasty, Nguyen Anh, and Monsignor Pigneau de Béhaine, his French political and military adviser. In 1782, Siam, hard-pressed by a Burmese invasion, recovered its former glory under the founder of the present dynasty of Thai kings, General Chao Phraya Chakri, or Rama I. However, Rama I favored basically restricting commercial exchanges to China, thus reviving the policy which had led to the expulsion of the French in 1688. The only exception was Malaya. There, the Bugi Sultans, originating in Celebes, encouraged free trade, particularly on the island of Riau, off present-day Singapore. However, a certain amount of piracy went along with the trade, leading the Dutch fleet to shell the trading station in order to reestablish the authority of the Amsterdam company.

Except for Japan, all these countries fell within the Chinese orbit. The "Middle Kingdom"—the only superpower in the Far East at the end of the eighteenth century—was at its peak under the reign of Qianlong (1736–1796), of the Qing (or Manchu) Dynasty. On the one hand, foreigners could not fail to be impressed by the power of this expanding country. It had been able to subdue the nomadic tribes of Central Asia and force the surrounding vassal monarchies to become its tributaries, while refusing to establish any official diplomatic contacts with the outside world. It was enjoying an apparently remarkable economic prosperity which allowed its population to grow by 0.8 percent per annum during the 1780s. But a few cracks were beginning to appear in the edifice. Behind the stability of the "immobile Empire" an experienced analyst might have detected some early indications of decline. The population was growing faster than per capita production, leading to the gradual impoverishment of the masses. The heavy military expenditures required to deal with internal unrest caused by ethnic minorities and secret societies opposed to the Manchus were a drain on the budget. The Court was discredited by its corruption. The country was closing in on itself—Christians were being persecuted and books burned by the hundreds to ensure the triumph of a dogmatic neoconfucianism that forced all its subjects into a straitjacket of conformity. An intolerant despotism was leading to dangerous intellectual stagnation, just when the Enlightenment was making the West open to new areas of knowledge and new technology. China as a superpower was in fact fast becoming a paper tiger, though it was still too strong for any Western nation to dare challenge it. As a country, it was capable of self-sufficiency—or such at least was the official doctrine expounded by the Peking Court to English Ambassador MacCartney, in 1793, in rejecting any establishment of formal diplomatic relations. Nevertheless, trade with the West, in whose favor the balance stood, was far from negligible. It supplied the silver needed to mint coinage and also provided a means for the mandarins to enrich themselves, with minimal effort, by pocketing more or less arbitrarily what were sometimes considered exorbitant sums of money to reward them for turning a blind eye, or for influence-peddling.

When the first Americans arrived in China in 1784, they had to adapt to the already established set of rules governing foreign trade. Since 1757, commerce had been restricted to a single port, namely Canton (Guangzhou). Western traders were not allowed access to the city itself, and for no more than five months, from November to March, they were granted the use of a narrow strip 300 yards long along the waterfront and nine miles downstream, at Huangpu, where they were subjected to permanent surveillance. When their business was done, they had to leave, spending the remainder of the year at Macao. They were unable to communicate freely, being required to use a group of Chinese merchants with monopolistic powers, the hong merchants of the *cohong*, as intermediaries. The firms officially permitted to belong to the *cohong* were few in number—thirteen at the most. They fulfilled a complex role as agents and as tax collectors—a function which brought in considerable income but also required substantial expenditures in bribes to corrupt mandarins, such as the *hoppo*, or supervisors of customs and excise, leading to their indebtedness to Western merchant houses. As a result, periodic debt crises arose, which were supposed to be absorbed collectively by the hong merchants, who were reestablished in 1780 after a nine-year spell of open competition. In 1784, then, trade with China was subject to rules entirely different from those prevailing in liberal economics. Selling prices in Canton were inflated by the mandatory middlemen, while foreign businessmen were treated like lepers, and barely tolerated. In spite of this, China was attractive to businessmen in search of profit, for it could offer commodities not available anywhere else, such as tea, and had no interest in shipping them itself, having renounced all long-distance maritime ventures since the fifteenth century. China offered a bonanza to daring seamen who were frustrated elsewhere by colonial exclusions and the policies of mercantilist governments.[5]

Between America and Asia, neither the scattered islands with their populations of Melanesians, Micronesians, and Polynesians, nor the immense area of Australia—which the British government would take over in 1788 only for use as a penal colony—were as yet integrated into the international economy. At best they were seen as an anthropological laboratory that could provide new knowledge of mankind. The cannibalism observed in certain groups of islands demonstrated that the state of nature so highly praised by Enlightenment philosophers was not always so benign, even if for some Western Christians the apparent absence of repression, particularly in sexual matters, conjured up the image of a Dionysian paradise. This lack of any communications with the mid-ocean archipelagoes, and the extreme tenuousness of the shipping link between the two continents provided by the Manila galleon, make it impossible to consider the Pacific as a unified whole at the end of the eighteenth century. Louis Dermigny refers to "the empty space of the Pacific, an uninhabited wilderness." Pierre Chaunu ends his analysis of the Philippine economy by rejecting any view of the Pacific as "an autonomous economic space, endowed with its own dynamic, capable of either arresting or furthering the destiny of Atlantic Europe." At that time it was no more than "a simple extension of the Hispano-American Atlantic ruled from Seville, as far as trade is concerned."[6] In 1784, while the "Atlantic world" already had a long history behind it, one looked in vain for the first indications that there would one day be a "Pacific world." But this was precisely what was about to change, most notably with the coming of Americans, attracted by the magnet of China.

DIRECT SINO-AMERICAN TRADE

At the end of the eighteenth century and during the first half of the nineteenth, China, in fact, played a central role in trade between the United States and the Pacific area. It is true that official American statistics do not provide a continuous record until after 1820–1821, for prior to this date the *American State Papers* contain only incomplete information, which can, however, be partially supplemented from other sources, such as the trade figures for Canton.[7] Nevertheless, we do know that trade with China was more significant early in this period than later on, if only because of the relative impregnability of Latin American markets before their liberation from Spanish rule between 1817 and 1825. If we omit the period of the Revolutionary and Napoleonic Wars (whose consequences for the tea trade we shall examine later on), the major features of economic relations between the United States and the Pacific emerge quite distinctly, in respect to both the average levels and the major trends (cf. Appendices, Table 2).

Expressed in gold dollars, imports grew by 3.3 percent per annum on average, and exports by 4.2 percent. At first sight, these rates appear quite high, but they must be seen in context. First of all, the United States population was growing by 3 percent per annum during the same period, so the per capita rate actually amounts to 0.3 percent for imports and 1.2 percent for exports. On the other hand, price fluctuations should also be taken into account. Between 1821 and 1867 prices declined by 0.8 percent in the case of imports and 0.2 percent in that of exports.[8] By volume, in constant dollars, American purchases in the Pacific therefore increased by an average of 4.1 percent per annum and sales by 4.4 percent per annum, or 1.1 percent and 1.4 percent, respectively, on a per capita basis. Where the current account balance is concerned, however, the important figure is the amount in gold dollars. In that respect, the United States regularly accumulated a trade deficit with the Pacific region, to which it sold on average two times less than it bought— although this proportion did improve, thanks to a rapid increase in exports.[9] For American protectionists, this gave cause for either lamentation or rejoicing, depending on whether more importance was placed on the prevailing situation or on the way things were evolving. In any case, the Pacific played a minor role in U.S. trade overall, and there was little change in the share of the total which it accounted for. Between 1821 and 1867, for example, the Pacific contributed an average of 7.8 percent of total imports, with a slight decline from year to year. Exports, on the other hand, accounted for an even smaller proportion of the total, at 4.2 percent, but this figure was increasing annually, though by only a small amount. As suggested by our outline of the frontier, American trade with the Pacific was still of marginal importance, even when the area accounted for an increasing proportion of it, as was the case for exports between 1820 and 1850.

Of all the countries of the Pacific rim, China made the most substantial contribution to the overall economic picture, particularly where imports were concerned, where it accounted on average for 64 percent of Pacific trade. Even if the Chinese magnet was gradually losing its strength, as is suggested by the average decline of 0.85 percent per annum in its share of American purchases from the region, it was not until 1857 that its share fell below 50 percent. The very high figures we find for the early 1820s (over 80 percent), which were probably comparable to those prevailing during the Revolutionary and Napoleonic Wars, could not be maintained once both the western and eastern shores of the ocean were opened to trade, but they do illustrate the powerful pull exerted by the Cantonese market until China as a whole was opened up in 1842. But where exports are concerned, it was a different story, for on average China accounted for only 27 percent of American sales in the Pacific region—a figure which furthermore was declining by 0.22 percent per annum, though this was a much less marked trend than in the case of imports. Between 1784 and 1867, because of the accumulating deficits, China represented the major problem faced by the Americans where trade in the Pacific was concerned. The nagging question was how to pay for the desirable goods which only China had to offer, among which tea had a special place.

The Age of Tea

Under English influence tea had become a fashionable drink in the eighteenth century, adopted first by the wealthy, with its popularity later spreading to the lower classes.[10] The mother country's intention to tax tea had provoked unrest and rioting, the best known incident, the Boston Tea Party in 1773, leading directly to the American Revolution. It was only to be expected, therefore, that when their country became independent, American merchants would want to avoid buying tea in London. Since they owned the necessary ships, they preferred to import their own stocks directly from Canton. This also meant that the Asian trade would partly make up for the lost markets in the British West Indies, from which Britain's mercantilist policy had excluded its former colonies. The first ship flying the American flag to set sail for the Pacific to bring home the precious infusion was the *Harriet*, but her captain never got farther than the Cape of Good Hope, where he encountered an Englishman prepared to pay him a good price for the ginseng he was carrying—most likely to rid himself of a dangerous competitor. The venture was soon pursued by other investors possessing the information necessary for such a long voyage in waters which as yet had never been sailed by Americans. A partnership was formed, including, on the one hand, Daniel Parker of New England, William Duer of New York, John Holker, a Frenchman, who had been the official purchasing officer for the Royal Navy during the War of Independence, and on the other, in for a half share, the most important merchant in the United States, Robert Morris of Philadelphia, one of the financiers of the struggle for independence. These four men raised the $120,000 needed to fund the venture—a considerable sum for the time, well out of reach for an average businessman. Properly fitted out, bearing letters of accreditation from Congress, under the command of Captain John Green and with a crew of forty-five, the *Empress of China*, a three-master of 360 tons, set sail from New York on February 22, 1784. Its departure was hailed by the poet Philip Freneau as a symbol of liberation from British economic domination:

> With clearance from Bellona won
> She spreads her wings to meet the Sun,
> Those golden regions to explore
> Where George forbade her sail before.
> . . . .
> To countries placed in burning climes
> And islands of remotest times
> She now her eager course explores,
> And soon shall reach Chinesian shores.
>
> . . . .

From thence their fragrant teas to bring
Without the leave of Britain's kin;
And Porcelain ware, enchased in gold,
The product of that finer mould.[11]

What did it carry + bring back?

On August 23, the *Empress of China* reached Macao, after an un-
eventful crossing lasting 183 days. Four months later, her business done,
she left Huangpu on December 28, arriving back in her home port on
May 10, 1785, having taken sixty days fewer than for the outward
voyage. This venture brought her promoters (who had fallen out in the
meantime) a profit amounting to 30 percent of their original investment.[12] It was, all things considered, an average profit for a venture of
this nature, but still substantial enough to encourage Morris to send the
Empress of China back to Canton the following January. The French
Consul in New York concluded: "Whatever pains the interested parties
took to conceal the outcome of this venture, various circumstances,
and their preparation of the vessel for a second voyage, gives the impression they have made some money. They have told their friends that
the profit was no more than encouraging, and that the major benefit
accruing from the endeavour would be the information acquired for
future use."[13]

In the United States, where economic liberalism held total sway, it
was difficult to prevent competitors from entering any market which
promised a decent return. Unlike in England or on the continent of
Europe, there was no company with a monopoly authorizing it and it
alone to trade with a particular overseas region. This being the case, a
host of other vessels flying the Stars and Stripes set out for Canton in the
wake of the *Empress of China*. Between 1784 and 1833, a total of 1,164
arriving ships were recorded, which comes to an average of twenty-
three per annum.[14] The increase in traffic became most apparent after
1795, when the Americans took advantage of their neutrality during
the European wars to take over business abandoned by European companies, and played an active role as middlemen, an activity interrupted
only by Jefferson's embargo in 1808 and by the conflict between the
United States and Britain between 1812 and 1815. In some years, as was
the case in 1801, 1809, and 1819, one ship out of every two arriving in
Canton was flying the American flag. When the tonnage is taken into
account, however, the performance was less impressive, since American
ships made up no more than a quarter, for the average American vessel

involved in the China trade was of between 300 and 400 tons, while the huge vessels of the British East Indies Company were of over 1,000 tons (cf. Appendices, Graph no. 1). But the Americans, the last of the nations to sail to Canton, had already brilliantly succeeded in putting themselves in second place, with only the peerless English doing better.

Tea was the crucial commodity which allowed the Americans to make their mark.[15] Its role grew ever more significant in trade between the United States and China. Between 1821 and 1867 it accounted on average for 66 percent of the value of American imports from the "Middle Kingdom." This share grew by 0.52 percent per annum, with a particularly noticeable growth trend between 1833 and 1848, when it frequently made up three-quarters of the total or more. In overall trade between the United States and the Pacific, tea alone amounted to 41 percent of imports, though this proportion was decreasing, even if, it is true, by the insignificant amount of 0.16 percent per annum. During the period of the great frontier, then, a single product dominated trade as our profile predicts. However, tea is not a homogeneous commodity. The Americans, who initially mostly shipped home black teas, such as Bohea, Souchong, and Congou, rapidly acquired a taste for the hyson, or imperial, green teas, which they came to consume in great quantities in preference to other varieties.

The Yankee traders were not content to supply their home market, and reexported a portion of their cargoes. In some cases they even shipped the tea directly to Europe from Canton—a fact which creates a disparity between the figures recorded by the Chinese customs and those compiled by the fiscal authorities in American ports. Between 1800 and 1811, for example, at a time when the function of middleman was at its peak, a third of the tea imported into the United States was reshipped abroad. After peace returned to Europe, almost one-fifth of the tea was sold to other countries.[16] Still, the major share of an import which was growing at a rate of 4.1 percent per annum by weight was destined for the domestic market. Such rapid growth can basically be explained by population growth (3 percent per annum), but a combination of changes in price and income also had an effect. In fact, if we ignore occasional sharp fluctuations caused by military and political events, and particularly by the War of 1812 and the embargo of 1808, as well as by the size of the inventories held by speculators or merchants, the per capita consumption of tea in the United States remained stable between 1790 and 1832, at around 8½ ounces per annum.[17] It rose sharply after 1833, remaining at a peak of around 12⅓ ounces until the Civil War. This increase of almost 50 percent, which explains the increasing share of tea in imports from China, can be more easily explained by trends in prices rather than in incomes. In estimating the rise in per capita productivity in the United States between 1800 and 1860, economic historians can only make conjectures which, depending on the premises on which they are based, give rather different results. Thomas Weiss, puts the rate at 1.12 percent per annum, rising sharply after 1840.[18] Be that as it may, the evident growth in productivity, since it does not necessarily correspond to an equivalent increase in disposable income, does not explain the sharp increase which took place in 1833, even if it does show that American households did enjoy quite substantial purchasing power. The average import price of tea, calculated by adding the buying price in Canton and the shipping costs, fell by 0.4 percent per annum between 1821 and 1867. This stimulated consumption among a population whose purchasing power was increasing in other ways, particularly since the elasticity of demand in relation to price was rather high, suggesting that the brew was considered a luxury product at the time.[19] The best proof of this is provided by the much greater quantities sold in 1833, when the federal government decided to abolish all import duties on tea, allowing it to be imported tax free, even though it had made a substantial contribution to the public coffers since the creation of the Republic.[20] Like coffee, it was a favorite drink among Washington politicians, who would only be brought to reimpose a duty on tea to help defray the cost of the Civil War—although the resulting price increase came too late to influence what was by then a deeply ingrained habit.

Trading Silks for Cottons and Dollars

Apart from tea—and even it is a manufactured agricultural product—direct trade between the United States and China concentrated on manufactured goods rather than on raw materials. Among imports, cassia, a variety of senna used medicinally as a laxative, was a regular import, while sugar was not brought in significant quantities until shortly before the Civil War. Bills of lading mostly listed Chinese industrial products. Porcelain, to which Freneau's poem refers, was used as ship ballast, but the quantities were quite small—though certain pieces were of a high artistic standard, and specially decorated for the

American market with designs inspired by the symbols of the new Republic. Chinese cottons, known by the name of nankeen, after Nanking, the town where they were manufactured, were popular with American consumers until around 1830, when sales collapsed as a result of New England entrepreneurs adopting a successful strategy of providing substitutes for imported goods. Chinese silks, on the other hand, had no American competition to fear, but they were unable to avoid being gradually replaced by French products from Lyons, more in tune with fashion and closer to the transatlantic market.[21] This is why, after 1823, silks no longer surpassed tea imports in value, yielding to them the predominant role in Sino-American trade.

During the first half-century of their trade with Canton, the Yankees had almost no products of their own to sell in return. Only their ginseng was appreciated because of its aphrodisiac qualities, but Chinese demand quickly reached its limit, and any excess in supply only brought about a collapse in prices. In spite of some reexporting, the trade deficit with China remained substantial until American cotton goods made their appearance in the East, beginning in the 1830s. As soon as China was opened to trade, Yankee manufacturers, now quite able to compete on their own market with British goods, took advantage of a cheap raw material which was available domestically, and of their high level of productivity, to supply cloth, unbleached or printed, to merchants in the ports which had been opened to trade by treaties signed in 1842–1844. This made China their principal customer in the Pacific, for it imported almost $4 million worth of cottons in the years just prior to the Civil War—an untimely event which obliged the Americans to withdraw abruptly from this promising market (cf. Appendices, Table 2).

For lack of goods to trade, commercial deficits with China had to be paid for in hard cash. But the United States, until gold was discovered in California, was poor in precious metals. The drain imposed on its reserves by trade with China was becoming a serious problem. The Mexican dollars earned from commerce with Latin America were used, until the 1820s, to purchase most of the luxury goods (tea, silks, and nankeen). Beginning in 1828, however, things took a turn for the better. It so happened that China was by this time importing increasing amounts of opium from British India, resulting in a payments deficit with Britain. The Cantonese traders and dealers were obliged to sign bills of exchange which the Americans could then buy back and use as payment. This three-way financial arrangement resulted in considerable cash savings. In addition, at the same time, cotton exports to Europe were expanding rapidly. The United States was considered a sound risk for capital investment, and American merchants had the advantage of ready access to the London financial and exchange markets. British credit enabled them to pay off their Chinese debts with bills of exchange drawn on English banks—which alone enjoyed a good reputation in Canton, where the financial institutions of New York or Boston had not yet managed to make any inroads. Henceforth, silver dollars would no longer be sucked up by China, and that is one of the causes—but not the only one—of the rampant inflation which occurred during the Jackson era.[22]

INDIRECT RELATIONS

The Opium Trade

Up to 1828, lacking British credit and wishing to reduce the outflow of hard currency, the Americans had to find imaginative ways to pay for Chinese goods. One solution was to act as middleman. In spite of the imperial ban, Canton was becoming a major center for the opium trade at the beginning of the nineteenth century. The drug came mostly from India, a colony of Great Britain covered by the monopoly of the East India Company. Americans were therefore disqualified from participating in the direct trade between Calcutta or Bombay and Canton. However, this exclusion was not entirely to their disadvantage, for it also excluded private British or Indian dealers, with the result that, since the company had little desire to draw attention to itself by participating directly in an illicit trade, it often found it most convenient to use Yankee ships in Cantonese waters as drug repositories, from where Chinese smugglers could collect their supplies. Apart from Olyphant, who on principle had always refused to take part in such an unsavory commercial activity, the other American firms in Canton, Russell and Augustine Heard, had never scrupled to make profits by acting as middlemen. Earnings were even higher if they could manage to control the transportation of the drug between its source and the market. This was why, in 1805, Thomas H. Perkins of Boston had the bright idea of sending his ships to Smyrna, in Turkey, to purchase opium for sale in China. The idea was quickly copied by his competitors, so that a regular trade route from the Mediterranean to Canton was established entirely under American control, until 1833 when the East India Company monopoly came to an end. The volume was never very great compared with the amounts originating in Bengal or Malwa, but it helped pay for some of the tea and textiles in such great demand in the United States. However, the opium trade never played as important a role for the United States as it did for England. The government in Washington refused to support illegal activities by its merchants, who, with their interests less directly at stake than those of their counterparts from Calcutta or Bombay, were more willing to submit to the edicts of the Chinese administration once it began to enforce firm measures. It was, therefore, inconceivable that Americans should take part in an opium war against China, even if they were not unhappy with the consequences of China's defeat in its confrontation with the British lion.[23]

The drug trade was certainly the most profitable form of commercial activity, in spite of the risks involved before 1842, when China was opened. Nevertheless, at the end of the eighteenth century, the Yankee traders were largely excluded from it. This made it necessary for them to scour all the coasts of the Pacific in search of goods for which there might be a market in Canton. Thanks to the Chinese magnet, the ocean was finally opened up to long-range international commerce, at the cost of destroying certain extensively exploited natural resources in the absence of any concern for conservation.

The Sea Otters of the Northwest

The voyages of Cook and Lapérouse had made the Western world aware that the northwest coasts of the American continent, and the deep fjords which cut into them, were the habitat of sea otters—superb creatures whose furs were highly prized by wealthy Chinese.[24] A year before the official publication of the account of Cook's third voyage in 1784, one of the Americans who had taken part in it, John Ledyard, lost no time in arranging for his own version to be printed by a Hartford publisher.[25] It contained the information that the coastline of Vancouver Island, explored in the spring of 1778, was rich in fur-bearing animals of all kinds—foxes, martens, sable, ermine, wolverines, beavers, and otters—whose pelts could be acquired from the Indians in exchange for objects of absurdly low value, to be sold in Canton at a fabulous profit. Having failed to sell his idea to either Morris or Parker, both much too absorbed by the planned voyage of the *Empress of China*, John Ledyard set off for Europe where he could count on Jefferson, who was in Paris at the time, to support his attempt to reach Nootka Sound by way of Siberia. However, the Czarina Catherine II refused him passage. Undeterred, the adventurous Yankee tried his persuasive skills on the British in 1786, but his efforts only resulted in his being charged with violating the monopoly held by two British companies. So he returned to his Siberian solution. At the second attempt he got as far as Irkutsk, but no farther, for the czarina had him arrested and deported. Perhaps he would have found a way to realize his ambition if his attention had not been temporarily distracted from the Pacific by an English invitation to take part in an expedition to the African interior, in the course of which he died in Cairo, in January 1789.

Other Americans, with the requisite financial means, were to succeed where John Ledyard had failed. Six investors—four of them from the Boston area, including the architect Charles Bulfinch, a New York merchant, and a Salem outfitter, John Derby—invested $50,000 to send two vessels to the Northwest coast: the three-master *Columbia Rediviva*, of 212 tons, and the sloop *Lady Washington*, of 90 tons. This little squadron, setting out from Boston on September 30, 1787, under the command of John Kendrick and Robert Gray, was the first to round Cape Horn under the American flag. A year later, it reached Nootka Sound (which had been a haunt of the English since 1785) just before a serious international crisis broke out between the Spanish and the British over their claims to these shores lying at the outer edge of the known world. The year 1789 was spent acquiring furs in exchange for chisels, which the natives were particularly eager to own, since they provided them with a better tool for wood carving. In May and June, while the *Lady Washington* was cruising along the shores of the Queen Charlotte Islands, Robert Haswell noted in his log:

> At 6 PM a vast number of Natives men Women and Children came off and brought with them several sea otter skins we understood of them that their was a large tribe not far off the weather was very thick hazey and we were but little distance from the land We soon saw their village from which they launched twenty or thurty very large canoes and came off in great perade padleing off

swiftly and singing a very agreable air. [. . .] of those people were purchaced to the amount of two hundred skins in a very fue moments for one chizle each we bought all the skins they appeared to have by 10 in the evening when they returned to their Village for the night no doubt intending to bring off more in the morning but we did not stop but stood on to the southward.[26]

On June 11, another deal was made in the Houston Stewart Channel, but in this case the chisels had little appeal for the natives, who preferred clothing. It seems that British traders had already passed through, saturating the market and teaching the Indians the relative value of goods. Overall, the voyage was not a great success, for John Kendrick turned out to be somewhat lacking in zeal. In fact, he decided to remain in the Northwest, and, in the fall, dispatched the *Columbia* to Canton under Gray's command, with instructions to sell the otter skins and use the income to buy tea and cloth which he was to convey back to Boston, thus accomplishing the first circumnavigation of the globe by an American.

The *Columbia* had scarcely docked, on August 9, 1790, when, in spite of the disappointing financial results, its Bostonian owners dispatched it for a second voyage, which was also to last almost three years from September 22, 1790, to July 29, 1793. It was in the course of this voyage that, on May 12, 1792, Captain Gray discovered the mouth of the river to which he gave the name of his ship. The event is recorded in John Boit's journal:

> N Latt. 46° 7' W. Long. 122° 47'. This day saw an appearance of a spacious harbour abrest the Ship, haul'd our wind for itt, observ'd two sand bars making off, with a passage between them to a fine river. Out pinnace and sent her in ahead and followed with the Ship under short sail, carried in from ½ three to 7 f[atho]m, and when over the bar had 10 f[atho]m Water quite fresh. the River extended to the NE as far as eye cou'd reach, and water fit to drink as far down as the *Bars*, at the entrance. we directed our course up this noble *river* in search of a Village. the beach was lin'd with Natives, who ran along shore following the Ship. Soon after above 20 Canoes came off, and brought a good lot of Furs and Salmon, which last they sold two of for a board Nail. the furs we likewise bought cheap, for Copper and Cloth. they appear'd to view the

Ship with the greatest astonishment and no doubt we was the first civilized people that they ever saw.[27]

The Columbia River appeared to be the long-sought-for route leading from the Pacific into the interior of the American continent. Its discovery would enable the United States to make a legitimate claim that its territory extended as far as the ocean. In this way, Robert Gray prepared the way for future claims over what are now the States of Washington and Oregon.

In the short term, however, the east coast shipowners had few imperialist ambitions. They dispatched their ships with the sole objective of trading cottons, metal objects, firearms, alcohol, or slaves in exchange for skins. Taking advantage of the English involvement in the interminable struggle against Revolutionary France and Napoleon, as well as of the East India Company's bitter defense of its monopoly against any incursions by British competitors, the Americans were the only Westerners present on the Northwest coast at the end of the eighteenth century (cf. Appendices, Table 3). They did not do any of the hunting themselves, leaving it to the Indians. Initially, from spring until the early fall, their boats cruised from inlet to inlet and through the labyrinth of islands, around Vancouver Island and the Queen Charlottes. Trading conditions varied from season to season. Unscrupulous captains did not hesitate to use violence to force the natives to sell the produce of their trapping, caring nothing for reprisals that might be taken against the crews of other ships. Trade still remained a dangerous business requiring constant vigilance, being careful not to allow too many Indians on board ship. In the early days, they would winter in the Hawaiian islands, but later, to save time, the off-season was spent sheltering behind wooden fortifications on the northwest coast itself, using the time to repair the damage the ships suffered during their long voyages, to construct longboats, and, eventually, to trade all through the year. A voyage would generally last three years, though in some cases somewhat longer during the War of 1812, lasting as long as eight years in the case of the *O'Cain*'s third voyage, between May 1809 and October 1817. From a financial point of view, the results were very variable. With luck, it was possible to make a killing, but it was not long before the stock of sea otters was severely depleted, for there was no authority to manage the resource, and investors cared only for immediate profits.[28]

Other avenues therefore had to be found. Some Americans turned from trading to hunting. Between 1803 and 1812, following the lead of Captain Joseph O'Cain, they reached an agreement with Baranov, the agent for the Russian monopoly company based in Sitka, to poach otter in Spanish California. The Russian provided teams of hunters, the American the ship, and the profits were divided equally between them. Then, since the Russians had no ships available and had been granted no rights to trade in Canton, the Americans undertook the shipping and marketing on their behalf, just as they would do between 1817 and 1822 on behalf of the fur companies of British Canada, which had also been excluded from China by the official monopolies. The New York investor John Jacob Astor attempted to put this commerce on a permanent footing by sending out two expeditions, one by sea and the other over land, to found Astoria, at the mouth of the Columbia. Unfortunately, the undertaking came to nothing because of the War of 1812, and Astor decided to sell out to his British competitor, the North West Company, for $58,000. By about 1825 the prime years of the sea otter hunt were over. Most of the New England investors who had finally come to dominate the trade—Perkins, Dorr, Lyman, and Lamb—got out, and what remained was controlled by three firms: Bryant and Sturgis, Marshall and Wildes, and Boardman and Pope. In any case, the skins had only accounted for a small proportion of the trade deficit with China. However, they had enabled the Americans to gain a footing in Hawaii, to become well acquainted with the waters of the North Pacific, and to familiarize themselves with the southern shipping routes along the coast of Chile, or even as far as the Marquesas (whose northernmost islands were discovered in 1791, two months ahead of the Frenchman Marchand, by Joseph Ingraham, the captain of the *Hope*, who claimed them for the United States.[29])

Sealskin, which was not as much in demand in Canton—being worth only $1 or $2 apiece, compared with $20 to $30 for an otter pelt—played only a minor part in Sino-American trade. Many colonies of these aquatic mammals inhabited the coasts of Tasmania, the uninhabited shores of Chile and Peru, and the islands of Juan Fernandez, San Felix, and San Ambrosio. Captain Stewart, reaching Canton in March 1793 from Mas Afuera, was unfortunate enough to hit on a period when prices were particularly low, so that his sales earned him only $16,000, instead of the $38,000 he expected. But that was not enough to discourage others. According to Amasa Delano (the "good"

captain and hero of Melville's *Benito Cereno*), who had himself organized two expeditions to hunt seals in 1799–1802 and 1803–1807, from
1797 to 1804 between two and three million animals were bludgeoned
to death, skinned, cured, and dispatched to Canton. The Chilean historian Pereira Salas suggests a figure of three and one-half million skins
for all of Chile up to 1809, with a value of seven million dollars on the
Canton market. At this rate, the resource, too, was soon exhausted, and
other hunting grounds had to be sought in colder latitudes. This would
become the specialty of the shipowners of Stonington (Connecticut),
like the famous Edmund Fanning, nicknamed the "pathfinder of the
Pacific."[30]

Sandalwood and Sea Cucumber

As the supply of skins declined, it became necessary to seek out other
exotic products likely to appeal to the Chinese. Sandalwood was used
mostly in temples, because of its incense-like perfume. Traditionally, it
was imported from India, but was also to be found on tropical Pacific
islands. There was no question of the Americans harvesting it themselves, since the natives were much too protective of land ownership.
The only solution was to make agreements with local chieftains able to
mobilize their subjects to cut logs and transport them to the coast. Favorable terms were offered in the Hawaiian islands where King Kamehameha I had extended his rule to the entire archipelago. An initial
shipment of sandalwood to Canton in 1789 turned out a failure because
of the poor quality of the wood. It was not until 1811 that the Winship
brothers recognized the advantages of supplementing cargoes of furs
with extremely marketable sandalwood logs. They obtained exclusive
rights from the king, but were unable to take advantage of them because of the war against England. When peace returned, the new
Boston merchant houses, who henceforth would control the major
share of American trade in the North Pacific, took over exports to
China. If Kamehameha I, jealously protecting his exclusive ownership
of the wood, was prepared to sell it in only modest quantities, his successor, Liholiho (1819–1824), a weak and easily influenced individual,
awarded his favorites the right to participate in the trade, with the result
that the sandalwood was harvested without any restraint whatsoever. It
was used as currency to purchase American ships at exorbitant prices,
resulting in debt for the Hawaiian nobility, who, consequently, imposed
even greater woodcutting duties on their subjects. By 1821, exports

peaked at 1,800 tons; and then, after one last surge between 1827 and
1830 (with an average of 850 tons per annum), they collapsed. Only a
decade and a half were needed for this natural resource to be wiped out.
However, this was time enough to allow American traders to become
established in the islands. The same rapid depletion of the resource occurred in the Marquesas and in Fiji, where harvesting had begun earlier,
often in collaboration with Australian investors.[31]

The Chinese, always lovers of good food, found the sea cucumber,
a type of holothurian native to tropical waters, particularly delectable.
It was collected by native divers from among the rocks at a depth of
between six and twelve feet and was then dried over fire until its
weight was reduced by two-thirds and it was ready to be shipped. A
large thatched building, in which large quantities of wood bought
from local chiefs was burned, was constructed on the beach for this
purpose. Several months of work would be required to fill a ship's
hold, during which time an attack was always possible. This is what
happened to Captain Benjamin Morrell, who lost fourteen men in Fiji
in 1830, and to the *Charles Doggett*, six members of whose crew were
slaughtered four years later. But it took more than this to discourage
the doughty Yankees, for sea cucumber was worth from $13 to $25 a
picul in Canton, and considerable profits could be made. In five or six
voyages, for example, Captain John H. Eagleston conveyed 265 tons,
earning him $80,000, all in return for an outlay of at most $10,500—
in other words, a profit of almost eight times the initial investment.[32]

Otter and seal skins, sandalwood, and sea cucumber: these commodities
harvested from the Pacific and destined for the Chinese market were not
in plentiful enough supply to long survive the reckless exploitation to
which they were subjected. Their share in solving the problem of the
United States trade deficit with China was only marginal. However,
they are worth mentioning as illustrations of the enterprising spirit of
the Yankees, always on the lookout for any opportunity to make a short-
term profit, in quest of which they would ply the Pacific in greater
numbers than any other maritime nation of the time, England included.

OTHER TRADE DIRECTIONS

However central it was, trade with China was far from the whole story
of American commercial activity in the Pacific. Other trade channels

developed, such as Indonesia and the Philippines in the western part of the ocean. By the end of the eighteenth century, ships from Salem had discovered the pepper-growing region in northern Sumatra. They took advantage of their country's neutrality and of the flexibility of their free enterprise system to cut out the Dutch middlemen and make themselves the main suppliers to Europe until around 1830, when relations with the natives deteriorated with the fall in prices.[33] Similarly, between 1797 and 1809, the Dutch East India Company carried on a certain amount of trade with the Japanese island of Deshima using American vessels, as a way of evading the British blockade. In 1800, the *Franklin* was the first Yankee ship to return to Boston carrying goods from Japan. But this was to last for only a short time, since the Tokugawa Shogunate enforced a closed door policy until Perry's expedition in 1854, and the Dutch no longer had any need to make use of American ships now that Europe was at peace once more.[34] During this entire period, Java became much more than a port of call on the way to Canton, for it was the final destination for many ships from New England and New York, in search not only of the spices and coffee the Americans were beginning to consume in considerable quantities, but also of sugar and tin.

The Spanish Philippines, for their part, accounted for an increasing share of United States imports from the Pacific. Indeed, Manila had two commodities to offer which the United States itself was unable to produce in sufficient quantity, namely, sugar and, above all, hemp—indispensable for the manufacture of rope and ships' rigging. The amount of hemp produced in Kentucky was far from sufficient to satisfy the demands of a merchant fleet undergoing rapid expansion in the 1850s, and by turning to the Philippines, America could avoid an excessive dependence on Russia, which had previously been the main source of supply. This explains why, on the eve of the Civil War, hemp from Manila alone accounted for between 7 percent and 11 percent of American imports from the Pacific.[35]

Australia, on the other hand, played virtually no role in American trading relations during the first half of the nineteenth century. Ships flying the Stars and Stripes were actually excluded from trading there as long as the British Navigation Acts were still in force. The Jay Treaty (1794) granted them only the right to trade directly with the East Indies, and cabotage, or coastal shipping, which was very important in Australia, was forbidden. In spite of this, during the wars with France, making a virtue of necessity, the Sydney authorities allowed fifty-eight American "peddlers of the seas" into port between 1792 and 1811—twenty-two of them between 1804 and 1807—even though they were accused of selling rum for all too scarce currency. At the time, Australia was a port of call for ships heading to the seal colonies, the whaling grounds, or the sandalwood forests, rather than a final destination. Captain Folger, captain of the *Topaz,* was sailing on from there when, in 1808, he came across the single English survivor of the mutiny on the *Bounty* on Pitcairn Island. As was the case in the Dutch East Indies, the American commercial presence was short-lived. When the British East India Company lost its monopoly in the Pacific in 1813, the Navigation Acts were strictly applied once again, so that trade between America and Australia reverted to its former insignificant level.[36] It was only because of the happy coincidence of two events in the 1850s—the introduction of free trade in the English colonies and the simultaneous discovery of gold in California and in Australia—that it came to life once more, with the result that between 1853 and 1860 Australia would account for 30 percent of American exports to the Pacific, and for a major increase in sales. The balance of trade would be heavily in favor of the Americans—so much so in fact that regular shipping links could not be established. This imbalance arose because the wool manufacturers of New England used scarcely any of the raw material which was the major contributor to Australia's wealth, still preferring to fulfill their requirements in London. For this reason, Australia, like New Zealand, remained marginal to American commercial interests in the Pacific.[37]

On the opposite shore, in the direction of Latin America, trade got off to a slow start, for the Spanish empire maintained a policy of jealous protectionism. It was wary of American vessels sailing along its coasts and putting into port, ostensibly to take on food and water, but actually for the purpose of smuggling. The wars of independence and the conflicts between the new states made trading along these coasts a risky business. Seizures of vessels and confiscation of their cargoes fueled the dispute. On the other hand, if everything went well, juicy profits could be made. Richard Cleveland, for example, carried out two voyages along the western coast of Latin America. The first, in 1801–1804, sailed from Hamburg under somewhat unfavorable auspices, since the Treaty of Amiens brought about renewed trading relations between Spain and its colonies, so that in the future smuggling would be dealt with mercilessly. The *Lelia Byrd* was delayed for more than two months in

Valaparaiso by overscrupulous and suspicious authorities. From there, except for a short stop in the Galapagos to catch tortoises, Cleveland had to sail to the Mexican port of San Blas. There, it was the same story. One of the partners was sent to Mexico City to obtain trading rights, but in the meantime the ship was obliged to leave Mexican territorial waters and take refuge among the offshore islands, where she waited for a cargo of otter skins to arrive from California. To fill the remaining space in her hold, she put in for a third time to San Diego, attacked the Spanish fort, and sold on credit to the Franciscan missionary fathers. In May 1803 she sailed for Hawaii, having been able to dispose of only half the goods she had brought from Europe in the space of fifteen months. Fortunately, she found a good market for skins in Canton. With his profits, Cleveland bought tea and silks, which he loaded on the *Alert* to be shipped home to Boston, while his partner, Shaler, returned to California to attempt to sell the remaining half of the cargo and collect the debt from the missionaries. This venture would end in disaster. Overall, however, the outcome was positive, for in 1804 Cleveland was left with a fortune of $70,000 net of debt. In 1806, Cleveland and Shaler organized a second voyage, which would get no farther than Rio de Janeiro, bringing them total ruin when their vessels were seized by the English fleet. Early in 1817, it was learned in New York that Chile had risen against Spanish rule. Sensing that handsome profits were to be made, with the ports now open, Cleveland proposed to Astor that the latter entrust him with one of his ships, the *Beaver*, carrying a cargo of European-manufactured goods worth $140,000—a relatively substantial investment at the time. Unfortunately for him, things would not turn out as he had planned. On putting in to Talcahuano, in October 1817, he found the port in the hands of the royalists, who confiscated the ship and its cargo, on the pretext that the latter was contraband. After whiling away seven months without making any headway, Cleveland set off for Lima to plead his case before the viceroy. The snail's pace of the colonial administration gave him ample time, taking advantage of his neutral status, to indulge in some cabotage between Peru and Chile, dodging the Chilean fleets ineffective blockade. Finally, after purchasing a cargo of cocoa in Guayaquil, he sailed home from Callao in June 1820. He had spent almost three adventure- and misadventure-filled years along the coast—years which, despite the confiscation, actually turned out to be quite profitable for both owners and the captain, who came out of the venture with a fortune of $75,000. This, however, Cleveland soon lost again in a foolish investment.[38]

In trade relations between the United States and the Pacific, the western coast of Latin America was far from devoid of interest, for the Americans managed to develop a favorable trade balance with it which partially made up the deficit with China, Java, and the Philippines. It provided them with the silver dollars which were common currency on the far side of the ocean. Indeed, before China was opened to trade in 1842, Chile and Peru were the principal market in the Pacific for New England cottons (cf. Appendices, Table 2). Imports, on the other hand, were restricted to raw materials, and even these in insignificant quantities. One commodity was predominant: Chilean copper, indispensable until the mines in northern Michigan came into operation, which was only made possible by the opening of the Sault Sainte Marie Canal in 1855. Some Chilean copper was also shipped directly to China. The guano taken from the uninhabited islands off Peru was low in value, but benefited for several years from the campaign to regenerate the soil exhausted by tobacco growing in the southern United States, which created a short-lived demand for this type of fertilizer. It reached a peak of 150,000 tons in 1848, attracting a good number of ships and some capable businessmen like the Grace brothers, originally from Ireland, but it was already falling into decline by 1854–1855. This activity finally collapsed for lack of a market when interrupted by the Civil War.[39]

A third noteworthy commodity, cattle hides, became another important commodity because the New England footwear industry, in the process of mechanization, being unable to find adequate supplies of leather in the United States, turned abroad, to the Pacific region, for somewhere between about 10 percent to 20 percent of its requirements. The most famous source, even though a marginal one, was California, thanks to the vivid picture drawn of it in 1840 by Richard Henry Dana in his autobiography, *Two Years before the Mast*. Obliged to cut short his studies at Harvard by an eye ailment, the young Dana signed up in August 1834, at the age of nineteen, to join the crew of the *Pilgrim*, owned by Bryant and Sturgis. The objective of the voyage was to round Cape Horn and sail north to California to trade manufactured goods for hides collected by the Franciscan missions, then in the process of being secularized. In 1835, this kind of trade had become routine, and, since 1829, the Americans practiced it regularly. A ship could hold as many as 40,000 hides, but to assemble such a quantity it had to spend a long time on the coast, sailing up and down between San Francisco and San Diego, where the curing and storage sheds were located. Loading took time, for there was no port equipped for it in California at the time. The hides had to be conveyed by rowboat from shore to ship, through the surf, and often in windy conditions. Where the coastline was rocky, the hides would be thrown down on the beach from the cliff tops. After sixteen months, spent at times in exhausting labor and at others in total idleness, the young man had exhausted the joys of travel, and felt a pressing desire to return to Boston—which he did on board the *Alert*, in September 1836. From an economic point of view, the voyage to California was profitable only because hides were cheap there—$2 each—and could fetch over ten cents a pound in Boston. When prices collapsed after 1844 it dealt a deathblow to this business, for it became more profitable to import hides from Argentina or Uruguay, where there was a more plentiful supply and from where the shipping costs were lower.[40]

All these different trade directions give a complex picture of American commercial involvement in the Pacific. However, because they were divided between a great number of ports, none was able to play a role comparable with that of the Chinese magnet, centered first on Canton, and then, after 1850, on Shanghai. It was from there, on the outer edge of China, that traders could hope to penetrate the most immense market in the region—a continuing dream for the most visionary among American entrepreneurs.

AMERICAN COMMERCIAL ACTIVITY IN THE PACIFIC

Well-adapted to a low-density economy, trading activities were practically the only ones to gain the attention of American businessmen and entrepreneurs during this initial phase of the exploitation of Pacific resources. To acquire land in Latin America, it was necessary to be a citizen and a Catholic—a status most Protestant Yankees were reluctant to embrace. Before the treaties of 1858–1860, China was inaccessible. Only the Kingdom of Hawaii, with reluctance at first, was prepared to liberalize the laws governing land acquisition in order to encourage the development of sugarcane plantations. In 1835 it leased land to Ladd and Company to carry out agricultural experiments, and six years later it offered fifty-year leases to foreigners, much to the frustration of Americans eager to enjoy the same kind of total ownership as in their own country. Under the intense pressure exerted by the latter, it did not take long for the feudal system to collapse. In 1846, new property legislation, inspired by the desire to increase government revenue, was a sign that Western economic principles had triumphed. It also enabled the rapid development of the first sugar plantations, which would soon see the markets of California and Oregon open up to them. This set in motion a process which would lead to the annexation of Hawaii a half century later.[41] Industrial investment, however, was still nonexistent.

Yankee Traders

All those Americans who saw the Pacific as a vast domain where their instinct for profit could have free rein devoted their entire energy to trade. First came the seamen: captains and supercargoes, at a time when the businessman was often the same individual as the shipowner, and when the latter would not hesitate to leave his office to try his luck on the ocean wave. Usually they began young, hardening themselves to the tough discipline of life on a sailing ship, and acquiring business experience by trading with the native peoples. As long as he had a head for business, was not afraid of hard work, and was lucky enough to survive shipwrecks, typhoons, epidemics, and skirmishes with the natives, a young fellow could move rapidly up the ladder to positions of responsibility, and be able to retire, in the prime of life, as the owner of an honest fortune, to reside in his beloved United States, which from far across the ocean had always seemed the best country in the world—as long as there was money in the bank. As he sums it up himself, the career of Robert Bennet Forbes (1804–1889) provides an exemplary portrait of success:

> Beginning in 1817, with a capital consisting of a Testament, a "Bowditch," a quadrant, a chest of sea clothes, and a mother's blessing, I left the paternal mansion full of hope and good resolutions, and the promise of support from my uncles. At the age of sixteen, I filled a man's place as third mate; at the age of twenty, I was promoted to a command; at the age of twenty-six, I commanded my own ship; at twenty-eight, I abandoned the sea as a profession; at thirty-six, I was at the head of the largest American house in China.[42]

All this was despite a very minimal formal education, consisting of one year in France and two and a half years in Massachusetts. It was also true, as he says himself, that his family connections made many things easier for him. His uncles were none other than James and Thomas Handasyd Perkins, the richest American merchants engaged in trade with Canton. For a relatively long time, Robert Forbes pursued a career as a sailor, for which he had a liking, but also because he preferred to leave the business to his elder brother Thomas, who would perish in a storm at Macao in 1829. At thirteen he began his life of adventure on board the *Canton Packet*, on which he served for six years. During his first voyage to China, he was put up by his cousin, John P. Cushing, who had himself taken charge of the Perkins firm in Canton at the age of sixteen. He was initiated into the tea and silk trades. On the return journey, supplies ran low and the "bread became locomotive by reason of the weevils and worms."[43] In 1819 he made a second voyage to China, where he invested the $500 his uncles had given him by purchasing cloth, which it was up to him to sell at a good profit, and did so in Hamburg. Promoted to third officer at a salary of $10 per month, he sailed to Gibraltar, from where he made his way back to Canton via Batavia. On his return voyage he was promoted to second officer, at $25 per month. He took just enough time to spend a few weeks with his family and make two other voyages. In 1824 he was made captain of the *Levant*, without ever having held the rank of first lieutenant, but did not remain long on this vessel, which was destined to be used as a repository in China. He was entrusted with the *Nile*, a brig of 250 tons, at a salary of $50 per month, with, in addition, an entitlement to carry six tons of merchandise worth between $180 and $240. He set off on a trading expedition to the Pacific, not returning to Boston until October 1827, after a three-year absence. In the meantime, he had been given good reason to complain of his supercargo, who proved inept at trading along the American coast between California and Peru, being too parsimonious to bribe the high local officials and so incompetent that he allowed his smuggled goods to be confiscated in Guayaquil.[44] On the return voyage, in Hawaii Forbes collected a cargo of sandalwood destined for Canton, on his return from where he intended to sell merchandise in Buenos Aires. Unfortunately, he found this port under blockade by the Brazilian fleet, and it was only by bribing the enemy that he was able to escape confiscation. His last voyage to the Pacific was undertaken on others' behalf, in 1828–1829, with a detour to Smyrna to pick up a cargo of opium. This left him wealthy enough to purchase his own ship, the three-masted bark *Lintin*. On her, he sailed once again for Canton, with his brother, John Murray, and Augustine Heard on board, remaining there until April 1832, using his vessel as a repository for drugs and other contraband. His profits enabled him to retire from the sea at the age of twenty-eight, after plying the ocean for fifteen years. He could take a wife at last, buy a home in the state where he was born, and live a quiet businessman's life, until the 1837 crisis left him no alternative but to return to China to rebuild his fortune. This time he went as a partner in the firm Russell and Company, thanks to the support of the Perkinses, his brother John, and Houqua, the hong merchant. He assumed his duties right in the middle of the opium war, but refused to leave Canton as the English did, for the American government refused to cover any losses suffered by its merchants. Taking advantage of the situation, he served as an intermediary for British businesses until Her Majesty's Fleet enforced a blockade in June 1840. His health failing, his only option was to return home, after remaining long enough not just to pay off all his debts, but actually to accumulate a sizable nest egg. Back in Boston, he devoted himself to his business as consignee of oriental merchandise, and as an investor in a number of steamship enterprises, such as the *Midas*, the first American propeller-driven ship to round the Cape of Good Hope and to operate—albeit unsuccessfully—in Chinese waters. When, in 1849, his cousin Paul S. Forbes wanted to come home, Robert set off one last time for Canton, where he managed Russell and Company for a further nineteen months, after which it was time to bid farewell to the Pacific and leave the field to younger men.

The career of Robert B. Forbes was hardly typical of the thousands of Americans who sailed the Pacific between the end of the eighteenth century and the Civil War. The great majority of them never became anything more than ordinary crewmen, and failed to make their fortunes. Nevertheless, in all the important ports it was quite usual to find Yankee merchants living in relative affluence, whether their stay abroad was a long or a short one. In the course of his voyage on the *Dale*, Philo White mentions their presence all along the Latin American coast. They often supplemented their strictly commercial pursuits by acting as United States consul, and often married into the best families in the country, like Stanhope Prévost in Lima, Samuel Haviland in Coquimbo, and Seth Sweeter in Guayaquil.[45] In Mexican California, American businessmen—Henry Delano Finch, Jacob Leese, and William Sturgis Hinckley—made matches in the best ranching families, and dominated business in this region at a safe distance from the central authorities. The consul in Monterrey, Thomas O. Larkin, laid the groundwork for California's annexation by the United States, carrying on a profitable business all the while.[46] William Wheelwright, a former consul in Guayquil, developed steamship traffic between Ecuador and Chile, though he was obliged to do so under the British ensign, for he was unable to find any American ready to invest in this new mode of transportation in the Pacific.

It is true that Yankee capital preferred China. It was there that the most substantial deals were to be made, and where the main trading routes converged. As trade became more and more intensive, captains and supercargoes became increasingly willing to entrust the local management of their affairs to specialized agency houses. Initially, these were commission houses serving as middlemen for principals in the eastern United States or Europe. Thanks to their knowledge of China and to their close ties with the Canton hongs, they reduced the risk to their clients, providing them with price information and advice on the profitability of whatever commercial venture they had in mind. They bought and sold commodities which were in demand or on offer, and made arrangements to ship them. They were, in other words, genuine entrepreneurs.[47] Nevertheless, it was not their concern to find buyers for the merchandise, so that the risks they took were minimal, and did not require a large capital investment, while their profits could be fabulous. John Cushing, a nephew of the Perkinses, arrived in 1803, and was able to retire as a millionaire a quarter of a century later. His cousin, John Murray Forbes, accumulated $150,000 in two years, between 1834 and 1836. When China was opened to foreign trade, they were obliged to diversify by providing various services which were lacking locally, so that the functions of ship's chandlers, banking, insurance, and foreign exchange were added to their activities as mercantile agents. When they had to, they engaged in trading on their own account, though not always with success, for it took a long time for information about Western markets to reach them in those days of sail. The years during which the Chinese Empire was threatened by the Taiping Rebellion (1851–1864) provided some excellent opportunities for profit, in spite of the temporary suspension of trading links, since Western vessels, which were faster and able to obtain insurance coverage, provided much more security than the traditional Chinese junks. The American agency houses provided wealthy Chinese businessmen with an opportunity to protect their capital from extortion by the mandarins, and Houqua, using Russell and Company as an intermediary, invested in railroad construction in the United States. The firm of Augustine Heard and Company recorded the best net revenues between 1854 and 1862. At Russell and Company, in 1858–1860, profits reached an average of $240,000 per annum, providing $45,000 per annum for the two senior partners, who each held a ³⁄₁₆ share, and $15,000 for the junior partners. When R. W. Spooner retired in 1861, after a long illness which had prevented him from playing an active role managing the firm, his former partners refused to pay him a pension of $3,333 per annum for three years, on the grounds that he had already accumulated a fortune of between $70,000 and $80,000. He was given to understand that:

> . . . altho in the eyes of some may appear a modest fortune, is in the estimation of no man a contemptible sum—and is quite sufficient to maintain any gentleman in abundant comfort—even luxury—while out of business in the United States, or elsewhere, while it affords a fair capital upon which to embark in business when his health or inclinations had [sic] him that way.[48]

However, clouds were gathering on the horizon for the commission houses. With the disappearance of the *Cohong*, they became dependent on Chinese compradors, unless they were themselves prepared to learn the language and travel in the interior. This amounted to encouraging future competition. The transport revolution decreased dependency on intermediaries. The sections of the continuous telegraph link between the West and Hong Kong, which were finally connected in 1871, resulted in a trend of awarding business to smaller agency houses, for size was no longer such a prime advantage in avoiding risk. The practice of buying tea by sample was a step in the same direction. Specialized banks were established in China, ending the monopoly over foreign exchange. Unless they could adapt, the commission houses were destined to disappear. Such was indeed the fate of the American companies, Heard declaring bankruptcy in 1875, and Russell, the most prestigious of all, in 1891.

The Heyday of the United States Merchant Navy

Apart from the trading communities residing in the various ports, the most visible sign of the American commercial presence in the Pacific was the Stars and Stripes fluttering the masts of ships. According to official statistics, between 1824–1828 and 1856–1860 the tonnage entering and leaving port increased dramatically, the annual growth rate being close to 8 percent—much more than the increase in value of the merchandise conveyed.[49] However, in total shipping as in trade, the Pacific was still of only marginal interest, although it did account for a larger share after the discovery of gold in California.[50] During the decade preceding the Civil War it accounted for 9 percent of incoming and 13 percent of outgoing shipping through American seaports. The difference between the two figures is striking, for it seems to be in contradiction with the negative trade balance. There seem to be two explanations. On the one hand, the triangular trade routes (leading from the United States to the Pacific, then on to India or Europe, and back to the United States) were numerous enough to account for the difference, and, on the other, many vessels which sailed to the Pacific may never have returned, ending up as hulks in San Francisco, being lost in shipwreck or other mishaps, or sold in Latin America, China, or Hawaii.

At the end of the eighteenth century and at the beginning of the nineteenth, all American ships trading with the Pacific left from, and returned to, Northeastern ports. At the time, Philadelphia and New York, from which the original expeditions had set out, were very soon outstripped by New England. Salem, under the impetus of men like Darby and Crowninshields, grew wealthy from trade with China and Sumatra, while those archetypal Americans, the Bostonians, showed the flag from India to the northwest coast.[51] During the next half-century, shipping activity became increasingly concentrated on Boston, and above all on New York, which benefited from an influx of New England investors and shipowners. All that changed after the annexation of California, and the Gold Rush. In 1856, the first year for which we have port-by-port statistics, San Francisco recorded almost half of the ships arriving, and almost three-quarters of the ships departing. It was natural for it to dominate traffic along the Pacific coast of Central America and Colombia. Indeed, after 1849 the isthmus of Panama, which was controlled by Colombia, became the main transit route between the Atlantic and Pacific coasts of the United States. Passengers in a hurry to reach the goldfields boarded steamships in New York, disembarked in Colón, and after the short trip across land—shortened even more in 1855 by the construction of a railroad—arrived in Panama City to find the boats of the Pacific Mail Steamship Company waiting to convey them to the Golden Gate.[52] In other directions, San

Francisco's supremacy was less marked. Even where her strong point, which was in sailings, was concerned, she had strong competition from Boston and New York for trade with Chile and Peru, the Dutch East Indies, and even Australia. In ships arriving, Boston and New York had a healthy lead where ships from Java and the Philippines were concerned. New York would from this point on be the focal point for shipping from China, while Norfolk and Baltimore became the principal destinations for ships with cargoes of guano. However, in San Francisco the Americans had acquired direct, if still limited, access to the Pacific—an advantage their European competitors could only envy.

If the United States held such a relatively strong commercial position in the Pacific between independence and the Civil War, it owed it to the fact that this was the golden age of its merchant navy. It had the best sailors, the most experienced captains, and the most innovative builders of wooden sailing ships. This explains why it needed no Navigation Acts to protect its fleet. In 1850, for example, only a quarter of the tonnage involved in trade between the United States and the Pacific flew the flag of other countries.[53] What is more, American vessels captured a major share of the shipping between other countries. They carried Chinese laborers to Peru, and linked the opposite shores of the ocean. As a supreme affront to the Red Ensign of Britain, they were able to carry tea to London in record times.[54] At the time, the Pacific was a vast testing ground, where sailing ships were improved by leaps and bounds. The average tonnage of vessels continued to increase at an ever more rapid pace. On the China routes, it lingered between 300 and 400 tons until around 1840, and then began to grow, rising to over 850 tons just before the Civil War, with a growth rate of 3.7 percent per annum between 1834 and 1860. As they grew larger, ships also grew faster and faster. During the 1850s the Americans adored the long-distance races between the great tea-clippers, the "greyhounds of the seas." Bets were laid on which ship would be the first to arrive in San Francisco, around Cape Horn. Between 1850 and 1857 the average duration of the voyage changed very little, remaining at between 125 and 130 days, with a stable coefficient of variation of around 15 percent. A new record brought enormous prestige. In the summer of 1851, Captain Josiah P. Creesy pushed his *Flying Cloud* to the limit, covering the distance between New York and the Golden Gate in 89 days and 21 hours. He did even better in 1854, breaking the record by thirteen hours—a feat which would be equaled, and perhaps even surpassed, by

Captain John E. Williams, of the *Andrew Jackson*, in March 1860. But these were the last few glory days before the onset of a decline already beginning to become evident well before the Civil War. The average time of voyages increased between 1858 and 1859, for it was becoming obvious that greater speed was being achieved at the expense of profitability. The great clippers were no longer properly maintained. When they began to carry guano their demise was no longer in doubt. Shipowners found fewer and fewer young Yankees prepared to lead such a hard, degrading life for miserable pay when industrial growth was beginning to provide more secure and better-paid employment. The high point of the "great frontier" coincided with the records set by the American clippers for the voyage across the Pacific. Only an extremely efficient merchant navy could maintain a level of activity so widely dispersed over the ocean. With the decline of the merchant navy, which was becoming noticeable by the end of the 1850s and was hastened by the two disasters of the Civil War and the technical innovation of metal-hulled ships, a new era began, one which would be typified by shipping routes that were shorter and contained within a narrower perimeter.

TWO

The Whalers' Pacific

If it were possible to plot all the courses followed by merchant ships on a single map, it would be seen that they plied the Pacific along relatively well-defined routes. For them, the ocean was merely an expanse to be crossed as quickly as possible. The clusters of islands strung out between the opposing coastlines were of little interest to them. Whaling ships, on the other hand, wandered about the entire ocean. In search of their quarry, they sailed from north to south, from east to west, moving slowly in all latitudes, on the lookout for islands and atolls. It was they who, for the first time, treated the Pacific as a single entity, with all parts of it serving the same purpose. The immense majority of these whalers were American. As in the case of the northwest coast, it was the English who paved the way, starting in 1788, but the Yankees lost no time in following their example, and soon were more numerous. In 1791, five ships—four out of Nantucket and one out of New Bedford— sailed for the Pacific. On rounding Cape Horn, the *Beaver*, commanded by Captain Paul Worth, became the first American whaling ship to enter this new hunting ground, although on the return voyage she was a month behind the *Rebecca*, which docked in New Bedford on February 21, 1793.[1] In the meantime, other Nantucket men were working on British and French boats, owned by William Rotch, of Dunkirk. From the outset, the Americans dominated the whale hunt, although they had to wait until after the Revolutionary and Napoleonic Wars to assert their superiority. 1815 saw the dawn of the golden age of whaling, which reached its peak between 1839 and 1857, after which the inevitable decline set in (cf. Appendices, Table 5).

HUNTING THE LEVIATHAN

Whaling underwent an extraordinary expansion in the Atlantic in the eighteenth century. To find their quarry, captains were obliged to sail their vessels farther and farther into the southern hemisphere, until eventually an apparent scarcity of whales obliged them to leave Atlantic waters in search of new hunting grounds, either in the Indian Ocean or beyond Cape Horn, in the eastern Pacific along the Chilean and Peruvian coasts. It was difficult to say how many whales there were at that time in the Pacific as a whole. But we can make an estimate, using the information we possess on present-day stocks, catches, and breeding rates. Before the start of intensive whaling there were between 1.8 and 2.4 million sperm whales in the southern hemisphere and the northern Pacific. Right whales were much less numerous, at between 370,000 and 400,000, and this was particularly the case for the bowhead, or Arctic right whale, of which there were no more than 30,000.[2] The entire Pacific nevertheless represented a fabulous hunting ground for anyone daring enough to confront the hazards of the sea and the fury of the wounded leviathans.

In the nineteenth century, American whalers hunted only two species of whale. The great rorquals of Antarctica were beyond their reach, for they could swim too fast for the technical capabilities of the time. They would only take humpbacked whales if nothing else was available, for this species tends to dive as soon as it is harpooned. Until 1845 the most highly prized catch was the sperm whale found in tropical waters. A carnivorous animal with horrendous jaws, it swims either in pods led by a polygamous old male, or alone. Its blubber yields an oil of high quality but, above all, its huge head contains the spermaceti, which was the source of the best lamp oil, and of a lubricating oil essential to the machinery of the industrial revolution.[3] In addition, there might be ambergris to add to the value of the catch. The second species, the right whale, of which the bowhead is a subspecies, is larger than the sperm whale (reaching 115 feet as opposed to 65 feet). It inhabits the colder regions of the ocean, in the north and in the south, keeping away from tropical waters (cf. Appendices, Map 1). It feeds on plankton, which it filters through its enormous plates of whalebone, or baleen. It is monogamous, and has a faster rate of reproduction than the sperm whale. It is also much less aggressive. For the whaler, it provided two marketable commodities: its very thick blubber which could be refined to make lamp or lubricating oil, and its supple, elastic whalebone. The right whale would be increasingly sought after by American whalers, and after 1865 seems to have been the most profitable product of the hunt (cf. Appendices, Graph 2).

Since hunting techniques were very much the same for all the species, it was quite easy to switch from one to the other in response to the state of the market and the likelihood of profit. In the course of his peregrinations across the Pacific, a captain would move between the tropical and temperate zones. When the sperm whale seemed to be the more desirable quarry, he would not waste any time in the southerly regions of the ocean which is the habitat of the right whales, preferring the warmer waters where he would certainly be able to fill his barrels with spermaceti. This happy state of affairs lasted only a few years, however, reaching a climax between 1835 and 1845, when it was fueled by the higher price for sperm oil (cf. Appendices, Graph 3). As was required by the low-density mode of exploitation characteristic of the frontier, to maintain such a high yield it became necessary to move on continuously to new hunting grounds, and, if possible, to arrive there before the competition. Information did not remain secret for long, for the data compiled in logbooks would appear in newspapers, or be passed along by word of mouth on the docks, and even from vessel to vessel during midocean encounters. For almost three decades, the whaling captains of Nantucket or New Bedford would be content to sail the southern hemisphere, along the coast of Latin America ("onshore") or off Australia. Then, abruptly, as the nature of the demand changed, they began, in 1818, to hunt out across the ocean ("offshore"), westward along the same latitudes, from the Galapagos to the Marquesas. In September 1819, the first American whalers arrived in Hawaii. In the following year, the *Maro* discovered the so-called Japan Grounds, which actually extends from Midway to the Bonin Islands. Within two decades little remained to be discovered about the regions of the ocean frequented by the whales in their migrations, or about the areas where they were to be found at each time of year. Charles Wilkes, at the end of his account of his explorations, defined three belts, each from 20° to 25° of latitude in width, lying between the equator, the tropics, and 50° N and S:

> Within the tropics, whales are almost always to be met with. There are, however, particular places within this zone where they chiefly congregate. Whales are found in the first belt on the north side of the equator, to the southward of the Sandwich Islands, and thence westward as far as the Mulgrave island, for the greater part of the year; but the only spot or space they are known to abound at any particular season within this belt, is to the westward of the Gallipagos; they pass and repass over the rest of this space in their migrations, and may generally be found near to, or around the small islands.
>
> . . . .
>
> The third belt comprises the ocean from the coast of South America to the Kingsmill Group, including the Marquesas, Society, and Friendly Islands, the Samoan and Feejee Groups. Within these are the spaces and known as the "on shore and off-shore grounds." The latter the whalers frequent from November to February, and along this third belt they are found until the months of July and August, by which time they reach the Kingsmill and Feejee groups. There are, however, stragglers to be met with in this space during all seasons.[4]

In 1850, Matthew Maury was able to chart all the seasonal hunting grounds with even greater precision. But these advances in knowledge were unable to arrest the declining size of the harvest, provoking several controversies among the interested parties.

The apparent decline in the sperm whale stock was not a great disaster for the whalers, since the right whale was available to make up the difference. Previously, the right whale had been hunted in the southern part of the ocean, off Australia and New Zealand, or along the Chilean coast, but now the northern Pacific was discovered to be even richer in whales, yielding often 100 and sometimes as many as 200 barrels of oil. The Northwest coast of America, which had been neglected since the end of the sea otter hunt, suddenly became busy again. It appears to have been a French whaler, Captain Chaudière, of the *Gange*, who first braved the inhospitable mists of these northern latitudes in 1835.[5] The Americans wasted no time in following suit and, in 1840, the number of ships heading there increased sharply. Indeed, it was quite easy to fill the holds, for the animals seemed to lack any sense of danger. Conditions were right for a massive expansion: demand exceeded supply and prices were rising, despite the greater number of whales being taken. Whalebone, long considered to be without value, found a market at last, for it was used to manufacture whips, gentlemen's collars, umbrellas, and, above all, corset stays.[6] Feminine fashion had decreed a new standard of beauty, requiring a wasplike waist in order to show off the breasts. Beginning in 1820, the corset industry flourished in Paris, and many patents were taken out on what was a "device rather than an article of clothing,"[7] growing in number from two to sixty-four between 1828 and 1848. To shape the busts of European and North American ladies, right whales would soon be hunted throughout all the waters of the North Pacific. They were sought from east to west, off Alaska, around Kodiak, then along the Aleutians, and finally as far distant as Kamchatka and in the Sea of Okhotsk. It seems to have been there that in 1843 the *Hercules* and the *Janus* caught the first bowhead whales. If Americans were to be found in all parts of the Pacific, it only remained for them to enter the Bering Sea and venture north into the Arctic Ocean. This is exactly what happened in July 1848, when Thomas Royce of Sag Harbor, disregarding his owners' instructions and the protests of his crew, sailed through the Bering Strait and took fifteen whales in thirty-four days, enabling him to sail home at the end of August with a full hold, containing 1,700 barrels.[8] The following year the rush of whaling ships to this new, frozen Eldorado began. In 1852, 220 whalers slaughtered 2,682 whales—a record. At such a rate, this source of riches could not last much longer, and just about ten years later catches of the right whale suffered the same kind of decline as the sperm whale. So even before the Civil War it became obvious that the glory days were over. A rise in prices seemed to bring a temporary reprieve, pushing sales over the $10 million mark on four occasions between 1851 and 1857, but by 1860 they had fallen again to $6.5 million—the level they had stood at twenty-five years earlier. But Pacific whaling was not on the way out just yet, for Americans continued the hunt until early in the twentieth century, though in smaller numbers, it is true. Whaling had become a marginal activity, the mere shadow of what had been the greatest adventure in the ocean's history.

The question remains: Why did the whale hunt decline in this way? Should it be blamed on a decline in the stocks, or on a fall in demand?[9]

In the nineteenth century, people often tended to attribute it to the over-exploitation of natural resources. It was claimed that American whalers had exterminated the whales in greater numbers than would have been the case if a sound policy for resource management had been in place. A less extreme version of this argument suggests that the whales, which until then had only been hunted by the coastal Japanese and Indians when they came close to shore, quickly learned to fear a much more deadly enemy than the killer whale. They became "shy," it was said, avoiding man as soon as they became aware of his presence. It appears most unlikely that the Americans were the cause of ecological disaster for the sperm whale. Indeed, to judge from the quantities of oil brought back, and counting the animals which dived when harpooned, between 180,000 and 300,000 whales were probably eliminated in the nineteenth century, which represents between 8 percent and 18 percent of the initial stock.[10] It is true that the species has a low breeding rate, and it seems unlikely that captains would have systematically selected the adult males who yielded the most oil, but rather harpooned all the sperm whales they could, whether young, medium-sized, or large. The *Mariner*, out of Nantucket, for example, during a voyage lasting from 1836 to 1840, took 82 sperm whales, yielding a total of 1,944 barrels, i.e., an average of 23.7 barrels per whale. Seven animals alone yielded more than 50 barrels each, for a total of almost 500 barrels, whereas 55 supplied fewer than 20 barrels each, for a total of 750.[11] These statistics seem to indicate that Captain George W. Gardner took large males whenever possible, but that in general he did not scruple to harvest females and young whales. But in spite of this, there is no reason to think that there was any reckless extermination of sperm whales by the Americans in the nineteenth century. On the other hand, it seems probable that the toll of right whales was a heavier one. The same number were taken as of sperm whales (180,000), despite the fact that the population of right whales was only one-fifth the size. Also, whalers preferred females, who yielded much more oil than the males. This slowed down the rate of reproduction, in spite of a natural breeding rate higher than the sperm whale's. All in all, it seems that only two subspecies, the gray whale and the bowhead whale, suffered a considerable diminution of their stock, while for the others there was no disaster to speak of.

It would seem then that the reason for the decline was not a lack of supply, caused by an apparently questionable overdepletion of the available stock. Sometimes the declining standard of the equipment was

blamed. This certainly did not apply to the ships, which were constantly undergoing improvement by their owners, who never hesitated
to rerig three-masters to make them into more maneuverable barks,
better able to catch the bowhead whale.[12] Perhaps it was true in respect of the labor force, for, it was said, the hardy young New England
Yankees were gradually being replaced by unsavory individuals and
outsiders from tropical latitudes. However, this, too, is a questionable
argument, given that from quite early on the Nantucket whaling ships
depended on the skills of colored seamen.[13] It is, rather, to the demand
that we must look for the main explanation for the decline. The discovery of petroleum in Pennsylvania in 1859, and the rapid spread of
kerosene for lighting purposes dealt a deathblow to the much more expensive whale oil. Excellent industrial lubricants could also be produced from the "black gold." This meant that the principal product of
whaling could no longer compete. Following the period of inflation
during the Civil War, prices collapsed, with the exception of whalebone, which climbed to dizzy heights, but which was after all just a by-
product, not important enough to make it worthwhile for owners to
send out as many whaling ships as in the past. If the world had not entered the age of kerosene in 1860, whalers would probably have continued to exploit the resources of the Pacific Ocean at a level slightly
below the peak attained during the preceding three decades. Contretemps such as the depredations of the Confederate cruiser *Shenandoah* in 1865, or the losses suffered by the fleet when caught in the
Arctic ice in 1871 and 1875, would never have discouraged investment
in such profitable ventures. In any case, as Davis, Gallman, and Hutchins
have shown, profits from the Pacific were never as high as between
1876 and 1885, years when the harvest of right whale oil and of whalebone per ships' tonnage and per whaling season reached record levels,
thanks to a fortunate combination of excellent prices and a plentiful supply, as a result of decreasing competition among whalers.[14] The
golden age of whaling was indeed over, but a few nuggets still remained for the hardy souls willing to carry on.

THE VOYAGES OF THE *LAGODA*

Of all the hundreds of American whaling ships which sailed the Pacific, the *Lagoda* has been made famous by the scale model on display in
the main gallery of the New Bedford Museum. While not necessarily

typical, its story provides an excellent illustration of the different aspects of a life spent hunting the leviathan of the seas.[15] The *Lagoda* plied the oceans from 1826 to 1890, as a cargo ship initially, before becoming part of the fleet owned by Jonathan Bourne, a New Bedford agent. Dana describes the three-master in the port of San Diego in March 1835, ready to sail for Boston with a cargo of between 31 and 32 thousand hides, its "large, high forecastle, well lighted," contrasting with the dark, cramped crews quarters of the brig aboard which the author had made the voyage out. On docking, the *Pilgrim*, under its mediocre captain, Frank Thompson, had bumped the *Lagoda*, fortunately not causing much damage. Bearing no grudge, Captain Bradshaw invited his colleague to dinner on board that same evening. When the first officer announced: "Captain Thompson has come on board, sir!" he inquired, "Has he brought his brig with him?" in a tone loud enough to be heard all over the ship—a jest which became a standing joke among the crew for the remainder of the voyage.[16]

The Twelve Voyages (1841–1886)

After twelve years spent carrying freight, the *Lagoda*, like many other ships, was converted for whaling. Boilers for melting down blubber were constructed on the deck, she was equipped with five longboats for chasing whales, and a block and tackle for hoisting the catch on board. The new owner, Jonathan Bourne (1811–1889), who acquired her for $7,500, had started out in the grocery business, but at the age of twenty-five he saw he could make more money by organizing whaling expeditions, on his own account as well as for other shipowners. In this he was following the lead of certain wealthy New Bedford Quakers, originally from Nantucket, who, despite their pacifist convictions, had been making their fortunes from the slaughter of these marine mammals since the end of the eighteenth century. The superb residences along Country Street, with their view over the mouth of the Acushnet, built by William Rotch Jr., James Arnold, William R. Rodman, and Joseph Grinnell, still stand today as a reminder of the opulent lifestyle of these cultured, charitable Yankee families, with their talent for making dollars grow.[17] During more than a half-century, Bourne was to own twenty-four ships and hold a stake in twenty-two others. Counting his agents' commissions, they would earn him $2,100,000, or almost $40,000 per annum, a truly magnificent income at the time.

The *Lagoda* alone earned him almost $350,000 from her twelve voyages, and was the jewel of his fleet, despite the rivalry of other, equally profitable vessels, such as the *Hunter* and the *Northern Light*.[18]

Whaling expeditions were so risky that the owner invariably shared the costs with other willing investors. This was a temporary arrangement, involving at the most eight individuals in the case of the Lagoda, for Bourne almost always retained at least a relative majority of the shares for himself. After the voyage, the profits would be distributed in proportion to the amounts invested, or losses would be shared in the same manner. The arrangement was that initially the funds would be advanced, and remain frozen for increasingly long periods of time, as the duration of voyages increased (cf. Appendices, Table 6). Before sailing, the ship had to be fitted out with everything needed for the hunt, with food, and with articles to be sold to the crew or traded with the natives for pork, flour, or fruit (to guard against scurvy). On signing up, the crew generally asked for an advance, which they would have no difficulty in obtaining, since a heavy rate of interest would be charged. During the expedition itself, the principal expenditures would result from mishaps and storms, with damage being partly covered by maritime insurance. The owners bemoaned the increase in their initial costs as the nineteenth century progressed—not without justification during periods of inflation and of general price increases (cf. the *Lagoda*'s voyages nos. 5, 8, and 9), although it must also be remembered that the selling price of oil and whalebone usually increased at a similar rate. When the voyage was completed, the books would be balanced. On the debit side would be the drafts drawn during the voyage, the shares owing the captain and crew, and freight costs involved in sending home part of the catch on board another vessel. In the case of the *Lagoda*, these expenses were relatively low in proportion to the income, leaving quite respectable profits, with the exception of the tenth and twelfth voyages. The profits included, in addition to the proceeds generated by the hunt and the freight shipped home in advance, the numerous deductions from the amounts due to the crew. The sailors were supplied with clothing and tobacco at a 100 percent markup, and the advances made by the outfitters and recruiting agents often carried interest amounting to 40 percent for a four-year voyage. Deserters received nothing. For many of those who had dreamed of making a fortune from the whale hunt in the Pacific, the financial rewards were often most disappointing. Earnings were tiny considering the risk involved. It would have been better to have worked for a regular wage—on a merchant ship, for instance—than place all one's hopes in a share in profits which often turned out to be illusory, if not indeed a swindle.

The profit margin, as it was calculated in the nineteenth century—i.e., as the ratio of the net payment to the owners and the capital advanced before the ship's departure—seems very high in the *Lagoda*'s case. Whether real or nominal, it averaged around 100 percent per voyage, with some large variations. The most profitable years were during the Civil War and at the beginning of Reconstruction (1860–1873), when whaling went into decline. From 1873 on, the financial results were outright mediocre, while during the two decades prior to the war they were about average.[19] If general price trends are taken into account, a somewhat different ranking is obtained. The nominal profits from the seventh voyage were artificially boosted by the inflation during the Civil War, while those from the eighth voyage seem to have been hurt by the 25 percent fall in prices between July 1864 and May 1868. A better idea of whaling's real profitability can be obtained from a comparison of the real net income per day spent at sea. The longer voyages, from 1856 on (the ninth lasted almost five years), meant a larger total catch, at the expense of a lower yield per day. Using this criterion, the most profitable voyages were those preceding the Civil War, in 1841–1843, which earned $68 per day, and the two from 1850 to 1856 ($54 and $66, respectively), while the two most vaunted voyages, in 1860–1868, fell considerably short of this ($43.50 and $52). The last three expeditions still bring up the rear, with the twelfth earning a net real daily revenue of $11, only a sixth as much as the first voyage. After 1868, the decline in productivity per day shows that it was not merely a matter of bad luck in an activity always vulnerable to chance, but of an actual decline, as is confirmed by the reduction of the American whaling fleet from 230,000 tons in 1846 to 38,000 tons thirty years later.

The Seventh Voyage (August 27, 1860–April 18, 1864)

Although, as we have just seen, the *Lagoda*'s seventh expedition scarcely deserves the plaudits heaped upon it by Benjamin Baker, the historian of the Bourne Company, it was well enough documented to allow us to follow all its ups and downs, typical of a Pacific whaling ship.[20]

Two months after docking from its preceding voyage, the *Lagoda*, which had just been rerigged as a three-masted bark, was ready to put to sea again. Jonathan Bourne renewed his partnership with some old New Bedford friends: Edmund Maxfield, who had captained her on the first voyage, in for a $^2/_{16}$ share, and Lemuel M. Kollock, for a $^1/_{16}$ share. Bourne kept a majority share ($^9/_{16}$) for himself.[21] The firm of Taber, Read and Company, the whaling agents, acquired the $^2/_{16}$ share previously taken by Otis and Edward N. Seabury.[22] Finally, the captain, Zebedee A. Devoll, also took a $^2/_{16}$ share—an incentive for him to make sure the hunt would be profitable. In July and August 1860, Bourne, acting as agent in addition to being the owner, purchased the material and equipment required for a lengthy voyage of over three years, skimming off a commission of 2.5 percent in the process. He recruited a fresh crew of 26 men, with whom he discussed the share of the proceeds they would receive after the voyage. In all, 32.3 percent would go to them, the other two-thirds being reserved for the shareholders. If the captain and officers were well provided for, the harpooners had to content themselves with $^1/_{90}$ of the net proceeds, and the inexperienced sailors, the "green hands," with $^1/_{190}$, or 0.52 percent.[23] By 1860, indeed, the days were over when the majority of hands came from Nantucket, an island so poor in natural resources that its male inhabitants had no employment opportunities apart from whaling. But neither did the *Lagoda*'s crew resemble the Tower of Babel that Melville describes in *Moby Dick*. The list indicating the origins of twenty-one of the crew shows a strong representation (ten) from the north of New York State, a substantial minority of New Englanders (eight), and two from Mississippi.[24] There were very few Portuguese from the Azores or the Cape Verde islands on board when the ship left port, since it was preferred to hire them en route, while natives of the Pacific—all called Kanakas, whatever their origin—were likewise added to the crew after the ship entered that ocean. Another characteristic of this crew was its youth, for, excluding the officers, the average age was twenty-one. Taking their geographical origins into account, this means that a number of those attracted by the venture, or persuaded to sign on, lacked any experience of the sea, not to speak of the risks involved in whaling. When the modest sum of $65 was paid as an advance to these "green hands" it must have seemed to them a foretaste of the fabulous profits and marvelous adventures they would have on the other side of the world, with free board and food and lodging thrown in—a dream from which they would have a rude awakening after a few days of seasickness and living in discomfort before the mast.

In her trip around the globe, the *Lagoda* first headed for the Cape of Good Hope, stopping only in the Azores to sign on four Portuguese, whose experience was considered extremely valuable. In the South Atlantic, fortune smiled on her. In the latitudes of the cape she captured three sperm whales and nine right whales in the space of six weeks, between November 17 and December 31, 1860. It was a very busy time, for the carcasses had to be cut up, the blubber boiled down, stored in barrels, and stowed in the hold, after which the whalebone had to be cleaned and tied up in bundles. Once the area had been combed, it only remained for the *Lagoda* to let herself be carried along by the westerly winds, along the roaring forties, to cross the Indian Ocean in a single stretch, reaching the western coast of Australia on February 21. At Bunbury, some of the crew took French leave, or attempted to desert—a sign that they were finding it difficult to cope with the long months at sea, out of sight of land, with alternating periods of intensive labor and idleness.[25] Captain Devoll had no alternative but to put out to sea as soon as possible and sail along the Australian coast. Then bad luck intervened, for in five months only a single whale was captured, and all Devoll could do was watch the rorquals and humpback whales, which were too fast for the longboats. Intending to take on fresh provisions and water, he reached the southern extremity of Java in August 1861. Five men took advantage of the opportunity to make off in a rowboat under cover of night. Devoll's efforts to bring them back were in vain. On shore, he contracted a dangerous fever. The logbook entry for September 27 contains these simple words:

> This day commences with light winds from ESE heading South middle part the wind and weather much the same some time during the night Capt ZA Devoll died as he was found dead in his berth in the morning. Latter part light winds heading SSW. So ends. Lat. 21°53' S Long 98°46' E.[26]

The first lieutenant, E. H. Cranston, took command, backtracking across the Indian Ocean to Mauritius to unload the whalebone and Devoll's remains, which would be sent home via England. In port, two more men jumped ship. On the way back to Australia, on November 28, Cranston chanced on a pod of sperm whales, and two were taken, yielding 155 barrels, while two others, hit by the harpoon-gun, were lost. Then came a lengthy period during which nothing was caught. It was time to move on. The *Lagoda* set off for New Zealand. At Monganui, near the famous Bay of Islands, Cranston took on food and water. Four new deserters were handed over by the natives, but the captain preferred to leave them in prison and take on three new crew.

After a stay of a month, on March 11, 1862, the whaling ship set out for the northern hemisphere under full sail. She took two months to reach Kamchatka, after a short stop in Samoa to take on supplies and put ashore a black member of the crew who refused to work, in spite of the lash. From May 19 to October 20 she cruised in the Sea of Okhotsk, in search of right whales. The Bay of Shantar was a prime hunting ground, but there was a lot of competition, and the longboats were lowered into the fog and ice for nothing. In the log, Second Officer Vanderipe noted: "whales scarce and hard to strike." It seemed easier to catch a deserter than a whale. At these extreme latitudes, the news from home was already a year old, so it was there that they learned about the outbreak of the Civil War.[27] The season was a total disaster: just three whales, one a dead one, and a rorqual yielding only eight barrels. Winter fell early, and there was much sickness. They were obliged to sail to Hawaii, where all the officers were replaced—probably for incompetence—as well as part of the crew. The new captain, Abraham W. Pierce, shipped the oil home, got the vessel back into shape, and set sail, on Boxing Day, for the new 1863 season, spending the winter in the tropics and the summer in the Arctic. Along the equator not a whale was to be seen, but as they headed for the Mariana Islands things improved somewhat, for they took a female and a humpback. At Yokohama, which was now open to foreigners, Pierce made preparations for the summer hunt, which would decide the success or failure of the voyage. Putting to sea on April 24, he headed north toward Kamchatka. May 13 to September 20 was spent in the Bering Sea and the Arctic Ocean, in the company of several other American vessels, cruising along the edge of the pack ice as it shrank back above 71° N. It was possible to hunt all day long, in sunlight or in the midnight fog. After a disappointing beginning, fortune finally smiled. On July 15 the first whale was taken, and ten more were harpooned later, making these northern summer months a time of exhausting labor. With the hold well filled, they were able to make a relatively early return to Hawaii, avoiding the risks brought by the early onset of winter, which would destroy twenty-nine ships in 1871, and thirteen in 1876.[28]

The time had come to think of returning to home port. The Hawaiians, who had signed on for a year, were let go. A new captain—the fourth—took command of a crew of twenty-one, four of them new.[29]

The voyage would be uninterrupted, with the *Lagoda* taking five months to cover the distance between Honolulu and New Bedford, rounding Cape Horn. The voyage was completed on April 18, 1864, having lasted 1,331 days, or more than three and one-half years, and providing the owners with some cause for satisfaction.[30] On the debit side, the working expenses amounted to $44,153, the principal items being the captains' expenditures during the voyage ($13,527), the shares due to Captains Pierce ($9,661, thanks to his profitable time in the Arctic), and Devoll ($3,897, due to his success in the Atlantic), and the share of the profits due the officers and those members of the crew who had not deserted ($13,160). Out of the crew of twenty-three who had set out, only eight were entitled to anything at all,[31] the best provided-for being the ship's cooper, with $1,802, although he was actually paid only $1,217 after deductions were made for the commission owing to the agent who had hired him (with added interest of 37 percent), for the sums of money he had sent home, and for goods purchased from the captain ($145). So, when he disembarked after three years and eight months of service, he was left with $1,217 in his pocket—and heavily devalued dollars at that, for they had lost 45 percent of their purchasing power since the *Lagoda* had left port. When the voyage was over, he had earned the equivalent of $186 per annum in 1860 dollars, while, by comparison, the average pay of a nonagricultural worker at the same period would have been $363 per annum.[32] What then can be said about the treatment of the unfortunate William Barker, described, it is true, as a "good-for-nothing," who saw his share of $586 reduced to a payment of $383, or $58 per annum, in 1860 dollars? At that rate, he would have done much better just to have joined up for the return journey, like the Portuguese, avoiding the risks involved in the hunt, and being paid a fixed wage of $12 per month, amounting to $32 (in 1860 dollars) for less than five months work![33]

If there was no bonus for accepting the physical risk of the whale hunt, the financial risk, in the *Lagoda*'s case, was rewarded handsomely. The catch was quite respectable, and prices had reached previously unknown levels, even when expressed in devalued dollars: $1.67 for a gallon of sperm oil, $1.15 for a gallon for ordinary whale oil, and $1.44 a pound for whalebone.[34] In the credit column, the total came to $141,312, leaving a net profit of $97,159, of which Jonathan Bourne alone would pocket $54,652, thanks to his majority share—i.e., almost one hundred times more than the unfortunate Barker or his fellow ordinary seamen. The *Lagoda* was able to sail again two months later for a new expedition under the orders of Captain Charles W. Fisher.[35]

The social inequalities so evident in the large cities of the United States just prior to the Civil War were even more obvious in the Pacific whaling business. In addition to the disparities between the forecastle and the quarterdeck—social and economic differences, to which was added the crew's subjection to the captain's sometimes tyrannical authority—there was an equally marked contrast between the shipowner's office and the ship. The Pacific whale hunt was indeed an element of the capitalist mosaic beginning to spread over the entire surface of the globe by the middle of the nineteenth century.

AMERICAN WHALERS AND THE PACIFIC

The American presence throughout the entire Pacific was never more considerable than in the glory days of the whale hunt. Usually, between 1839 and 1857, more than one hundred ships set sail from northeastern ports each year. In 1848 and 1851, the actual number was over two hundred (cf. Appendices, Table 5). As the expeditions lasted longer and longer, particularly after the discovery of the rich stocks in the North Pacific, it was in the Pacific that each whaling ship spent an ever-increasing proportion of its time.[36] The *Lagoda*, for example, during its voyage in 1860–1864, spent exactly two years there, or 55 percent of the total time at sea. As a rough estimate, we can consider that in any one of the years when the hunt was in its prime the number of whaling ships in the Pacific was equal to the number leaving port during two consecutive years. The record was set in 1844–1847, with more than three hundred vessels, and in 1852. This would have made a total of about 10,000 men if we estimate a crew of thirty on each vessel. Of course, not all of these were Americans. According to the ships' articles from the Pacific, there were numbers of Portuguese and Polynesians. However, until the Civil War, United States nationals made up the majority— sons of New England and youths from the north of New York State, but also African-Americans, who much preferred the tough discipline of shipboard life, where everyone was treated the same, to slavery on some southern plantation. It seems to have been typical that, in the famous story of the *Essex,* of Nantucket, sunk by a sperm whale in 1820, six of the crew of twenty were black, all of them fated to be eaten by the white survivors, thus turning the stereotypes of cannibals current at the time on their heads!

What part was played by these thousands of men? Did they merely ply the ocean without leaving any more lasting traces than the wake of their three-masted ships? Or did they have some influence on the course of Pacific history?

The first thing worth mentioning was the contribution they made to knowledge. Before the American whaling ships began to penetrate every region of the ocean looking for their prey, hundreds of islands and atolls had remained unknown to Western mapmakers. Thanks to the information registered day after day in ships' logs, these gaps were filled in, and errors were corrected. Even mishaps and shipwrecks contributed to a more detailed knowledge of shorelines and of the ocean's depth. Even today some islands still bear the names of whaling captains from Nantucket and New Bedford, such as Johnston, Jarvis, Starbuck, Baker, and Howland, while in other cases the ancient Polynesian names have been restored. These discoveries would serve to justify the American government's claim to many tiny atolls in the central Pacific. In 1856 the guano law laid claim to forty-eight islands from which, for the most part, no guano was ever harvested.[37] At the Naval Observatory in Washington, from 1842 on, Matthew Maury, himself prevented from going to sea by an injury, compiled all the information required to make charts indicating the winds and currents in all the oceans. Thanks to him it became possible for ships' masters, instead of navigating rather blindly, to assess the likelihood of gaining some time by taking a certain route at a particular time of year. One achievement led to another. Maury collected more and more information during the 1850s, laying the foundations of the new science of oceanography. The clipper ships owed him a considerable debt when setting their records. The Pacific, about which people had until then only the vaguest of information, lost much of its mystery.[38] Without the American whaling ships—the only vessels to venture outside the major commercial shipping lanes—it would have taken much longer to fill in so many gaps in knowledge.

More than any other commercial activity, whaling left many sailors behind it, all along the shores of the Pacific. Some were deserters, while others, falling ill, were abandoned by captains wanting to unburden themselves of useless mouths to feed. Nearly every ship suffered a loss of crew. When Captain Peter Cromwell, of the *Cortes*, returned from two successive voyages with his crew virtually intact, having lost only two men on the first voyage and three on the second, he was hailed as a hero.[39] There were many reasons for desertion. Sometimes the ordinary crew members complained of the excessive demands officers made on them, and of their brutality. One, Stephen Curtis, made a detailed catalog of the abuses committed by Captain Fordyce D. Haskell, of the *Mercury*, and which, according to him, were worthy of the Spanish Inquisition. They included striking, a sadistic use of the lash, throwing heavy objects at men's heads, and hurling insults intolerable to Christian ears. It was hardly surprising that many of the crew jumped ship in each port of call, and that the writer was overjoyed to return to the United States, "that land, where we could once more taste the sweets of liberty—liberty dear indeed to us, having tasted, to the very dregs, the bitterness of voluntary slavery for a period of three years and more."[40] Other chroniclers, weary of lengthy "off-shore" voyages, speak, like Melville, of the boredom of life on board:

> Six months at sea! Yes, reader, as I live, six months out of sight of land; cruising after the sperm-whale beneath the scorching sun of the Line, and tossed on the billows of the wide-rolling Pacific—the sky above, the sea around, and nothing else! Weeks and weeks ago our fresh provisions were all exhausted. There is not a sweet potatoe left; not a single yam. . . .
>
> Oh! For a refreshing glimpse of one blade of grass—for a snuff at the fragrance of a handful of the loamy earth! Is there nothing fresh around us? Is there no green thing to be seen?[41]

Even more than the tyranny of the *Dolly's* (alias the *Achushnet's*) "Lord of the Plank," it was the perspective of a lengthy voyage—one which would actually last almost four and one-half years—which persuaded the young Melville to take his chances among the cannibals of the Marquesas rather than die of boredom of further long months before the mast, waiting for a sperm whale to be considerate enough to offer itself up to its hunters.[42]

The whaling ships abandoned their human jetsam by the hundreds along the shores of the Pacific, not only in all the ports of call—Talcahuano, Valparaiso, Paita, Tumbes, or in the ports where they would put in to allow the crew some liberty time ashore, such as the Bay of Islands, Tahiti, or Hawaii—but also in every archipelago or island scattered across the vast ocean. They were to be found in the most unlikely spots in Polynesia and in Micronesia—even in uninhabited island groups like the Galapagos, and, more rarely, in Melanesia. Some were quite prepared to return to sea again if it meant they would see their native land once again, while others made themselves at home in the native societies, where their technical know-how and their usefulness as interpreters endowed them with a prestige they could never have achieved in democratic America. But their presence would break asunder the old, traditional worlds of the Pacific societies. The diseases they often carried, against which the natives had no immunity, decimated entire populations, initiating the decline of civilizations which had flourished since the migrations of the first millennium of the modern era. These Pacific castaways—the original "beachcombers"—whose presence helped to bring about cultural change and destabilize the native societies, were, for better or for worse, part of the litter left behind by the eternal restlessness of modern society, a litter which did have its effect on the environment, though in many cases a delayed one.

Without the whalers, the Hawaiian islands might never have come under American domination. When the brief sandalwood era came to an end, it might have marked the end of any active relationship between the little Polynesian kingdom and the Western world, if the rich resources of the northern Pacific waters had not been discovered at precisely that moment. When, in the 1840s, the main focus of the whale hunt shifted to this region, Honolulu, Lahaino, and Hilo became indispensable as the only ports in the northern hemisphere where sailors could spend some time relaxing between the summer season in northern waters and the winter season spent around the equator. In the wake of the whaling ships there soon came entrepreneurs specializing in ship's supplies and even building dockyards for repairs. The Hawaiian economy became one more cog in the machinery of the international market and, since the great majority of companies, such as Eliah Grimes and Company or Pierce and Brewer, had come there from the United States, ever closer links were forged with America. These were even further reinforced when, in 1846–1848, the United States finally acquired its own Pacific seaboard, in Oregon and California.

It was because of the whaling industry, just as much as by reason of its other commercial interests, that the United States Navy was led to show the flag in the Pacific, to provide protection for its citizens. Lagging somewhat behind in economic involvement, but as an inevitable consequence of it, the United States was shortly to make its appearance on the Pacific stage as a political power.

THREE

Ambition and Modesty

— At the end of the eighteenth century the Pacific held no immediate interest for the American government. The country lacked a Pacific seaboard, and only an insignificant number of vessels flying the star-spangled banner sailed the Pacific, heading to Canton, to the Northwest coast, or hunting sperm whales "onshore." In this marginal zone situated on the farthest fringe from the Atlantic center, the federal authorities gave a free hand to private entrepreneurs ready to take a risk. The most they were prepared to do was appoint a consul in China, and the first to occupy this post, Samuel Shaw, during his second term of duty between August 1786 and January 1789, did little but look after his own business interests and devoted very little time to representing his country. During the nineteenth century the situation gradually changed. Economic interests—trade and whaling—assumed increasing importance. In purchasing Louisiana from Napoleon Bonaparte on April 30, 1803, President Jefferson was opening a window to the future, though its limits were ill-defined and liable to be contested by the Spanish, the British, and even the Russians. Before signing the agreement, he suggested to Congress, on January 18, that an exploratory expedition should be sent out, "even to the Western Ocean," to parlay with the Indian tribes and lay a foundation for trade. The suggestion was supported by the argument that it was necessary to offer some compensation to the fur traders who, under the new Indian policy, would henceforth have to compete with government trading posts.[1] Once Louisiana had been acquired, the project took on greater importance. At the head of fifty or so men, Meriwether Lewis and George Clark left St. Louis in the spring of 1804, traveling up the Missouri, crossing the Continental Divide, and made their way down toward the Columbia, to reach "this Ocean, the object of all our labours, the reward of all our anxieties," in mid-November 1805.[2] After spending the winter in the rain and fog of the coast, they started out on the return trip in March 1806, arriving in Saint Louis in September. The Americans were now in a good position to lay claim to the Pacific Coast, which until then they had had to be content with approaching from the sea. But could such a modest beginning justify high hopes for the future? For these to be realized, six decades of sporadic efforts would be required, during which the visible hand of the "manifest destiny," more than market forces, would finally give history its direction.

THE PRINCIPLES OF AMERICAN POLICY

The almost complete absence, until the end of the eighteenth century, of any links between the opposite shores of the ocean had made the Pacific one of the most peaceful regions on the planet, spared anything resembling the wars continually breaking out on the continent of Europe, from where it spread to the areas of the globe colonized by the great naval powers. Only a few rare military or naval clashes marked the history of what, in the absence of any combatants, appeared to be a zone of international peace. In the nineteenth century, on the other hand, the Pacific would emerge from its isolation, to be gradually drawn into the maelstrom of rivalries among the world powers. When this happened, the danger was to come less from the original inhabitants than from the appetites of the nation-states coming into being on its shores. In attempting to impose some order on this emerging chaos, the United States, while reluctant to conclude alliances with foreign countries, was led to sign an increasing number of treaties with the republics or kingdoms situated along the shores of the ocean (cf. Appendices, Table 7).

The primary objectives were to ensure the United States would have a gateway to the Pacific, and to settle any remaining doubt about the limits of the recent Louisiana Purchase. No way having been found to divide up the Oregon territory, the agreement of October 20, 1818, established a joint Anglo-American occupation between the 43rd and 54th parallels. Joint occupation based on a compromise between two dynamic powers rarely turns out to be a lasting solution. Fortunately, the heyday of the fur trade was already over, and the interests at stake were too insignificant to justify any confrontation. So, in 1827, the 1818 agreement was renewed for an indefinite period, each party having the right to rescind it on a year's notice. All went well until the beginning of the 1840s, when this makeshift solution—which in any case had only lasted because the population was so sparse—suddenly collapsed. Long columns of American pioneers traveled along the Oregon Trail, across the prairies and over the Rocky Mountains in search of new land to clear. They settled in the valleys, where they demanded democratic institutions similar to those in the states they had come from, and encroached on the commercial interests of the Hudson's Bay Company. The United States, where dislike of the English was never far below the surface, enthusiastically took sides with this new territory, which clearly seemed destined to advance American political ideals right to the shores of the Pacific. This was particularly true of the Democratic party, whose expansionist ideology was inherited from its agrarian roots. Early in 1846 Democratic newspapers were aggressively trumpeting the slogan: "Fifty-four forty or fight!"—an explicit reference to what it was felt should be the territory's northern frontier. Tension with Britain rose, but no blood would be shed, for both sides ultimately accepted a compromise. The English, who were not very numerous, gave up very little by ceding the southern part, which was relatively poor in furs, while it was out of the question for the Americans, who had been engaged in an open war over Texas with Mexico since April 25, to fight a war on a second front. So, a treaty signed on June 15, 1846, terminated the joint occupation of Oregon. Henceforth, the 49th parallel would separate the United States from British North America, leaving the former in possession of the Columbia Valley and Puget Sound, the two most important regions from an economic and strategic point of view.[3]

In the negotiations over Oregon, only the region lying between the Columbia River and the 49th parallel became the object of any real dispute, for the English made no claim to the zone lying south of the river. From there, southward to California, the United States would have to contend with Spain if it wished to establish a clearly defined frontier. The visionary imagination of John Quincy Adams, President Monroe's secretary of state, had led him, in January 1818, to press for a border extending all the way to the Pacific, so as to ensure an outlet for his country's ambitions. The invasion of Florida by General Jackson increased pressure on the recalcitrant Spanish: their obduracy would encourage the United States to recognize the independence of colonies which revolted against the mother country. By the end of the year Madrid had accepted the idea of a border running all the way across the continent, from the Gulf of Mexico to the Pacific. The signing of the Anglo-American agreement on Oregon gave Washington an even stronger hand. Adams's preference was to make the 41st parallel the frontier, but Monroe, who was more interested in Florida and what was to become Louisiana than by any unknown quantities in the West, agreed that the border should run through the source of the Arkansas River, i.e., along the 42nd parallel. For the first time, the treaty signed on February 22, 1819, gave the Americans undisputed and exclusive right of access to the Pacific seaboard, while Spain was obliged to abandon for once and for all its designs on the Columbia basin.[4]

Treaties with Latin America

At the very moment when these agreements with the British and Spanish rivals were being concluded, the Spanish empire in the Americas was beginning to crumble. New independent states were taking shape along the Pacific coast and gaining recognition by the United States, after a brief delay.[5] The "Monroe Doctrine," formulated by Adams, and expressed by the president in his speech to Congress on December 2, 1823, rested on two principles. The first was that of noncolonization, prohibiting the future creation of any new European colonies on the American continent, and the second the principle of nonintervention, based upon the fundamental difference between the political systems prevailing in the two hemispheres, one being essentially republican and the other monarchist. This meant that the United States was ruling out any involvement in European alliances, but in return expected the great transatlantic powers to give up any intention of imposing their principles on the newly independent nations of the New World. Furthermore, in a note addressed to the czar's representative, the American government dismissed the possibility that ownership of any part of the American continent could be transferred from one European colonial power to another—thus, for example, preventing Restoration France taking over the torch from an ailing Spain. As a unilateral American declaration, the Monroe Doctrine certainly had little effect when it was first formulated. This was because Adams, whether out of anti-British prejudice or presidential ambition, did not wish to appear to be "a cockboat in the wake of a British man-of-war," and was opposed to any joint declaration. But, in fact, the independence of the former Spanish colonies was better safeguarded by England's desire to block any initiatives by the Holy Alliance, and to reach an agreement to that effect with France, than by the forceful, peremptory terms of the Washington declaration. Nevertheless, the United States was laying down rules of conduct for the rest of the world which, although they would be forgotten until President Polk revived them in 1845, demonstrated its ambition to expand across the entire continent as far as the Pacific, and showed its legitimate concern for the future of its sister republics in Latin America.[6] Since it applied specifically to the "western hemisphere," the Monroe Doctrine implied that the Pacific was not viewed by the Americans as a single strategic unit. The eastern shoreline, from the Bering Strait to Tierra del Fuego, was seen as one distinct region, with its own political principles and a certain common destiny, in contrast with the archipelagoes and the western shoreline which, rather than a Far West, was in fact a Far East—the back door of the Old World. So the Pacific was viewed as a barrier, a political bulkhead between two different universes, each characterized by its own rules of behavior—a conception which is still alive today and which makes it impossible to speak of a Pacific policy in the way one can of an Atlantic one.

The contrast between the two shores of the ocean emerges clearly from an examination of the treaties signed. Those with Latin American countries contained reciprocal obligations, whereas the Asian empires generally made unilateral concessions to the Americans.

Among the new nations of Latin America, the United States signed agreements with Great Colombia and with the United Provinces of Central America shortly after they won their independence. Mexico and Chile had to wait a little longer, since the American envoy, Joel Poinsett, had upset the Mexicans by some untimely interference in their domestic politics, while in Chile the conservatives had little sympathy for the *gringos'* democratic ideals. The breakup of the former political entities made it necessary to sign numerous "treaties of peace, amity, commerce, and navigation," but the same basic provisions appeared in all of them, many taken from the 1778 treaty with France.[7] The "most favored nation" clause, allowing the signatory of a treaty to benefit from all the advantages granted a third party, was an encouragement to trade and a curb on mercantilist temptations. It also helped create a homogeneous trading region in an area where closed national borders had hampered the free movement of trade. The principle of freedom of trade and navigation forbade any discrimination between foreigners and nationals, except where cabotage was concerned. It did not necessarily imply the establishment of free trade, and indeed was quite compatible with protectionism on condition it was enforced in the same way by each country, the emphasis being put essentially on equality. An American doing business in a Latin American country was entitled to the same legal rights as any citizen of that state. Furthermore, the recognition of freedom of conscience was an invaluable step forward for American Protestants resident in Catholic countries. It is hardly surprising that the most numerous provisions dealt with shipping, for at that time trade was exclusively carried on by sea, and Yankee sea captains were frequent visitors to the ports along the western seaboard of the hemisphere. Ships seeking shelter, and shipwreck victims, were to be given assistance. Goods taken by piracy were to be returned to their legitimate owners. In case of a war with a third power in which one of the signatories remained neutral, the principle already formulated in 1778, according to which "free Ships shall also give freedom to goods," was to remain in effect: the national flag also covered the property on board, and the freedom of the seas was guaranteed. The right of inspection at sea was strictly regulated. In case of war between the two signatories, their nationals were to be given the opportunity to settle their affairs without any fear of confiscation. Consuls with diplomatic immunity were to be located in the principal ports and political centers, to ensure that these provisions were properly applied, and the authorities of the country were to provide them with assistance in arresting deserters.

All of these stipulations intended to promote trade worked much more to the advantage of North Americans than to the countries of Central and South America, for they alone possessed a merchant fleet capable of operating in both the Pacific and the Atlantic. This kind of international agreement to eliminate discriminatory practices always favored the more technically advanced partner, although it was not necessarily a hindrance to the development of more backward nations as long as they had the will to reduce internal obstacles to progress, and if it was feasible for them to do so. In certain cases, behind the apparent even-handedness which seems to have been characteristic of the language of these treaties, the imbalance of power is obvious. For example, Article 35 of the Bidlack-Mallarino Treaty with Colombia (December 12, 1846), dealing with the right of way across the isthmus of Panama, gave equal status to Americans and Colombians, and, in return, the United States, while recognizing the sovereignty of the country (called New Grenada at the time), guaranteed the complete neutrality of the isthmus and gave a pledge that freedom of passage would not be denied. This clause became even more advantageous after the annexation of California, so that it appeared not only in the 1853 Treaty with Mexico concerning the Isthmus of Tehuantepec but also in the agreement with Honduras in 1864 dealing with the possible construction of a railroad linking the two oceans. It was essential to the United States to avoid the long voyage to San Francisco around Cape Horn, and it was also most certainly to the advantage of the English, who had important trade interests in Latin America and who had access to the Pacific through what would later become British Columbia. Wishing to forestall any imperialist intrusion by perfidious Albion, the Americans succeeded in negotiating the Clayton-Bulwer Treaty on April 19, 1850, under which neither of the two powers would "maintain for itself any exclusive control" over a ship canal, nor "erect or maintain any fortifications commanding the same," nor "exercise any dominion over Nicaragua, Costa Rica, the Mosquito coast, or any part of Central America." Once the canal was constructed, the United States and Great Britain were to guarantee its neutrality and uninterrupted operation, without any discrimination in tolls, "for the benefit of mankind on equal terms to all." The two powers also under-took to support any private company prepared to make a capital invest-ment in such a worthy undertaking. By tying England's hands in this way, the United States was assured the right to exert control over the future of the Central American isthmus, across which there were several possible routes. A half-century later, when the balance of power between the two nations had changed, they in turn would find their hands tied, so that if they were to build the Panama Canal on their own they first had to enter into the sensitive negotiations resulting in the Hay-Pauncefote Treaty of November 18, 1901, which nullified the Clayton-Bulwer agreement.[8]

Treaties with Far Eastern Empires

The United States signed its first treaty on the far shore of the Pacific with the kingdom of Siam. In 1833, nine years before China was opened to trade, Edmund Roberts won from Bangkok the freedom to trade without prices being set by officials. Only the import of arms and opium, and the export of rice, were forbidden. Tonnage duties would no longer be arbitrarily fixed, nor could American debtors be reduced to slavery to pay off what they owed. The marked difference from the Latin American treaties is quite obvious, in that the Asian kingdom had not obtained any reciprocal concessions.

China followed Siam's example when, on July 3, 1844, it signed the Treaty of Wanxia with President Tyler's envoy, Caleb Cushing. This was not as a result of China's defeat by the English, but because it still wanted to open its doors no more than was absolutely necessary. The Peking government, as it had done in the past, felt no need to trade with foreigners, and saw no advantage for its subjects in gaining free access to the American market. In its negotiations with China, the United States was following in British footsteps, and was given an even greater incentive to accept the model of the Treaty of Nanjing [Nanking] when Qiying, the Chinese negotiator, wishing to prevent the formation of a Western coalition, offered it favored nation status, of his own accord, as a way to put the "barbarians" in competition with one another. The rules governing trade were modified profoundly. The authorization to trade freely through five open ports, the application of new uniform customs duties, the granting of the right to employ interpreters and compradors, and to acquire land to construct storage facilities put an end to the hongs' long-standing monopoly and to the mandarins' extortionate practices. In return, in the only concession made to reciprocity, the Chinese government would no longer be held any more responsible than the American for debts incurred by their respective subjects. The appointment of consuls enjoying equal status with the top Peking officials at last put an end to the illusion of superiority cultivated by China. The Treaty of Wanxia included several new provisions which make Chinese historians consider it even more unfavorable. There would be no change in customs duties without prior consultation with the United States. Americans were given the right to build hospitals, churches, and cemeteries in the open ports, to hire Chinese language teachers, and to buy books. Above all, Cushing obtained the right to extraterritoriality, meaning that Chinese found guilty of crimes against Americans would be punished under Chinese law, while American citizens committing a crime in China would only have to answer for their actions to the American consul. This was an undue privilege, for which the Yankee traders in Canton had never asked, and which there could never be any question of extending to Chinese nationals living in America. There was, however, one important exception, in that Americans involved in the opium trade were to lose all right to protection from their government and be subject to the full rigor of Chinese law, while the United States was to take steps to prevent the misuse of their flag by British subjects trafficking in drugs. In this, the moral influence of Protestantism tempered the overweening self-righteousness of the Westerners, all too confident of the superiority of their own laws.[9]

However, these good intentions did not last very long. In reaching a second agreement with China, the Treaty of Tianjin (June 18, 1858), President Buchanan's envoy, William B. Reed, managed to eliminate any reference to the opium trade, taking advantage of the Peking Courts' eagerness to win the good graces of the United States, for it hoped that they would act as mediator in its dispute with Great Britain and France. The Shanghai Convention, signed five months later, permitted the unhindered import of narcotics, henceforth considered a commodity like any other, subject only to a duty of thirty taels per picul. China gave up its freedom to set tariffs, for customs duties were covered by the treaty. Also, the Americans acquired additional privileges. They were allowed to reside in the capital, Peking, for only short periods at first, but from 1860 on were able to do so permanently. Additional ports were opened to trade. Christianity was characterized as a religion teaching men to do good and to behave toward others as they would have others behave toward them, and any person, whether American or Chinese, who professed these principles in an orderly way would be protected from harassment or persecution.

During the decade preceding the Civil War, the empires of the Far East were subjected to increasingly strong pressure to renounce their traditional policy of a partially or entirely closed door. The United States was no longer satisfied to tread in England's footsteps. In its turn, it took the initiative toward Japan, whose place in the northern Pacific was too important to allow it to be overlooked, particularly at a time when it was the principal hunting ground of American whaling ships and when trade was developing in Hawaii. The motives invoked to encourage opening Japan were humanitarian, economic, and religious. There was a need to ensure the protection of shipwreck victims from the harsh treatment inflicted on them by the shogun's regime, to include a population of almost 40 million in the sphere of world trade, and to resume the activities of the Christian missions, which had been interrupted in the seventeenth century. After several abortive attempts of a private and public nature (by the Olyphant Company of Canton in 1837 and by Commodore James Biddle in 1846) to make contact with the Japanese in Tokyo Bay, Commodore Matthew C. Perry was put in command of a formidable armada, considered the only thing likely to force compliance from those who held power under the shogun. In the course of two voyages, in 1853 and 1854, during which he displayed both firmness and diplomacy, Perry managed to make a definitive breach in Japan's defenses, without needing to fire a single shot—unlike the English in China a few years earlier. This is also why the very succinct Treaty of Kanagawa, signed on March 31, 1854, contains very few concessions to the Americans. Two ports were opened, one, Shimoda, not far from Tokyo, but in a dead end, the other, Hakodate, useful only to whaling ships en route for the Sea of Okhotsk. It granted the right to travel within a radius of seventeen miles, and required that Japanese officials be used as intermediaries in making purchases. There was to be a consul at Shimoda, and assistance was to be given to shipwreck victims. All in all, this was a modest result, although significant when one considers the situation which had prevailed for the previous two centuries. But the Tokugawa shogunate would never recover from the shock, and Consul General Townshend Harris would have little difficulty adding to these early advantages. The Shimoda Convention (June 17, 1857) consolidated the rights of American citizens. Nagasaki was opened, Americans were allowed to take up permanent residence, and a more equitable rate of exchange was introduced. Extraterritoriality was granted, as in China. The treaty of friendship and trade signed at Edo in the following year, on July 29, 1858, granted religious freedom, allowed diplomats free access to the capital, provided for seven more ports to be opened before 1863, including Tokyo and Osaka, eliminated the requirement to trade through official intermediaries, and established a genuine freedom to trade with nondiscriminatory customs duties and clear rules. Only the opium trade was barred. For its part, Japan obtained some advantages: an exchange of consuls, the right to purchase warships from the United States, and to hire American military experts and engineers. All the Japanese concessions, which were immediately demanded by the European powers under the most favored nation clause, generated an outbreak of xenophobia throughout the country. The Shimonoseki incidents in 1864 resulted in Japan having to pay a heavy indemnity (the American share amounted to $785,000, handed back in 1883) and then, two years later, losing its freedom to set duties, with a 5 percent ad valorem tax being imposed. During these few last gasps by Japan, the actions of the United States were in no way different from those of its English, French, and Dutch colleagues, for its value system and their interests were much too similar to allow it to act independently. The spirit of the Monroe Doctrine applied to only one side of the Pacific.[10]

THE TWO NAVAL STATIONS

Whereas whaling ships did not hesitate to ply the entire ocean in search of higher profits thus bestowing a certain degree of economic unity on it in spite of the immense distances involved, diplomats and their secular arm, the United States Navy, had a more compartmentalized vision. For them, the ocean was divided into two zones, almost entirely distinct from one another. On the one hand, there was the so-called "Pacific," corresponding essentially to the Latin American waters, and, on the other, the East Indies or Asia, with its center of gravity on the Chinese coast.

This lack of unity arose primarily from the way the use of the fleet was envisaged. The high command saw no advantage in concentrating resources in the North Atlantic to defend the coast of America and be in a position to strike at the heart of an enemy's position. They followed England's example in gambling that there would be no more wars between them once the Treaty of Ghent (1814) had been signed. Consequently, the few naval vessels in sound condition they could call on were scattered over every corner of the globe. This was the policy of naval stations. It was rare for a commodore to have a sufficient number of vessels under his command to undertake a large-scale offensive operation. The essential objective was to show the flag in foreign ports, and ensure an intermittent presence sufficient to discourage any thoughts of aggression by the new states or the native population. The orders given

the principal warships—frigates or corvettes, rarely ships of the line—
were that they should alternate long courtesy visits to allow their
officers an opportunity to mix with local society, with spells at sea,
devoted to training exercises for the crew. Indeed, the different elements of the fleet were rarely assembled into a squadron and each ship
put into port individually. These general principles considerably diminished the effectiveness of the Navy in an ocean as vast as the Pacific, but
without the requisite funding it was the best that could be done, always
trusting that international unrest would not get out of control. And
anyway, if the worst came to the worst, they could always count on
Britannia, the ruler of the waves.[11]

The Campaign of the *Essex* (1813–1814)

And yet it was during the war against the former mother country that
the first incursion of the American navy into the Pacific occurred in
1813, almost thirty years after the *Empress of China*'s voyage.[12] Setting
out on the frigate, *Essex*, in October 1812 with a crew of 319 to cruise
in the South Atlantic off Brazil, Captain David Porter decided to make
his way into the Pacific—the only way, it seemed to him, to escape
capture by the Royal Navy, and to take on fresh supplies in a friendly
port, while dealing the enemy a few hard blows. On February 3, 1813,
in accents worthy of Napoleon about to cross the Alps into Italy, he
made the following appeal to the men serving under him:

> Sailors and Marines!
> A large increase of the enemy's force compels us to abandon a
> coast, that will neither afford us security nor supplies, nor are there
> any inducements for a longer continuance there. We will, there-
> fore, proceed to annoy them, where we are least expected. What
> was never performed, by a single ship, we will attempt. The Pacific
> ocean affords us many friendly ports. The unprotected British
> commerce, on the coast of Chili, Peru, and Mexico, will give you
> an abundant supply of wealth; and the girls of the Sandwich is-
> lands, shall reward you for your sufferings during the passage
> around Cape Horn.[13]

Rounding the Horn turned out not to be particularly perilous
compared with the storms which had battered Lord Anson's fleet in the

previous century. At Valparaiso, Porter found a country in the throes of revolution, having just cast off Spanish colonial rule, and where the patriots felt strong ties with their fellow republicans in the United States. He learned that armed English vessels were preparing to capture American whaling ships, which were still unaware of the state of war between the two countries. Countering this threat became his principal objective in the Pacific. Between April and September, Porter seized twelve British whaling ships near the Galapagos, five of which he kept with the *Essex* to form a little fleet, sending the others home with their cargoes to the United States. In calculating the results of his campaign, he put his contribution at $5,170,000 ($2,500,000 for the ships he had captured, and the same amount for the American whaling ships which would probably have been captured by the enemy if the *Essex* had not been there to protect them) and $250,000 for the expense of the ships the British Admiralty was obliged to send in pursuit of him—and all this in return for an outlay of only $80,000.[14] It is sometimes maintained that his campaign was enough to discourage English whaling ships from returning to the Pacific, leaving the field open to the Nantucket and New Bedford shipowners after 1815, but there seems to be little basis for this argument, since several of the captured vessels were recovered by the enemy, and, in any case, if the hunt for the sperm whale had seemed profitable to the London shipowners they would surely have had little difficulty raising the capital necessary for such a venture in England. The predominance of the American whaling ships in the Pacific was not a result of Porter's heroic exploits, but must rather be attributed to their superior productivity.

In October 1813, Porter's tiny fleet left the Galapagos to sail to the Marquesas in order to repair the ships in preparation for the return voyage and, since a visit to Hawaii was out of the question, to offer some agreeable relaxation to the crew. Putting in to Nuku-Hiva, Porter was unable to resist the temptation to take a hand in some intertribal quarrels, taking the field against the much-feared Taïpis, a "happy and heroic people," whose valley he laid waste, to his later regret:

> When I had reached the summit of the mountain, I stopped to contemplate that valley which, in the morning, we had viewed in all its beauty, the scene of abundance and happiness—a long line of smoking ruins now marked our traces from one end to the other; the opposite hills were covered with the unhappy fugitives, and the

whole presented a scene of desolation and horror. Unhappy and heroic people! the victims of your own courage and mistaken pride, while the instruments of your own fate, shed the tears of pity over your misfortunes, thousands of your countrymen (nay, brethren of the same family) triumphed in your distresses![15]

On November 18, 1813, Porter claimed Nuku-Hiva for the United States. In a reference to the earlier discovery by Ingraham in 1791, he renamed it Madison Island. This was the first American colonial conquest in the Pacific, but it was a short-lived one, since the government in Washington, which had other fish to fry in its war with England, failed to endorse it. In any case, Porter's heroic exploits were over. Returning to Valparaiso after leaving the Marquesas, he was blockaded by two powerful English warships. His attempt to run the blockade, on March 28, 1814, quickly turned to disaster when a last-ditch stand cost him half his crew. This defeat obliged the American envoy, Joel Poinsett, who had enjoyed considerable influence until then, to leave Chile. The little garrison Porter had left behind on Nuku-Hiva fared little better, for mutinies and native uprisings forced the survivors to leave, only to be taken prisoner on reaching Hawaii. In the end, the first armed American intervention in the Pacific would leave no lasting trace.

The Pacific Squadron

However, once peace returned, American economic interests became too important for Monroe's administration to leave his captains and whaling ships subject to the tender mercies of Latin American royalists and patriots without any ability to react. For this reason, in 1818 the first ship, the *Macedonian*, commanded by Captain John Downes, was sent to provide a permanent presence on the South American coast.[16] However, a single frigate could not suffice to cover such a vast area, with the result that in 1822, Commodore Charles Stewart, in command of two ships, became the nucleus of a squadron which was to grow slowly— numbering three vessels from 1824 to 1838, and normally at least six after 1840, with additional reinforcements during the Mexican War and the California Gold Rush.[17] For three years, beginning in 1867, the station was divided, with a base for the North Pacific at San Francisco, and one for the South Pacific at Panama.

During this entire period, the Pacific Squadron was involved in very few naval engagements. The United States maintained a policy of neutrality during the Latin American wars of independence, and avoided intervening in disputes between the newly created countries. To protect the lives and property of American citizens it was enough for ships to pay occasional visits to the major ports along the coast. Buccaneers no longer felt free to attack Yankee merchant vessels and whaling boats, while the Peruvian or Chilean navies had to show restraint in enforcing their blockades. Incidents did occur, as in the case of the arrogant Cochrane, an English mercenary in the employ of the Chilean republicans, but overall there were very few.[18] Callao was used as an outlying base of operations, and the corvette *Dale*, part of the squadron commanded by Commodore Thomas Catesby Jones, put in there on five occasions in 1841–1842, while, in comparison, she called only twice in Valparaiso and Lambayeque—though she did put in three times into Paita, a favorite port of whaling ships. In October 1842, she was at sea between Panama and the Galapagos when Commodore Jones, uneasy about British designs on California and believing a state of war existed between the United States and Mexico, appeared on the eighteenth of the month with two ships off Monterey, capital of the Mexican province, and demanded that Alvarado, the governor, surrender it to him. However, twenty-four hours later Jones became aware that his actions were without justification, so that he was obliged to apologize and hand back his illegitimate conquest.[19] But this was only postponing the inevitable, for before four years went by the Pacific squadron was to play an active part in the occupation of Upper California and help to mop up pockets of Mexican resistance. This was, however, a secondary field of operations compared to the eastern side of the country where the outcome of the war would be decided.

The Latin American coast remained the major preoccupation of the Pacific Squadron, but from time to time a ship would be dispatched to the islands—the Marquesas, Tahiti, and Hawaii—to remind the natives that any attack on American lives or property would be met with force, if necessary. This also provided an opportunity to pick up deserters such as Herman Melville, who was overjoyed at the prospect of sailing home on board a frigate after his nightmare experience on a whaling ship (though this turned out to be a considerable disappointment, if his antimilitarist novel *White Jacket* is anything to go by).

During the Civil War, the principal task of the Pacific Squadron (which had acquired a steam-powered vessel as its flagship shortly before, in 1858) was to protect the shipping route along which gold was conveyed from San Francisco to Panama, from where it was transported across the isthmus and onward to New York. Confederate piracy did not pose a real threat, but raids were always a possibility, and this route was vital in maintaining the solvency of the federal government. So, when the steamship *Shenandoah,* sailing from Liverpool in October 1864, entered the Pacific to attack Union shipping, there were no ships available to defend it—particularly given the difficulty of tracking down the *Shenandoah* in such a vast expanse of ocean. News of military operations only reached the Pacific with a considerable time lag, so it was there that the last hostile action of the Civil War took place. After destroying three whaling ships at Ponape in the Caroline Islands, on April 1, 1865, Captain James I. Waddell headed north toward the Bering Strait. Between June 22 and 28—i.e., more than two months after Lee's surrender at Appomattox—he seized any sailing vessels which failed to escape his clutches. Twenty were burned, two others lost, and four obliged to sign an admission of debt. A short time later, he learned through a chance encounter that the Civil War was over and, in spite of the vessels finally sent to hunt him down, returned safe and sound to Liverpool on November 5. He had provided ample evidence of the weakness of the American naval presence in the Pacific, for it had shown itself powerless to protect the Arctic whaling fleet against just one steam-powered warship.[20]

The Asiatic Squadron

It was not until twenty years after the deployment of the Latin American squadron that one was stationed in the Asian region. This was because the American merchants in Canton, like the British East Indies Company, preferred to make their own arrangements with the Chinese mandarins rather than antagonize the Chinese government by calling for the protection of their national Navy. They were far from happy when United States warships made one of their rare visits, for these seemed to suggest an element of confrontation which was bad for business. When Captain John H. Henley, commander of the frigate *Congress*, the first American Navy ship to enter Chinese waters, in 1819–1820, proposed escorting their sailing ships to protect them from piracy, they immediately rejected an offer they felt would endanger their good relations with the empire—particularly since Henley had not hesitated to approach Canton to obtain supplies, contrary to the rules in force at the time.[21] In any case, their rejection of the offer cost them little, since it was not until ten years later that a second warship came by, after putting in at Nuku-Hiva, Tahiti, and Hawaii—the corvette *Vincennes*, making a round-the-world voyage on its way home from the Pacific.[22]

During the 1830s, however, British traders, conscious of their country's power, were finding the extortion practiced by the mandarins increasingly intolerable. Little by little the Americans, in their turn, began to ask their government to send a warship on a regular basis, as long as it respected the understandings in place and remained at anchor outside the estuary of the Pearl River. Visits by Navy ships to Canton began to become more frequent. In 1832, the *Potomac*, commanded by Commodore John Downes, after inflicting severe punishment on the Malays who had pillaged Kuala-Batu, put in briefly to Macao before reaching her Pacific station. She was soon followed by the *Peacock*, with Edmund Roberts on board, on a mission to negotiate treaties in the Far East. He met with success in Siam, but failure in Vietnam. To complete his mission, the diplomat sailed a second time on the *Peacock*, accompanied by the *Enterprise*, under the command of Commodore Edmund P. Kennedy. It was in 1835 that the latter obtained permission from the Department of the Navy to hoist a commodore's ensign—which explains why this is generally held to be the year in which the Asiatic Squadron was created, although its existence was to become official only ten years later. In Vietnam, the mission again met with failure. Roberts's death, occurring at Macao in June 1836, made it necessary to cancel the planned visit to Japan and decided Kennedy to return home via Hawaii and the Latin American coast. In the same year, the *Vincennes*, coming from Callao as it had done in 1829, called in again, after collecting some deserters from Nuku-Hiva and Tahiti, putting down a mutiny in Fiji, punishing marauding natives in Samoa, and freeing two hostages on Palau. Then there was a hiatus of three years until the arrival of the squadron commanded by Commodore George C. Read, who once again had to carry out reprisals by destroying a village on the pepper coast of Sumatra, and who arrived in Macao in April 1839, a month after Lin, the imperial commissioner, had decided to suspend trading in Canton until the chests of opium were handed over to him. Despite the insistence of the American merchants, left on their own by the English evacuation, Read, during the very tense weeks which preceded the Opium War, refused to stay on to protect them for longer than three months, after which he returned home via Hawaii and Valaparaiso. So, between 1819 and 1839 a total of only nine warships had put in to Macao, and for relatively short calls in port, often in the course of a voyage round the world. The American naval presence was remarkable only for its insignificance, even though the commercial interests at stake amounted to millions of dollars— certainly a much greater amount than on the coast of Latin America, which, by contrast, had enjoyed protection of a permanent squadron for the previous twenty years.

All this changed when China became open for trade, even though the Americans did little more than follow in British footsteps, earning themselves a reputation as hangers-on. To protect the interests of his citizens, while observing the most scrupulous neutrality, President Tyler dispatched Commodore Lawrence Kearny to China, with two ships. When they docked in Macao in March 1842, the Opium War was almost over. Kearny was content to observe the situation, since he had not been empowered to negotiate any treaties. He did, however, succeed in having indemnities paid as compensation for the damage arising from the hostilities. In 1846, the first American commissioner, Alexander H. Everett, arrived in China as the permanent representative of his country. This meant that the commodore and the diplomat had to find a way to work together. Each reported to a different authority in Washington, one to the Department of the Navy, the other to the Department of State. The voyage out took so long that there was no way for the federal government to intervene in any contest of vanity or dispute over jurisdiction. At times, tension mounted between these two civil and military authorities. Perry, whose heart was set on opening Japan, refused to give up part of his fleet at the request of Commissioner Humphrey Marshall, who was concerned at the unrest in China resulting from the Taiping Rebellion. On the other hand, there were warm relations between Commodores James Armstrong and Josiah Tattnall and Commissioners William B. Reed and John E. Ward when the treaties were renegotiated in 1857–1859.

The American policy was to obtain concessions equivalent to those granted the English and French, but to achieve this as far as possible by peaceful means. During the Northern China campaigns in 1858–1860, vessels flying the Stars and Stripes followed on the heels of the flotilla sent out by London and Paris, while keeping well out of the way when hostilities broke out. There was just one exception, on June 25, 1859, when the Anglo-French forces suffered heavy losses while attempting to remove the barricade across the river at Taku. Tattnall did not hesitate to disregard his orders to remain neutral by towing the allied pinnaces out of range of Chinese guns. As far as he was concerned, he was only returning a favor by the British admiral who, the previous evening, had sent one of his gunboats to tow the American vessel off the mud bank where she had run aground—unless, that is, Anglo-Saxon solidarity played a part, as is suggested by Tattnall's celebrated remark that "blood is thicker than water." The Americans, in Chinese waters, developed ambiguous feelings toward the English. They were quite conscious of their inferior strength and were reluctant to allow themselves to be drawn into imperialist conduct, but they were aware of the similarities between them, and that the merchants and missionaries of both nations needed protection. They mounted joint actions against pirates, and at Shanghai in 1854 they stood side-by-side against the imperial troops. Two years later, on the other hand, during the hostilities over the *Arrow*, sparked by the seizure of a vessel manned by Chinese but registered in Hong Kong, Commodore Andrew H. Foote refused to cooperate with the British admiral in taking Canton. Nevertheless, when the Chinese forts downriver shelled his ship, he put a landing party ashore which, after a sharp engagement, destroyed the fortifications and the guns. The goodwill which the Americans displayed toward the Chinese authorities was often no more than an expression of their weakness in the region. Washington had neither the means, nor the desire, to conduct a policy of aggression.

The effort made to open Japan was quite exceptional. The American government had realized that it was not enough merely to invite the shogun to sign a commercial treaty, for that had been Biddle's mission, and was the reason for his failure. A show of force was required to ensure success. For this reason, Perry, on his second voyage in 1854, appeared in Tokyo Bay with a powerful fleet of nine ships, totaling almost 12,000 tons, three of them steam-powered frigates, with a total of 1,775 men on board, and over a hundred guns. The best the Japanese government could do was give in by half-opening a door which until then had been firmly shut. But the United States did not have the means to continue such a substantial commitment. At the outbreak of the Civil War, the Navy left only one ship in Asian waters, the *Wyoming*, which was far from enough to impose respect on the xenophobic *daimyos*, who were determined to close the Straits of Shimonoseki to shipping. On July 16, 1863, the *Wyoming* took heavy losses beneath the fortifications at Choshu. In the following year a Western naval coalition was put together to destroy these, but the United States could only contribute an ancient sailing corvette, so slow that it had to be replaced with a steamboat hired from a private shipowner to ensure that the American flag would not be absent from the expedition. This episode marked the nadir of the United States' naval presence in the Pacific, at a moment when all the nation's energies were devoted to overcoming the Confederate rebellion.[23]

The Role of the Navy in Pacific Exploration

For the Americans, whose presence throughout the Pacific was more considerable than that of any other Western nation, it was humiliating to have to depend on European navigation charts. President John Quincy Adams had this in mind when, on December 6, 1825, he argued that the United States should become actively involved in scientific progress, particularly in the sciences of geography and astronomy. While France and Great Britain—all considered backward monarchies—showed no reluctance to sponsor voyages of discovery in order to push forward the frontiers of knowledge, the democratic United States was content to feed parasitically on the disinterested scientific research coming out of Europe. However, prudently recognizing his fellow citizens' lack of appetite for any speculations for which an immediate application could not be found, Adams made it clear that it was not his "intention to recommend the equipment of an expedition for circumnavigating the globe for purposes of scientific research and inquiry." "We have," he said, "objects of useful information nearer home"—in particular the exploration of the entire northwest coast of the American continent.[24] And yet, more than anyone else, the whaling captain from New Bedford or Nantucket needed good charts if his ship was not to meet with disaster on poorly identified reefs. All one had to do for Washington to accept some involvement was demonstrate the practical application of a research program, in spite of Democratic dogma, which advocated limiting activity to within federal borders in the name of the state rights. So it was that between 1838 and 1842 the most substantial exploration ever undertaken in the Pacific was launched by Jackson's successor, President Van Buren.

It is true that the preparations for the expedition took a long time—a good twelve years, in fact. The initial inspiration was provided by the fantasies of the "Newton of the West," John Cleves Symmes, who, with all the contempt of a self-taught man for official (which is to say European) science, in 1818 proposed a new theory according to which the earth was hollow and habitable on the inside, which it was possible to reach by way of holes situated at the North and South Poles (which he thought to be regions of warmth and plenty). The publicist Jeremiah Reynolds became the proselyte for this novel geophysics and called for an expedition to explore the "open polar sea," the route to the still almost unknown Antarctic. To gain support for this scientific undertaking, he stressed its relevance to practical matters, such as the seal or whale hunts. Jackson's election in 1829 put paid to any publicly financed expedition, but since the attempt organized by Edmund Fanning was a miserable failure, there was no alternative but to call on a federal initiative. In the meantime, Jackson had acquired a better appreciation of the role of government in scientific research. The expedition was approved in principle in May 1836. But there remained no shortage of obstacles—the opposition of Dickerson, the secretary of the navy, the lack of an appropriate vessel, tension between the organizers, and above all, controversy about the role of scientists—so that it was more than two years before it sailed from Norfolk. The ideal model was Napoleon's expedition to Egypt, but the American officers, although themselves generally lacking in any scientific training, were reluctant to welcome on board ship civilians over whom they would have very little authority. On the other hand, the young scientists, wishing, with the help of the federal government, to demonstrate their professionalism and distinguish themselves from the host of self-taught amateurs by initiating the birth of modern science in the United States, wanted to enjoy total autonomy. The Pacific was at the time an unexplored domain, which it was reasonable to hope would provide answers to problems in botany, zoology, geology, and anthropology. Despite the difficulties, some of the best American scientists—in the same way as Darwin who at the same time was setting out on the *Beagle*—signed up for the expedition. They included the mineralogist James Dwight Dana, the zoologist Charles Pickering, the conchologist James P. Couthouy, and the philologist Horatio Hale. Finally, the trickiest matter was to find someone to take command. The various officers considered all backed out, and finally the position was given to a 40-year-old lieutenant, Charles Wilkes, a schemer and somewhat paranoiac, but well-versed in hydrography and astronomy. He was entrusted with six ships and 350 men. No country, not even Great Britain, had ever made such a substantial effort.[25]

In his instructions, Secretary of the Navy J. K. Paulding stressed the utility of the expedition to the important interests "embarked in the whale-fisheries, and other adventures in the great Southern Ocean." It was necessary, he said, to "determine the existence of all doubtful islands and shoals, as to discover and accurately fix the position of those which lie in or near the track of our vessels in that quarter, and may have escaped the observation of scientific navigators." Furthermore, a harbor would have to be selected in Fiji where American ships could allow their crews some relaxation, and an agreement would have to be made with the islanders for the supply of provisions. The exploration of the Antarctic and a possible study of Japan made the expedition's program one of the heaviest ever. Paulding stressed that the expedition was "not for conquest, but discovery," in the service of trade and navigation and, in the second place, of science.

With the exception of Japan, Wilkes respected the directions given him (cf. Appendices, Map 3). Not only did he venture into the Antarctic on two occasions, discovering at the same time as Dumont d'Urville that it was a continent, but also he carried out extensive research in the tropical Pacific. A number of features whose coordinates were previously known only approximately, if at all, could now be located with accuracy on navigational charts. The Fiji Islands, long a terror to navigators because of their dangerous reefs, could now be visited with safety. In the temperate latitudes of the northern hemisphere, painstaking surveys of the coast of Oregon and northern California were carried out. The labyrinth of Puget Sound was clearly mapped, Wilkes impressing on the government the usefulness of this region as the site for a port of easy access. The outcome was not entirely positive, however, for only two vessels returned to their home ports. Of the four others, one was lost in a storm near Tierra del Fuego, another broke apart on the bar at the mouth of the Columbia River, one was sold in Singapore, and the supply ship turned out to be useless because it was so slow. Almost half the original crew had deserted, had been sent home, or had died during the voyage, which had lasted almost four years. Wilkes, seeing plots against him on all sides, had created an atmosphere of tension, and indulged in an excessive use of the whip. For this he was court-martialed on his return, and sentenced to a public reprimand. Contacts with the native peoples were not always peaceful, particularly in Samoa and in Fiji, where, at Malolo, Wilkes destroyed several villages in reprisal for the killing of two of his sailors, but the treaty signed there provided whalers with a measure of security (cf. Appendices, Table 7). The most disappointed participants were the scientists, who considered that there was an excessive preoccupation with hydrography and not enough attention to collecting samples. Nevertheless, they brought back an enormous amount of documentation, most of it gathered in Tahiti, or while climbing the Hawaiian volcanoes, or in Antarctica, or from the various atolls they were able to explore.

Unfortunately, by the time the expedition returned in 1842, Americans had lost their interest in it. Congress, faithfully reflecting public opinion, and in the absence of any national scientific association to act as a lobby group in Washington, dragged its feet in approving the funding needed to publish the scientific findings. Some of the researchers finally tired of the delay, preferring quick publication in journals in order to win recognition by their peers. Wilkes, the principal organizer of the publication of the expedition's findings, which he set out to administer in shipboard style, preferred exhaustive investigations to be completed before they were made available to the public. Of the twenty-four volumes planned, five would never see the light of day. Spacing out publication over thirty years, from 1844 to 1874, lessened the impact of what had been a substantial scientific effort. It nevertheless contributed a host of new facts and descriptions of unknown species to botany as well as to the study of zoophytes, mollusks, shellfish, and crustaceans, and also to the fields of ethnography, philology, and anthropology. In carrying out this expedition to the Pacific, the Navy helped lay the foundations of American science and demonstrated that the democratic spirit is not necessarily anti-intellectual.

The northern Pacific was not part of the area Wilkes was supposed to explore. Whaling ships rarely ventured that far in 1838. This was no longer the case in the 1850s. By then, people had come to better appreciate the importance of a more complete knowledge of the ocean. Maury, at the National Naval Observatory, was making detailed charts of winds and currents. To obtain a more exact picture of the most northerly seas, Congress voted funding in 1852 for an expedition of five ships under the command of Cadwalader Ringgold, just as Perry was preparing to open Japan. However, it got off to a bad start: the supply ship was unsuitable, and another ship was lost in a typhoon. Ringgold, suffering from depression, was replaced by Lt. John Rodgers.

After surveying the coasts of Formosa and of the Ryukyu Islands, Rodgers made excellent charts of the waters around Japan, which had just become accessible, and then continued his work in the Sea of Okhotsk, in the Arctic Ocean, and around the Aleutians, before returning to San Francisco. Now the direct route to the north was no longer so dangerous, despite the fog and the storms. Thanks to the American Navy, the Pacific had been made a more welcoming place for mariners.[26]

Dreams of Empire

From early in the nineteenth century the Pacific exerted a strong fascination on the imagination of certain Americans. They were convinced that this was where their future would be decided. On the western side of the ocean the great empires of the Far East, with their masses of humanity, held out a promise of fabulous markets to complement those in Europe. At a time when publicity tended to portray the tiniest village as a future metropolis it was tempting for visionaries to point out how ideally the United States was situated. As the only noteworthy power with both Pacific and Atlantic seaboards, it would inevitably become a crossing- and meeting-point between Asia and Europe. This new Silk Road would bring activity to an expanse of ocean which until then had been deserted. Such an ambitious commercial geopolitics neglected the fact that, in conveying goods, transshipment from one mode of transportation to another is often more costly than mere distance, and that between Shanghai and London via San Francisco and New York two such transshipments were necessary, resulting in prohibitive costs for all but the most expensive commodities, in whose case the shorter delivery time might justify the additional cost. Europe had so many advantages, and had established such a lead, that before the Civil War no one could think it possible that the Pacific could overtake it so quickly. However, the laws of history did seem to show that no supremacy lasts forever, and that the torch was being passed from East to West, shifting from the Middle East and ancient Egypt to the Mediterranean, and then to the North Sea, before crossing the Atlantic and arriving on the east coast of America. Westerly expansion, driving the population to settle on the plains beyond the Appalachians, was blazing a trail to the future which led in the direction of the Pacific. So it seemed as if the time had come to pay some heed to the deeper forces controlling human history and for the United States to make its move into the ocean of the future.[27]

The messianic message seemed clear. As for the actions it would inspire, two attitudes were possible. One could either wait until these latent possibilities were realized in their own good time, or one could forge ahead and hasten the inevitable. On the one hand there was the modest confidence of those who believed that history was on their side and, on the other, the ambitions of the empire builders, always impatient to see their destiny fulfilled. Beginning in the second decade of the nineteenth century, but mainly in the 1840s and 1850s, speeches made by American politicians often expressed their belief in the Pacific's importance for their country's future preeminence. The development of the western ocean would enable them to loosen commercial ties with an aging Europe, now considered much too constricting. But we should not take these flights of rhetoric too literally, for they were primarily aimed at the electorate. The current pragmatism was more wont to take account of hard reality and give due priority to more urgent responsibilities, of which there were quite enough, before launching out on futuristic adventures. This explains why such flamboyant ambitions often gave rise to more modest successes.

First Objective: The Acquisition of Ports

In fact, this modesty only applied to the ocean itself. Where the continent was concerned, the ambitions were realized. As the historian Norman Graebner has shown, President Polk had one systematic aim, namely to endow the United States with a continuous Pacific coastline by adding to it the 1,200 miles stretching from the Columbia River to the 42nd parallel. His objective was to gain possession of the only three serviceable harbors along this forbidding and inaccessible coast—Puget Sound and the bays of San Francisco and San Diego—in order to shut out English and French competition and lay the groundwork for a magnificent future. The "manifest destiny" which had been responsible for United States expansion into Oregon and California was less an expression of the will of a people drawn by natural forces across the continent toward the Pacific coast than the implementation of a conscious policy developed in the corridors of power in Washington and based on the country's maritime interests. If the ideology had been exclusively

Nonsense!

continentalist and had focused on the western hemisphere alone, the Americans, once they had reached the western ocean, might have been content to turn their backs to it, as the Irish had done on the other side of the Atlantic. In fact, in the negotiations with England over Oregon, and in achieving the objectives of the war with Mexico, the main consideration was always to acquire potential sites for major ports. In this Polk was supremely successful, even if he appeared to fall short of the maximalist demands of the Democrats, with their insistence on claiming the territory lying between the 49th parallel and 54°40' N occupied by present-day British Columbia, or the total annexation of Mexico. His negotiator with the Mexicans, Nicholas P. Trist, brought his mission to a successful conclusion when he managed to have all of Upper California—even including San Diego Harbor, which the Mexicans considered part of Lower California—included in the Treaty of Guadalupe Hidalgo, which was signed after the American military victory on February 2, 1848. In spite of Polk's public expression of displeasure with a diplomat who had ignored his orders to return home and who, the president claimed, could have obtained further concessions from a defeated enemy, the Senate ratified the treaty as it stood, an indication that it found it quite acceptable and felt that the United States had no further need for frontage on the Pacific at that time.[28]

On the ocean itself, symptoms of imperialism are more difficult to find. Naval officers did propose annexations, as Porter did for Nuku-Hiva, or Perry for the Bonin Islands, and diplomats did push for the occupation of certain territories, as Peter Parker did in the case of Formosa, but Washington refused to be drawn into any such overseas adventures. The government wished to distinguish itself from the Europeans, going as far as to avoid using Hong Kong as a base for the Asiatic Squadron, at least until 1853. However, the American anticolonialist tradition was perfectly compatible with continental expansionism. The law of 1856 on the guano-producing islands represented the greatest extent of the imperialism that the United States as a nation was prepared to accept. Admittedly, these generally uninhabited atolls had little to offer.[29] This discretion was most evident where Hawaii was concerned, although it was there that the proportionately largest colony of Americans in the Pacific resided. The terms of the treaty signed in December 1826 by Captain Thomas Catesby Jones were similar to those negotiated with various Latin American countries. Nevertheless, the tone of relations between visiting American officers and the king are more reminiscent of those between a father and son than between two equal, sovereign states. In November 1829, on his departure, Captain William Finch addressed a message replete with worthy sentiments and condescension to King Kamehameha III:

> Soon after I arrived, strangers pointed out to me two taverns—one as belonging to the king and the other to Governor Boki. The impression made upon me, by this information, was unfavorable to you both. Such establishments are necessary in a port like Honolulu, but they ought to be conducted by private individuals. In the direct gains arising from them, you should not participate—nor ought you to frequent them. Very few should be allowed—and on such as are permitted, a high revenue or tax might be imposed; by which means its resources will be increased, and the burden, at present resting on the industrious, will be lightened. . . . [Y]our advancement in the acquisition of our language and letters persuades me to urge you to pursue English studies diligently. By a due division of your time between studies, business, and becoming pleasures, your life will be rendered delightful to yourself, and beneficial to those, who, by circumstances, are placed under your protection; and who must ultimately look up to you for example. You are yet young; now is the period to lay the foundation in your character for extensive usefulness.[30]

In spite of its close links with the Hawaiian monarchy, the United States had no wish to establish its protectorate over the archipelago, but in return expected Great Britain and France to show similar restraint. In December 1842, President Tyler, under pressure from Secretary of State Daniel Webster, formulated a doctrine applying to Hawaii which was very close in spirit to the Monroe Doctrine. The United States supported the independence of the Polynesian kingdom and implicitly guaranteed its protection as a way of retaining its advantageous position, declaring its opposition to the granting of any privileges to foreign powers. This was in reaction to some incursions by the French, as well as of the annexation attempted in 1843 by an Englishman, George Paulet, without the endorsement of his government. In December 1849 a treaty covering friendship, trade, and shipping, signed in Washington, reaffirmed the principles governing relations between the two countries and reasserted the Tyler doctrine. The development of California, beginning with the Gold Rush, naturally drew Hawaii into the American orbit, with the forging of increasingly close economic links between San Francisco and Honolulu. In the 1850s, sentiments favorable to annexation began to emerge among the American planters, who were eager to develop markets on the Pacific Coast. If not for opposition from France and England, Hawaii might well have become part of the United States at this time, for in 1854 a draft treaty proposing its annexation was sent to Washington. It got no further, however, since the Pierce administration had no wish to provoke a quarrel with the two great European maritime powers, and since the new king, Kamehameha IV, had come out in favor of independence—as did the majority of his subjects, for they feared for their land and distrusted the latent racism of many Americans. Congress, in any case, preoccupied by the perpetual disagreements between North and South over the spread of slavery, and not wishing to upset the fragile equilibrium between the two opposing factions, would probably never have found the majority required to approve annexation. It only remained to allow the right conditions to come into existence, in the course of time. Indeed, it was already clear by the time of the Civil War that if Hawaii was one day to lose its independence, it could only be to the United States. There was no strategic justification for hastening the inevitable.[31]

The Alaska Purchase

By 1848, no country was better placed than the United States to dominate the northern Pacific. The only power which might perhaps have stood in its way was the Russian Empire. When the Civil War broke out, the czar ruled over the entire area of land running from Vladivostok to the border with British Columbia, at 54°40' N. On a map, the North Pacific appeared to be a Russian lake, but this was not an accurate impression, since Americans were much more likely to be seen in Russian waters than were the czar's own subjects. Since the end of the eighteenth century relations between the two communities had been developing steadily. Some historians, such as Howard I. Kushner, emphasize the friction between them, while others put more stress on their collaboration. In fact, cases of conflict and collaboration often coexisted, until the Russians decided to put Alaska up for sale and the Americans to purchase it.

   In North America, Russia was represented by a monopoly company, controlled from Saint Petersburg. It had received its first charter in 1799, renewed until 1862. As a private company, and therefore expected to be profitable, the Russian-American Company nevertheless came under close scrutiny from the czarist administration. The government expected it to develop the colony and exploit its natural resources of furs, fisheries, and minerals. But there were too many obstacles for this Russian foothold in America to prosper. Its distance from the mother country made the cost of transportation prohibitive, settlers could not be supplied on a regular basis, and management was poor. To survive, Alaska was obliged to use American entrepreneurs and sea captains as middlemen, and very often they provided its principal outlet to the rest of the world. The czarist government was perfectly aware of all these problems, but did not consider them sufficient reason to abandon the region, if only because it wished to avoid the loss of prestige such a withdrawal would have meant. Even if Alaska was not vital to the empire, certain eminent Russians had sufficient interests there to oppose the Yankees on occasions, and on others to collaborate with them.

As we have seen, it was quite early on that the "Bostonians" entered Russian waters in search of the sea otter skins which could be traded for goods in Canton, thus avoiding the need to spend too much hard cash. When these resources were exhausted, the two parties joined forces in poaching along the California coast. Later, whalers took over, maintaining a constant American presence which was not entirely of benefit to the Russians, neglected as they were by both the Hudson's Bay Company and by their mother country. The Yankees were little concerned with maintaining the breeding stock, and traded directly with the Indian tribes, to whom they offered alcohol, arms, and munitions which was tempting to use against the colonists. The government in Washington, for its part, supported its traders in the name of maintaining existing advantages, or else claimed that it was not its affair to police its citizens. But far from seeking systematically to provoke conflict, the Americans were more inclined to collaborate, for this was more profitable for business. But now and then they did not hesitate to employ aggressive tactics in defending their commercial interests. In addition, we should not forget the part played in the balance of power by Great Britain, in the guise of the Hudson's Bay Company. The British lion viewed the Russian bear as an obstacle it often found barring its path, whether in Central Asia, on the Indian frontier, or in the Mediterranean. In the circumstances, it was tempting for the British to play off the Americans against the Russians, in the same way as the Saint Petersburg government was attempting to make an ally of Washington against London. So, strategic considerations tipped the balance toward collaboration rather than conflict.[32]

Made uneasy by the expansionist sentiments expressed by members of Congress, such as John Floyd from Virginia or Thomas H. Benton from Missouri, Czar Alexander I finally issued an *ukase*, on September 16, 1821, making the area between the Bering Strait and the 51st parallel off-limits to all trading, fishing, and hunting activities, and setting the limit of Russia's territorial waters at eighty nautical miles, any ship found within this area being liable to confiscation. These measures were aimed primarily at the Americans. The reaction of Monroe's government was to reject this unilateral decision. John Quincy Adams declared that commerce was "among the natural rights and duties of men." The Monroe Doctrine has at times been interpreted as a response to this Russian *ukase*, particularly with respect to the southern frontier, but this appears doubtful, since the United States, unlike England, seemed more preoccupied at the time with its commercial interests than with the precise location of the border. In any case, in June 1822 Alexander I suspended this measure, which had been in force for only a year. Henceforth, Russian ships would only patrol coastal waters north of 55° and arrest only smugglers. What could explain this change of strategy? Was it to leave the czar's hands free to help the Greeks in their revolt against Turkey? Or did the English and American protests have some effect? The czar's reasons seem to have been more intrinsic: as a believer in the freedom of the seas and in friendship with the United States, he saw that his *ukase* could not be enforced, that the monopoly company suffered from glaring inefficiency and corruption, and that it could only be made profitable again by improving its management rather than by shutting out the foreigners on whom it depended. So, in a change of policy, he initiated separate negotiations with the United States and Great Britain. The United States-Russia Treaty of April 17, 1824, provided for renewed cooperation for at least ten years. It granted freedom to trade along the northwest coast in any previously unoccupied areas. On the other hand, wherever a permanent settlement already existed, the administration's permission had to be obtained. Only trading in arms and alcohol was prohibited, though this was not considered a justification for the seizure of vessels.[33]

During four decades the Treaty of 1824 provided a framework for relations between the Russians and the Americans in Alaska. The ten-year limit was extended by tacit agreement. From time to time incidents did occur, though they were rarely serious. The Russian presence depended too much on the Americans to afford a quarrel with them. The annexation of California and the Gold Rush strengthened the ties between the two countries and led to a widening of trade. In 1852, the American-Russian Commercial Company of San Francisco was granted a license to import coal and ice. Henceforth, it would not only be Massachusetts that showed an interest in Alaska, for the Californians now did so as well. A visionary developer, Perry McDonough Collins, put forward the idea of a telegraph cable linking the United States to Europe across Alaska and the Bering Strait, and then running along the valley of the Amur, which he depicted as another Mississippi. He managed to win over the president of the Western Union Exchange Company, discouraged by the failure of the first attempt to lay a transatlantic cable in 1858. Only a charter from the Russians was required for work to begin, and this would have happened had the project not been dealt a fatal blow by the successful attempt to lay a second transatlantic cable in 1866.[34] Other entrepreneurs were attracted, some by the teeming salmon, others by the furs or gold mines. For the Russians, the death knell of the colony was beginning to toll, and in March 1853 the governor of Eastern Siberia, Nicolas Muraviev, wrote to the czar:

> The ultimate domination of North America by the United States is so inevitable that we shall have to withdraw sooner or later, but we should do so peacefully, obtaining other benefits from the Americans in return. Thanks to the amazing development of their railroad system, the United States will soon extend over all North America. We must accept that we shall have to yield our North American possessions to it.

Economic considerations did not play a predominant part in the czarist government's decision. Alaska represented no great financial burden to Russia. But, strategically, the colony was difficult to defend. During the Crimean War the Russian fleet proved quite unable to protect the colony, and if England chose not to attack it, it was to avoid the Russians selling it to the Americans—a solution Saint Petersburg had indeed briefly considered.[35] It was ultimately for political reasons that the czar decided to sell. Instead of overextending his power over an immense area which could only be controlled by whoever had mastery of the seas, he chose to concentrate on the Euro-Asian landmass. This made particular sense since the treaty of Aigun (1858) and the Peking Convention (1860) had allowed his empire to push its frontier with

China in the south as far as the Amur, a navigable river, and to Vladivostok, an ice-free port. In the circumstances, Alaska was little more than an impediment to Russia's imperialist ambitions, and a weakness in its armor, given the new geopolitics taking shape in the northern Pacific.

There was little doubt as to whom the purchaser would be. If the Russians much preferred the Americans to the English, it was not for sentimental reasons, but because competition between the two Anglo-Saxon powers could only be to their advantage. In the United States, a powerful pressure group was pushing for the acquisition of Alaska. It was headed by Secretary of State William Seward from 1861 to 1869, always a proponent of expansion overseas, who saw that the Pacific would ensure "the political and commercial control of the world," and become "the theater of our greatest triumphs," where the United States would "soon have no formidable rival."[36] The sale might well have taken place before 1860 if any new territorial acquisitions had not been made impossible by the dispute between North and South. Interrupted by the Civil War, the project was revived as soon as peace returned. In December 1866, Alexander II gave his approval, and henceforth things quickly came to a head. The negotiations held in Washington between Seward and Baron von Stoeckl resulted in a treaty under which the territory was ceded to the United States for the price of $7.2 million. With the support of most of the press, it was ratified by the Senate on April 10, 1867, by thirty-seven votes to two. Only the House of Representatives, upset at being presented with a fait accompli, and hostile to President Andrew Johnson, balked at approving the necessary funds. Final approval was not given until July 14, 1868, by which time the United States had already been in possession of Alaska since the previous October 18. What a sensation it would have caused if the United States had found itself obliged to hand back Alaska because the purchase price had not been paid! The acquisition of this immense stretch of coastline along the North Pacific seemed to prove the success of the United States' overt ambitions, but the conditions under which it took place are, rather, indicative of how modest their objectives actually were. In 1867, despite the Alaska Purchase, the northern Pacific was still not an "American lake," and this was even less true of the ocean as a whole.[37]

FOUR

Cultures in Collision

— The great majority of Americans who sailed the Pacific prior to the Civil War came from New England or the states partially colonized by the Yankees. Scions of an Anglo-Saxon, Protestant America, confident in itself, in its rights and in its faith, they came into contact with cultures which they considered, in varying degrees, to be inferior. On this scale of cultures, the Catholic Latin American peoples were placed above the heathen Asians, with the "savages" of the Pacific islands on the bottom rung. The new anthropology took little account of the biblical view according to which mankind was a single species, and Eve the first mother of the whole human race. Charles Pickering, a member of the Wilkes expedition, concluded that there were eleven distinct races of man. In the Pacific region, apart from the whites, he listed the Mongols, the Malays (with whom he classified the Californian Indians), the Australian Aborigines, the Papuans (on Fiji and New Guinea), the Negrillos (in the New Hebrides, Luzon, and the Solomon Islands).[1] If we add the blacks in Chile and Peru, and the Indians of Singapore, the Pacific was home to an astonishing racial and, consequently, cultural mosaic, whose different elements were now coming into sharp collision with one another for the first time.

DISCOVERING THE OTHER

The Motivations

In 1803, at the age of forty, Amasa Delano analyzed the reasons which had led him to undertake a third voyage to the Pacific, and concluded that they were three:

> The first that I know of was an ambition to excel others in achievements; the next as, to satisfy my own curiosity in a knowledge of the world, and particularly to know how far myself and others were imposed upon with exaggerated accounts of the world, and false statements of things a great way from home; the third, and last, and for many of my latter years by far the greatest, was, honestly and honourably to obtain a competency sufficient to support myself and family, through old age, should I live to see it; to do a benevolent act now and then, and to leave an unblemished character behind me.[2]

The economic motive was certainly predominant for Delano, who had hunted seals in Tasmania and Chile, as it was for the outfitters, ship's captains, supercargoes, and others. A ship would be sent on a distant voyage because high profits could be expected. The risk was recognized that it might never see its home port again, but it was only a minor one, as is shown by the longevity of most whalers. It was not even necessary to come back with a full hold to cover costs, which were relatively low. A single successful venture could bring sufficient profits to allow Delano and his compatriots to live the life of honest prosperity they dreamed of. And, in the vast ocean expanse, there was always the hope of discovering some new source of riches. In Fenimore Cooper's novel *The Sea Lions* (1849), in which the action is set around 1820, two ships of this name set sail, on the strength of an old sailor's deathbed story, one from Oyster Pond, the other from Martha's Vineyard, to hunt seals in the Antarctic around Cape Horn. The second vessel, not knowing the location of the fabulous hunting ground, refuses to let the first *Sea Lion* out of her sight, and when on one occasion she is shaken off, finds her again and by dint of moral suasion obliges her to spend the winter in the glacial cold, to allow her the time she needs to fill her hold. Cooper, who detested the Yankees, paints a pitiless portrait of Captain Daggett. "This man," he writes, "clung like a leech to the remotest chance of obtaining property. There is a bull-dog tenacity on this subject among a certain portion of the great American family—the god-like Anglo-Saxon—that certainly leads to great results in one respect; but which it is often painful to regard, and never agreeable to any but themselves to be subject to."[3] Not all the American captains who sail the Pacific are as crazed as this Daggett, who would rather die than sacrifice his ship to the common good. Nevertheless, they have sailed to the ends of the earth with the object of making money, imagining that a few lucky throws of the dice will spare them years of future toil earning a living and providing for their old age.

But the dream of a quick fortune might not have been sufficient to attract so many young Americans to the Pacific had motives other than a thirst for riches, both conscious and unconscious, not played a part. The first motive Delano lists—his desire to outdo his rivals—makes it clear that, in Nantucket, if a young man wanted to marry, he was well advised to make a name for himself in whaling first, for a coward had no hope of making a good match. Tales of his past adventures told at dockside or on board ship earned a man prestige. If he had sailed the Pacific, it meant that, for a few years at least, he had had the courage to break away from his local community to range beyond the limited horizons of the landlubbers, and that he preferred the ocean and its perils to the monotonous drudgery of farm or factory. It meant he had won some freedom from family and village life, some room in which to assert his individuality and prove himself capable of something different, in an America which was still predominantly rural. All this was largely an illusion, of course, for he would find out soon enough that life was actually much more monotonous at sea than on land. For every exciting whale hunt, how many weeks had to be spent cruising along at a snail's pace, performing the regular daily chores! On merchant vessels, too, it was only when the wind changed direction or a storm came up that there was some relief from the tediousness of shipboard duties. The freedom was even more illusory in that the seaman was always subject to the captain's supreme authority. He only had to chance on a tyrannical and sadistic master for the ship to become little better than a prison, like the *Pilgrim* and the *Alert* under Frank Thomson, whose brutality is described in Richard H. Dana's account of his voyage to

California. A member of the crew called Sam, who failed to answer a question promptly enough, was sentenced to the lash, while the remainder of the crew was subjected to a stream of abuse, in which the supreme insult was the comparison with a herd of "nigger slaves."[4] On the other hand, it was hardly possible to relax discipline on those American ships in the Pacific, for the captain and officers often had to make do with a heterogeneous crew, a veritable Tower of Babel which was made up of Yankees, Portuguese, Polynesians, and Chinese, with plenty of troublesome and degenerate individuals among them. Commentators at the time saw this as one of the reasons why whaling ships' productivity went into decline. According to the Reverend Henry T. Cheever, around 1,850 crews were no longer what they had been in the days when most of the crew came from Nantucket:

> They are now made up to a great degree (of course with some honourable exceptions) of the very refuse of humanity, gathered from every quarter, escaped from poor-houses and prisons, or gleaned from the receptacle of vagrancy and lazar-house corruption, with a large admixture of foreigners of all languages, complexions, and character.
>
> Such constitute the experienced portions of the crew. To them you may add one third or one quarter part more of land lubbers or raw hands, made up of very heterogeneous materials. Here will be found the young roving adventurer, who pants for opportunity to see the world; he has heard marvellous stories of the facilities of foreign observation furnished in this service, while, at the same time, visions of easily-acquired wealth, golden harvests to be here reaped, have filled his mind, and he hurried from the interior to ship himself on board a whaleman. The reckless and impatient who spurn all salutary control, are also here, thinking this is just the place to indulge unbounded licence.
>
> Here you will also find the spoiled sons of ever-indulgent parents, who, having made themselves intolerable by their vicious propensities, and constantly in danger of bringing disgrace on themselves, and their connections also, by their intemperance, their fits of passion, or unbridled licentiousness, are sent on a whaling vessel as a school of reform![5]

From these dregs of humanity came the "beachcombers" who made life so difficult for the missionaries. It must be said, however, that in general the Americans were less dangerous than the Australians, mostly escapees from the penal colonies, real brigands of the South Seas. Among the former were many fine young fellows who had blundered into a venture very different from what they had expected. Life at sea did not always allow them to fulfill their dreams of freedom.

The Pacific: A Sex Shop or a Brothel for Puritan America?

There is one form of freedom which was seldom referred to at the time, in an America where prudery was endemic, and this was sexual liberty. Everything in Puritan education aimed to teach control over the appetites, self-control, the repression of natural urges. The flesh was synonymous with sin. But however strong the moral censure may have been, it was unable to prevent Satan from gaining the upper hand from time to time—and to the Devil with morality!—or to escape the hypocrisy of voyeuristic pleasure or of self-congratulation for having withstood the most infernal seductions.[6] The demon of the flesh, exacerbated by the lengthy solitude of ocean voyages, and by the fantastic tales told by the crew to help pass the time or heard before signing on, always became active as soon as the ship dropped anchor off some Pacific island. In these tropical lands, the naked bodies of the native women were a revelation to the young Yankees, who were more accustomed to seeing the women in their own country covered up from head to foot. Every Pacific port was a gateway to the Garden of Eden— whence, as is well known, the depths of hell are never far removed. Confronted with such a spectacle, was a man to give in to sin, surrendering to the dictates of his fundamentally corrupt nature, or was he to reject such charming temptations in order to preserve the purity of his body and soul? A tempest must often have raged in the breasts of the young—and not so young—sailors, spending as long as two—and sometimes three or four—years so far from home.

When Captain David Porter with his little flotilla reached the Marquesas in 1813, fearing to go on shore directly, in spite of good relations with the native people, he offered them a small gift. "But," he wrote:

> . . . they had no article to offer in return but their women: and as two of them were not more than sixteen years of age, and both handsome, they no doubt considered them the most acceptable present they could offer us. The men repeatedly invited us to the shore and pointed to the women and the house near which they

were standing, accompanying their invitation with gestures which we could not misunderstand; and the girls themselves showed no disinclination to grant every favour we might be disposed to ask; and to render themselves the more attractive, they retired and soon appeared clad in clean, and no doubt, their best attire, which consisted of a white and thin paper cloth, which enveloped their whole persons, with the exception of one arm and breast: and this simple drapery, when contrasted with the nakedness of the men, gave them an appearance of grace and modesty, we had little expected to find among savages. Finding that all their allurements could not prevail on us to land, the old chief directed the young girls to swim off to us: but on the appearance of reluctance, the young men led them toward the water, where they were soon devested of every covering and conducted to the boat amid the loud plaudits of the spectators on the shore. On their entering the boat, the seamen threw them their handkerchiefs for a covering, with which they carefully concealed those parts which modesty teaches us should not be exposed. . . .[7]

Furthermore, much to the disgust of the Anglo-Saxon Puritans, Porter found nothing reprehensible in this behavior. After all, what could be more natural than this nudity of the island people, for was that not how "our great mother Eve is portrayed?" And he reflected that "dress is not always a proof of modesty and virtue, nor is nakedness that of disparity and want of shame." In their eyes, it was no sin to give oneself to the first visitor who came along, so that, "if there was any crime, the offence was ours, not theirs: they acted in compliance with the customs of their ancestors; we departed from those principles of virtue and morality which are so highly esteemed in civilization."[8] In Nuka-Hiva, Captain Porter, with his indulgent attitude toward the sexual freedom of the tropics, was no longer able to control his men. When a treaty with the Happah tribe was proclaimed, the sailors "ratified" it with the women, in their huts or among the vegetation. Before long, the *Essex Junior* was swarming with representatives of the fair sex, aged anything between ten and sixty, already familiar with visiting ships, and, as the author relates:

Indeed the ship was a perfect Bedlam from the time of their arrival until their departure, which was not until morning, when they were put on shore, not only with whatever was given them by all such as had shared their favours, but with whatever they could lay their hands on.[9]

When the whaling ships began to arrive in large numbers, they brought the island peoples customers ready to provide them with the products of Western civilization in exchange for the temporary services of their wives and daughters. This indigenous form of prostitution incurred the wrath of Protestant missionaries. At Ponape, in the Caroline Islands, the Reverend Luther H. Gulick noted that within five years, between 1852 and 1857, out of the seventy-four ships that had called in, only ten were not used as "the public residences of native females," some of whom lived in the cabin, others in the hold or in the forecastle. The clergyman's denunciation of these "public brothels" was aimed at the New England shipowners—men who were subsidizing the missionary effort while at the same time they allowed their captains to violate every principle of Christian morality.[10] His colleague, Alfred Sturges, described a similar situation which lasted until the Nahnken of Kiti, the most powerful chief in Ponape, declared, first, the ships and then the five brothels on shore, opened by whites acting as pimps, to be taboo:

> The natives seemed to have no sense of shame or wrong in the case—to them, virtue was one of the most lawful, as it was the most convenient articles of traffic. They would come from the ships directly to our house, bringing bundles of trade, & when asked where & how they got it, would reply, "I took my wife, my sister, my daughter on board such a ship (often naming the Capt., or officer, etc.) & here is the pay". One day I met a man in the road, bearing a bundle of trade, & to my question, "where did you get this", he said his wife had been stopping on the ship, & this is part payment, "another day I go again I get some more". When I told him it was wrong, he said "no, very good, Capts white men all say very good". On telling him that God was angry at them for this, & hence there were so few children born on the island, he thought for a moment at, then answered "no matter, me like tobacco, trade."[11]

For the puritan Americans, the responsibility for sexual license in the Pacific islands was shared equally, since it depended on the demand as much as on the supply. This caused them to view it with even greater horror. But not all Americans shared their gloomy view of human nature, a notable example being Herman Melville, with his hatred of missionaries. Deserting in the company of his friend Toby from the whaling ship which had brought them to Nuka-Hiva in 1842, he sought refuge in the valley inhabited by the formidable Typees. There, he discovered magnificent young women, whom he considered, together with the Tahitian women, the most beautiful in the whole Pacific, and spent his days leading an idyllic existence in the company of Fayaway, swimming in the waters of the pool, in an atmosphere of muted sensuality. His memories, as he recorded them in *Typee* on returning home, are filled with his sense of exaltation at living in such a state of nature, a thousand leagues removed from the staid sobriety of civilization. As he describes it, the as yet unspoiled Pacific was truly a place of bliss.[12]

The Pacific, Fragile Refuge of Utopia

Disillusioned with democratic America, Fenimore Cooper, unlike the socialists who attempted to found an ideal community in the second quarter of the nineteenth century, did not locate his Utopia in some region of the West, but in the very heart of the Pacific. Mark Woolston, the hero of *The Crater* (1847), is marooned on a tiny, rocky, desert island, with a companion, at the end of the eighteenth century. Like Robinson Crusoe, he manages to survive by making use of every single resource offered by this initially hostile environment. Gradually, he creates a garden, a little herd of livestock, and, when a storm separates him from his friend, resigns himself to growing old on this reef, with the thundering of its breakers in his ears. But, one day, an earthquake pushes the island up above sea level, enlarging its arable surface, with a volcano rising up not far away. As a result, he finds himself lord over a considerable amount of land, just as his friend returns, bringing with him a group of Americans in search of an earthly paradise. They set about creating a new society in which all the noble instincts of human nature are to flourish, for the existence of private property is considered no obstacle to an active sense of the common good. The islands become involved in commercial activity when sandalwood is traded to China. Hundreds of settlers are attracted from America and Europe to thrive under the enlightened despotism of the founding members, and of Mark in particular. The colony defends itself against marauding natives and pirates. But when the danger is past, the time for demagoguery arrives. The wealth and authority of the original settlers comes under attack, in an ultraegalitarian and ultrademocratic spirit. They are tried and condemned. It only remains for them to return to the United States, disillusioned by their utopian experiment. When, later, they return to the island, there is nothing to be found, for a second earthquake has plunged the recently emerged dry land back under the ocean and swallowed up the democratic society, as if to punish it for its ingratitude.[13] In this philosophical novel, Cooper uses the Pacific as a setting to settle a score with what he considered the excesses of Jacksonian democracy. In spite of its vast expanse, the ocean could afford no more than a temporary haven, destined to be soon invaded by civilization, bringing both advantages and disadvantages in its wake. In the long term there is no escape, for the spirit of the age will soon catch up.

Views of the Other

The inhabitants of the shores and islands of the Pacific had always inspired contradictory feelings and judgments in the Westerners who came into contact with them. Some of the foremost representatives of the Enlightenment saw them as proof of the degeneration of Western civilization. China was seen as a perfect empire, whose subjects were able to prosper under a wise and enlightened ruler. The Pacific islanders represented a state of nature which the vices born of modern society had not yet had time to corrupt. Such ill-founded notions generally stemmed from a blissful ignorance. For traders who had lived at Canton, like Constant of Geneva, China resembled a country of thieves at the mercy of their rulers' whims rather than the most benevolent regime on earth, and a number of seamen had already noticed that the inhabitants of the islands had a propensity for petty larceny and quite often for violence—if not indeed for cannibalism.

The Americans had adopted these contradictory views concurrently. Where China was concerned, for example, there was often a tendency to contrast the early decade, during which excellent relations had prevailed between the Yankees and the Chinese, still enjoying mutual esteem, with the period following 1870, when a wave of hysterical xenophobia originating in California is often thought to have led to the exclusion law of 1882. The close relations between American business and the Cantonese hong merchants, between the Perkinses and Houqua, have been cited in support of this thesis. The Boston traders apparently displayed much more patience, tolerance, and understanding toward their opposite numbers than did the British, often guilty of arrogance.[14] A strong taste for Chinese art, which created the nucleus of the superb collections in modern American galleries and museums, seems to show that there was a genuine communication between the two cultures. And yet historian Stuart Creighton Miller has amassed ample evidence to show that the three groups which were most directly involved in China—merchants, diplomats, and missionaries—were precisely those who, at the end of the eighteenth century, contributed most to creating a negative image of China and of its people, generally depicting the Chinese as superstitious, dishonest, cruel, xenophobic, and incapable of freeing themselves from an oppressive despotism.[15] According to this theory, it was in the eastern rather than the western United States that the negative perceptions began to build up, eventually leading to the rejection of the Chinese by American society. However, such categorical arguments need to be nuanced. China provoked varying reactions in the foreigner depending on his temperament, his existing prejudices, and his view of the world. In his book *The Middle Kingdom*, published in 1848, the missionary and sinologist S. Wells Williams endeavored to show "the better traits" of the Chinese national character as well as certain faults a Christian could only find repugnant. It hardly seems likely that the average American could have garnered either an entirely negative or entirely positive picture from all the newspaper articles and studies devoted to China—it must have been rather a mixed one.[16]

On the Latin American coast of the Pacific, it was a similar story. Yankee Protestants were unlikely to find very much to their liking in these countries which were not only Catholic but in which the Church played a central role in social and political life. They found the local customs and values disconcertingly different from those of New England. Dana describes Californians as "thriftless, proud, extravagant, and very much given to gaming." As for the womenfolk, they had very little education and their morals were suspect, though fortunately held in check by the fierce jealousy of their husbands.[17] Politics was carried on under rules that democratic North Americans found quite unacceptable. The widespread corruption of the administration made business risky. In short, Latin America still had a lot of ground to make up if it was to attain the level of civilization enjoyed by the great Anglo-Saxon sister republic. But, in spite of all this, some Americans were prepared to take up permanent residence there, founding families, and mixing in with the local aristocracy. There was also a sense of inhabiting the same republican hemisphere and standing up together against the Old World monarchies. Such things showed that there was no unbridgeable gulf between the two cultures.

Where the island peoples were concerned, the cultural conflict was stronger. There was so much to differentiate the average Yankee from the Polynesians or Micronesians—the level of technological development, the role of the individual in society, religion, and customs. In the cases of China and the Latin American republics, contacts were actually quite marginal, barely extending beyond the coastal fringes, and even then mostly confined to the ports, with the result that the bulk of the population remained unaffected, or nearly so. In the islands, on the other hand—which were often tiny, with a population of only a few thousand—the impact was immediately felt by all. Native society as a whole was shaken to its foundations, for no isolated havens were available farther inland to provide resistance to the intrusion of an alien culture. Epidemics would sweep rapidly through the entire population, as in the case of the smallpox carried by the whaling ship *Delta*, which killed off more than two thousand—more than one-third of the native population—on Ponape in 1854. The population of Kosrae, in the Caroline Islands, fell from more than two thousand to only seven hundred between 1830 and 1860.[18] These demographic disasters led to a loss of confidence in the traditional religious practices, weakened hierarchic social and political structures, and condemned all efforts to preserve autonomy to failure.

Some Americans, seduced by the attractions of the South Sea paradise, sadly observed the destruction of these enchanted islands where it was possible to escape the constraints of civilization. Even though he was kept a prisoner in the "Happy Valley" of the Typees, Melville became convinced that, despite certain disadvantages, the Polynesian savage, "surrounded by all the luxurious provisions of nature," enjoyed "an infinitely happier, though certainly a less intellectual existence than the self-complacent European." Melville saw him as free of all the evils engendered by civilization—jealousy, social rivalries, family discord, and poverty. And, above all, there was "no money! That 'root of all evil' was not to be found in the valley." Nor was there any disease. In short, "all was mirth, fun, and high good humor."[19] It is true that the Typees had a few faults, of which cannibalism was one, so that in the end our hero, enthusiastic disciple of Rousseau though he was, was not so reluctant to part company with them, to avoid ending up on the dinner table—though he did consider their cruelty to be no worse than that of British executioners in the eighteenth century. For a number of Americans, the Pacific adventure supplied a sense of cultural relativism. Western civilization lost some of its luster when it came in contact with other types of social organization. Nevertheless, even for the "outcasts of the islands," who had given up all hope of returning home and whose prestige among the natives was based on their role as intermediaries with the civilized world, its superiority was never in doubt. To the majority of Americans, Melville's views were mere utopianism. There were just too many tales of conflict between the natives, of continual tribal wars, of barbarous treatment of shipwreck victims, and of the massacres of too trusting crews for the image of the Pacific as a second Garden of Eden to remain credible. No doubt Amasa Delano and many others recognized that often sailors who were themselves blameless sometimes paid for the misdemeanors of their predecessors, who had slaughtered the natives or stolen their women. But violent behavior was by no means confined to civilized man. Collective guilt was ascribed by both sides—even by the American Navy when it failed to have the actual guilty parties handed over. It is symptomatic that, in many accounts and ships' logs, the natives were referred to as "Indians." This was the image of them brought back by Horatio Hale, the ethnographer who sailed with the Wilkes expedition. The Micronesians alone merited a positive report as a gentle, friendly, peaceful race, little inclined to licentiousness, and very intelligent—but then the expedition had only made brief contact with those islands. Wherever there was a prolonged and direct contact with the population, their negative traits were always stressed. The Polynesians were described as happy, ready to adopt new customs, but cruel and ferocious in war, inclined to debauchery, and displaying little love for their children. They were mendacious, hypocritical, thieving, covetous, and devoid of moral conscience, although not naturally deceitful, contrary to the Melanesians of Fiji, who were seen as bloodthirsty and rapacious, mistrustful of one another, and among whom it was unsafe to go about unarmed. For Hale, the Polynesian was to the Fijian as the dog to the wolf—a comparison which shows how formidable was the task confronting those Americans who wished to civilize the native peoples.[20] The Pacific was an extension of the American West, a frontier where savagery and civilization came face-to-face.

A Protestant Mission Field

For Americans in the nineteenth century, civilization went hand in hand with the Christian religion and, specifically, with its Protestant variety. The two were inseparable, and one was inconceivable without the other, in the medium term at least. At the outset, however, it was possible to debate the order of priorities. If the heathen natives were to be raised from a condition of barbarism to civilization, was it necessary to convert them first, or rather to give them the rudiments of a modern—which is to say American—education, to open their eyes to the true faith?

Urgency of the Protestant Missionary Effort

The sending of American missionaries to the Pacific originated with the spiritual crisis which shook New England at the end of the eighteenth and beginning of the nineteenth centuries. The Congregationalists and Presbyterians were forced on the defensive by the assault of the Enlightenment, whose most notable success was the triumph of Unitarianism at Harvard in 1805. The rise of rationalism was the first sign that the Calvinist beliefs of the first pioneers were being eroded. It was, however, in the West that Christianity first began to fight back, after 1799. The Second Great Awakening, launched by the Methodists and Baptists, set out to save souls from sin and from damnation by means of revival meetings that relied too much on emotionalism and on fear for the taste of the older eastern Calvinists. The latter believed that if conversions were to last, they had to be supported by an institutional framework, instead of leaving each sinner alone with his conscience. To resist the Unitarian enemy, the orthodox Calvinists built new churches.[21] The opening of the Andover Theological Seminary in 1808 put spiritual revival on a sound footing. Samuel Hopkins's New Divinity Movement, with its appeal to religious activism, to total dedication to serving the greater glory of God, to a "disinterested benevolence," awakened the enthusiasm of many young people. In an attempt to re-create their lost unity, and feeling threatened by a loss of prestige and by the attempts to disestablish the official churches, Congregationalist and Presbyterian ministers, brought together by the Plan of Union of 1801, turned to foreign missions as a way to enlist all the New England Protestants in a common cause, which it was hoped would heal the rift between the denominations and bridge their political differences. By turning overseas, New England Calvinists would rediscover the bonds which had united them. This was the thinking behind the creation of the American Board of Commissioners for Foreign Missions (ABCFM) in 1810.[22] The American foreign missions imitated the British model, as established by the London Missionary Society fifteen years earlier, in bringing the Gospel to heathens and saving them from eternal damnation.

The ABCFM wasted no time in getting to work. The first task was to recruit missionaries. Fortunately, there was no shortage of suitable candidates in the villages and small towns of inland New England—the Unitarian coast supplied very few. They were usually young people who would be prepared to go overseas shortly after getting married. College-trained, they were better educated than their British colleagues.[23] Secondly, to send them overseas and provide them with a decent living when they got there, a fund-raising effort would be required. To put this into effect, the populace was mobilized and convinced of the urgency of the cause, for there was little time to lose. Millenarians were depicting Napoleon as the Antichrist, and prophesying a thousand years of peace to follow the second coming of Christ. But before this new era came, the souls of the heathen must be saved. Despite this onslaught of religious propaganda, however, the New Englanders were slow to open their wallets, for they were still suffering, in the aftermath of the war of 1812, from the consequences of the blockade and the subsequent economic depression. The growth of missionary aid societies and of specialized fund-raising organizations finally overcame the reluctance to contribute. However, if the flow of financial support was to continue, positive results had to be shown. The opening of a school for heathen natives at Cornwall, Connecticut, in May 1817, was a key factor in the missionary campaign. Five of the first twelve pupils came from Hawaii, among them the young Henry Obookiah, who converted to Christianity, but unfortunately died in February 1818. His *Memoirs* would go through twelve editions and give a strong boost to enthusiasm for the missions.[24] However, the real test of the effectiveness of the Lord's work was not at home in the United States, but out in the field, among the native peoples. The first missions sent to India, though, were a failure. By 1820, some signs of success were eagerly awaited from the Pacific.

Success in Hawaii

American seamen had already been frequent visitors to the Hawaiian islands for thirty years when the ABCFM decided the population should be converted to Christianity, and brought the benefits of civilization. According to Hiram Bingham, who was to spend twenty-one years in Hawaii, the purpose was

> to honor God, by making known his will, and to benefit those heathen tribes, by making them acquainted with the way of life,— to turn them from their follies and crimes, idolatries and oppressions, to the service and enjoyment of the living God, and adorable Redeemer,—to give them the Bible in their own tongue with ability to read it for themselves,—to introduce and extend among them the more useful arts and usages of civilized and Christianized societies, and to fill the habitable parts of those important islands with schools and churches, fruitful fields, and pleasant dwellings.[25]

On October 23, 1819, the brig *Thaddeus* sailed from Boston for the Pacific with a little party of missionaries on board. It was made up of seven American couples (two clergy, Hiram Bingham and Asa Thurston, both recent graduates of Andover, a doctor, two teachers, a printer, and a farmer) and three Hawaiians. Five months later, this advance party of Christians landed on Hawaii, where, by a fortunate coincidence, they heard the happy news: "Kamehameha is dead—His son Liholiho is king—the tabus are abolished—the images are destroyed,—the *heiaus* of idolatrous worship are burned, and the party that attempted to restore them by force of arms has recently been vanquished!"[26] Actually, the powerful ruler who had unified the Hawaiian islands, Kamahameha I, had died in May 1819, leaving power jointly to his son, who was much more interested in drinking than in the oppressive traditional rites, and to his favorite widow, Kaahumanu, who wanted to abolish the taboos applying to women. After thirty years of contact with Westerners, the Hawaiian islands were ripe for a spiritual crisis. Polytheism, with its accompanying human sacrifices, was provoking disaffection among many of the chiefs and the downtrodden common people. Within a few months a vacuum had been created which would soon be filled by the arrival of the missionaries from Boston.

They did not meet with immediate success, but in a society as hier-archical as the one ruled over by the Hawaiian monarchy, the tactics they had adopted did not take long to bear fruit. The missionaries first turned their attention to the ruling class, the aristocracy of the chiefs, whose conversion would encourage the subordinate classes to opt for Christianity. If King Liholiho (1819–1824) showed little enthusiasm for the arrival of these missionaries bringing a new moral code, the women, who played a capital role in the political life of the archipelago, were very soon transformed into fervent Protestants. These included Kaahu-manu, who acted as regent during the early part of Kamahameha III's rule, from 1824 to 1832, and Kinau, "prime minister" from 1832 to 1839. They imposed respect for Puritanism. Hawaiian women were strongly urged not to indulge in prostitution on board visiting ships. Consump-tion of strong drink or failure to observe the Sabbath meant punish-ment for anyone caught in the act. The aim was to mold a new kind of Polynesia, based on Bible teaching. If this was to be achieved, the Word of God had to be made accessible to all. The missionaries quickly suc-ceeded in finding a way to write the Hawaiian language. Bingham wrote: "Aiming to avoid an ambiguous, erroneous, and inconvenient orthography, to assign to every character one certain sound, and thus represent with ease and exactness the true pronunciation of the Hawai-ian language, the following five vowels and seven consonants have been adopted: *a, e, i o, u, h, k, l, m, n, p,* and *w.* These twelve letters, and pos-sibly eleven, omitting either *u* or *w,* will express every sound in the pure Hawaiian dialect."[27] Such a simple alphabet made it an easy matter to school both adults and children, who quickly learned to read in their own language. As early as January 1822, the first pages printed in Hawaiian rolled off the presses the missionaries had brought with them. The Gospel message, in the true Protestant tradition, became easily ac-cessible to anyone who regularly attended school—which was the high road to conversion and Church membership. In 1828, eight years after they had landed, the Protestant missionaries could see the enormous progress that had been made:

> The heads of the nation and twenty subordinate chiefs being in
> favor of their work, and none to oppose their settlement; six con-

gregations, embracing about 12,000 hearers, 26,000 pupils, and 440 native school teachers, all needing their help, and nearly 100,000 of the population waiting for the means of competent instruction, while the spirit of renovating grace appeared to be hovering over the congregations where the gospel was regularly preached.[28]

But this progress was too rapid and poorly consolidated, for on Kaahumanu's death a reaction set in, and in 1832 and 1833 a number of converted or educated Hawaiians recanted, reverting to their traditional way of life. This setback only made the missionaries redouble their efforts. On the island of Hawaii, the Reverends Titus Coan and David B. Lyman began, in February 1837, to use the techniques of the Second Great Awakening. Gospel meetings would last for two weeks, during which sinners would be overcome with trembling and collapse in tears when presented with a vivid description of the torments which awaited them in the afterlife if they failed to repent in time, while salvation was assured for those who accepted Christ's word into their hearts. Between 1838 and 1840, the Great Awakening swept through the entire archipelago, making 20,000 new converts. Henceforth, the Hawaiian islands could be considered a Christian land—a remarkable success, achieved within only twenty years. The funding devoted to Hawaii could now be decreased, beginning in the late 1830s.[29]

Evangelization did not necessarily imply Westernization, nor, more specifically, Americanization. As the use of the native language indicates, the ABCFM did not pursue a colonialist agenda at the outset. Nor did it consider its role was to represent the United States, but merely to propagate the Christian faith. However, it is difficult to separate New England Congregationalism from the broader political and social context. It was often inseparable from a certain conception of the individual's place in society, of his rights, of his responsibilities, based upon democratic institutions. There was a considerable temptation for missionaries to turn themselves into counselors to the prince, in order to encourage him to set his subjects the example of a truly Christian life. There was a tendency toward "theocracy"—a State where the laws would be inspired by Gospel principles. Such, certainly, was the objection raised by the other foreigners, and by Americans in particular, whether they were bar owners, or whaling captains relying on the easy morals of the Hawaiian girls to provide an escape valve for the pent-up energy of their crews, or speculators who found it easier to extract concessions from the native chiefs by plying them with alcohol.[30] In 1826, Lieutenant Percival, of the United States Navy allowed his sailors, enraged by the taboo placed on the women going aboard ships putting in at Honolulu, to physically attack the Reverend Bingham. In 1839, Laplace, a French captain who had come to bring some Catholic priests to this Protestant dominion, declared himself willing to protect all foreigners, except for the American missionaries. In response to these accusations, the latter claimed that there was a genuine separation of church and state in Hawaii, for the natives in power were free to adopt whatever measures they deemed appropriate for the proper functioning of their society, so that if the puritan missionaries provided them with guidance this was indeed a reason to rejoice, but they did not play any active part in either the deliberations or the decision making. This argument carried little weight, particularly after 1838, when former missionaries became government ministers under Kamehameha III. Not only was moral legislation inspired by Gospel teaching and modeled on that of New England communities, but Western political structures were also introduced. In 1840, a written constitution restricted the powers of the traditional monarchy. The king yielded his legislative powers to a bicameral Parliament, composed of a Council of Chiefs and a House of Commons elected by universal suffrage, and was only allowed to carry out his executive functions with the approval of the prime minister. He sat on the bench of the Supreme Court alongside four judges chosen by the elected representatives. A mixture of distinctively American principles and British customs can be seen in all this, creating a system which was at the same time both conservative and democratic. Some would have liked to go further, to achieve complete Westernization, but, according to Rufus Anderson, secretary of the ABCFM, such a trend would have lost sight of the principal objective of the missionary effort, which was to evangelize, and not to modernize:

> It has seemed to be the mistake of some, wrote Anderson, that the great object of American missions should be *to reproduce our own religious civilization in heathen lands*, and just in the precise social and religious forms which that civilization has in this country. It may be that the Gospel will produce just these results, in process of time, all over the world; but that is not the proper object of Gospel missions. Their object is to proclaim salvation for immortal souls,

through repentance and faith in the Lord Jesus . . . It is no fault of the mission at the Sandwich Islands, that the social progress at those Islands, under the preaching of the Gospel, has resulted in a constitutional monarchy, rather than a republic.[31]

However, by 1840, the groundwork had already been laid for the Americanization of Hawaii. Apart from the massive, but temporary, presence of the whaling ships, with their whole commercial retinue, nothing tangible could have been accomplished without the sixty-five men and seventy women sent by the ABCFM between 1819 and 1847. It was they who turned the Hawaiian islands into a bastion of Protestantism, dominated by American cultural influences.[32] In 1848, the Boston organization felt that the conversion of the populace had gone far enough for them to envisage an independent Hawaiian church, where Americans, whether from the United States or born locally, and native Polynesians would worship side by side. Fifteen years later, it transferred its remaining responsibilities to the Hawaiian Evangelical Association. Nevertheless, it was the resident or naturalized Yankees who retained control. The indigenous culture, weakened by a declining population, was losing ground, contrary to the hopes of the first missionaries. In 1853, the government chose English as the language of instruction in Hawaiian schools. Conversion to Christianity was leading inexorably to Westernization.

Failure in China

Whereas in Hawaii the Americans were the pioneers and the sole representatives of the Protestant mission, in China they merely followed in the footsteps of their British colleagues. In 1807, the London Missionary Society had sent Robert Morrison to Canton, making a detour through the United States because of opposition from the British East Indies Company. It was not until more than twenty years later that the ABCFM decided to enter this new field. In 1829, the Olyphant company granted the Boston organization free passages for two of its clergy, David Abeel, who was put in charge of the spiritual well-being of the crew, and Elijah C. Bridgman, who was given the heavy responsibility, as a good soldier of the Lord, to "go to attack the prince of darkness in his most imposing fortress"—i.e., the Chinese world, still closed to foreigners at the time.[33] This first little group received reinforcements in

1833, with the arrival of Samuel Wells Williams, a printer destined for a fine career. In the following year came Peter Parker, a theologian and doctor, who opened an eye hospital which was soon to become famous. Other Protestant denominations followed the example of the Congregationalists—the Episcopalians in 1835, the Baptists in 1836, and the Presbyterians in 1838. Before the country became open, the missionaries had almost no possibility of converting the Chinese themselves, but spent the time among the emigrant communities of Southeast Asia, Siam, Singapore, and Borneo, learning the language.[34] They collected information on China, which they published in the best source of the day, the *Chinese Repository (1832–1851)*. For the Chinese public, Bridgman wrote *A Brief Description of the United Provinces of America* (1838), as an introduction to his country, its liberal and democratic institutions, its economic and technical advances, all in a slightly idealized form, and ending with the slogan: "a rich State, a strong army," which would later be an inspiration to the Chinese reformers. The missionaries also published Christian tracts, which would later find an echo in the ideas of the Taiping rebels.

While hostile to the opium trade, the American Protestants saw the hand of Providence in the Opium Wars between China and England, which finally opened China to the light of Christianity. For them, the end justified the means.[35] As early as 1844 they were granted the right to work in the open ports, and the treaty of 1858 allowed any American Christian or Chinese convert to teach and practice Christianity without fear of persecution. The various denominations soon took advantage of this to broaden the scope of their activity. The ABCFM sent its representatives to Xiamen in 1842, to Fuzhou and to Shanghai in 1847, the Presbyterians to Ningbo in 1844 and to Canton in 1845, the Episcopalians to Shanghai in 1845, the Northern Baptists to Hong Kong in 1843 and Ningbo in 1847, the Southern Baptists to Canton in 1845 and Shanghai in 1847, the Methodists to Fuzhou in 1847. After 1860 the ABCFM focused more and more on northern China, while the Presbyterians advanced into the Yangtze valley and Shandong, and the Baptists into the hinterlands of Canton and Shantou.[36] Nevertheless, this influx must be viewed in context, for by the time of the Civil War there were still only a few dozen American missionaries in China.

In the circumstances, the kind of success experienced in Hawaii was hardly to be expected. In Fuzhou, the missionaries encountered innumerable difficulties. In the first place, health problems were responsible for a substantial turnover of personnel. Furthermore, official restrictions prevented their finding a site for a house of worship in the center of town, forcing them to make do with a less prestigious location on the outskirts. Finally, there was the need to master the local dialect if they were to speak to the ordinary people, and not only to the educated class (or the "gentry," as an English historian might call it). The American government had no intention of indulging in gunboat diplomacy to support its nationals, and the missionary endeavor could at best count on nothing more than the goodwill of the local American consul. The general atmosphere in the 1850s was therefore not very favorable to missionary activity. In spite of this, the missionaries were not idle. They tried to spread the Gospel by distributing tracts, by preaching in the streets, by opening chapels and churches, welcoming young Chinese pupils into their day and boarding school, and practicing medicine. They embarked on a heroic translation effort into Mandarin and into the Fujian dialect, completing the so-called "Delegates' Version" of the New Testament in 1850, without resolving the dispute about the best way to translate the name of God, in the end allowing each denomination use the term it preferred—*Shang Ti* for the English, *Shen* for most of the Americans. In Ningbo as in Fuzhou, the use of the vernacular met with opposition. In spite of all these efforts, the results were disappointing. The Congregationalists had to wait nine years before recording their first baptism—an individual who three years later was expelled from the Church for unseemly behavior. By 1860 they had made only twelve converts in total, seven of them young boarding school pupils. Even the Methodists, with fifty-four converts, had little reason to boast. After 1860, the statistics improved, for the hinterland was opened up, so that places of worship could be built in the countryside and in small towns. With Chinese converts serving as lay ministers, these churches were more attractive to the local people, and particularly to the poor, the downtrodden, and the socially marginal. It was a modest success for the missions, but it provoked waves of xenophobia, like in 1871, for example, when the baseless rumor was put about that the Christians, enraged at their rejection by Chinese women, were poisoning the food and the water.[37]

Around 1870, in comparison with Hawaii, the efforts of American Protestant missionaries in China seemed to have been a failure. The reasons for this lack of success were many. The missionaries blamed the "literati," the mouthpiece of the official Confucian ideology, who rejected Christianity because of its threat to the social order they represented. The belief was that without the deceitful opposition of these people, the ordinary Chinese would have been much more open to evangelization. But there was a lot of self-deception in this, for the ordinary people themselves had an instinctive distrust of foreigners, and were quite able to see that Christianity went hand in hand with foreign penetration. The Chinese converts tended to take advantage of their newly acquired status to represent any sort of quarrel as a religious one, enabling them to ask for the support of the consular authorities in disputes with their neighbors.[38] In addition, Christianity was still considered a "barbarian" religion, and often an utterly incomprehensible one, the doctrine of the Trinity being a notable example. The Chinese were completely devoid of one trait fundamental to the Calvinist tradition: a sense of sin. Conversion required them to adopt a totally new lifestyle, to break with ancestor worship and with the still traditional family ties, and the missionaries would not allow any latitude in these matters. This meant that the convert might well become a social outcast, a pariah, constantly subject to threats. The very poorest, with nothing to lose, were less susceptible than others to such disadvantages. And it should be added that the example set by many foreign Christians—opium traders or frequent patrons of the brothel districts—did little to demonstrate the moral superiority of Christianity.[39] After forty years of activity on the threshold of China and in the coastal areas, American Protestantism still had to overcome such a multitude of obstacles that it had a very long way to go if it was to bring its crusade against heathen China to a successful conclusion. The two cultures were still as mutually impenetrable as they had been at the outset.[40]

Victory in Micronesia

In the minds of its promoters, the Hawaiian mission was to be a jumping-off point for the conquest of heathen souls throughout the Pacific archipelagoes. A first attempt in the Marquesas in 1833 ended in a hasty retreat after only a few months, when the missionaries, and particularly their wives, found the uninhibited sexual mores of the natives intolerable. By 1852, however, they felt ready to challenge Satan in his principal lair, in Micronesia, on Ponape and on Kosrae, where American whaling ships were putting in in ever-increasing numbers in search of water, wood, and women. Two missionary couples, the Sturgeses and the Gulicks, landed on Ponape from the schooner *Caroline*, while another, the Snows, accompanied by Hawaiian deacons, went on to Kosrae, to counteract the immoral depredations of their "unprincipled" compatriots, and save the souls of the ignorant, depraved, and superstitious native people. On Kosrae they received a warm welcome from King George, the single, all-powerful chief. On Ponape, which was divided into four rival chiefdoms, each governed by a sacred personage, the *Nahnmwarki*, and a *Nahnken*, responsible for practical matters, they decided to begin their activity in the south, at Kiti. There, the *Nahnken* was most welcoming and willing to learn, realizing this would enable him to be better able to trade with the whalers and thereby increase his political power. In the following year, Luther Gulick moved to the chiefdom of Madolenihmw, in the western part of the island, where his medical training stood him in good stead with the natives. On the other hand, eighteen months later, in 1855, the mission founded in the north, at Sokehs, turned out to be a failure. Doane, the missionary, was unwilling to comply with the custom of exchanging gifts which was an important underpinning of the power of the local chief. So, from the outset, the Protestant missions relied on the native authorities in undertaking their work of evangelization, with the result that the representatives of Christianity, seeking to subvert an existing social order based on a close association between the chiefs and the gods, were obliged to make an alliance with the very people who embodied these traditional values, and who therefore had no interest whatsoever in encouraging their people to renounce polytheism.[41]

On heathen Ponape, where whalers were numerous, the missionaries had their work cut out for them. First, they had to struggle against prostitution on board the ships. The *Nahnken* of Kiti assisted them in this, but his only reason for doing so was to prevent his subjects from trading directly with the sailors, a privilege he wanted to keep for himself. He therefore closed only the brothels on shore controlled by foreign drifters like the black Philadelphian, Edward Johnson, nicknamed "the Terror of the Pacific," when Johnson was killed in 1858 in a fight with some ruffian no better than himself. Second, an alphabet had to be devised for the Ponapean language, in order to translate the Bible and provide a basic education for the natives. The arrival of a printing press meant the pupils attending schools taught by the missionary wives could be supplied with textbooks. The terrible smallpox epidemic of

1854 provided the Protestants with an opportunity to demonstrate the effectiveness of vaccination, but this was of little help in encouraging the surviving inhabitants to convert. The traditional social structures, with their innumerable feast days celebrated by drinking kava, a narcotic beverage, resisted every assault. The missionaries had little to show for their efforts. In 1860, Albert Sturges wrote home to Boston that, after eight years on the island, the Church could still boast only twelve members. During the Civil War, when the mission was isolated from the outside world, except for a few rare appearances by ships, more rapid progress was made, in the face of fierce opposition from some chiefs, and in particular from the new *Nahnken* of Kiti. This was because a Christian party had been formed among the natives, was gaining in strength, and growing ever more able to resist attacks from the heathens. By 1867, the battle to convert Ponape had been won. On Kosrae, things evolved in similar fashion. During his ten years spent on the island, following King George's death in 1854, Benjamin Snow had to deal with the chiefs' hostility and the resistance of the old order, so that it was not until 1858 that he was able to welcome the first two members into his church. But after these modest beginnings, he made ever more rapid progress, so that by the time his tour of duty was finished he was able to leave behind a Christian party strong enough to withstand the last, dying resistance of tradition. In 1868, King George's son became the first native pastor in Micronesia, and, in the following year, King Sipe had no alternative but to accept democratic reform, surrounding himself with an elected council, made up mostly of Christians, which would depose him five years later and replace him with the first Christian King of Kosrae, a purely constitutional monarch.

Not content to erect these two bastions of Christianity in the Carolines, the American missionaries used them as a jumping-off point to the Marshall and Gilbert Islands.[42] Until that time, the natives of the Marshalls had been extremely feared by sailors because of their reputation for cruelty and greed. In spite of this, in 1857 the *Morning Star* put two missionaries, George Pierson and Edward Doane, ashore on Ebon, at the southern extremity of the archipelago, while Hiram Bingham Jr., the son of the Hawaiian apostle, accompanied by a Hawaiian schoolteacher, settled on Abaiang in the Gilbert Islands. The history of these missions is a repetition of those of Ponape and Kosrae. On Ebon, a pleasant surprise awaited the ABCFM missionaries, for the chief, Kaibuke, despite his ferocious reputation, gave them a warm welcome and continued to protect them until his death. With no need to struggle against white drifters, who had not yet reached these shores—dealers in copra oil would not arrive until the following year—the missionaries were free to spread the Gospel message throughout a very mobile population, by which it was carried from one island to the other. The Bible was translated into the local language, and schools were opened. Thanks to the energy of the Hawaiian auxiliaries, ten natives of the Marshall islands were considered worthy of admission into the Church in 1862, and things evolved smoothly from then on. As happened on Ponape and Kosrae, the death of the missionaries' initial protector—Kaibuke in this case—encouraged the traditional chiefs to persecute the Christians, in an effort to preserve their power. The converts, encouraged by Snow, who had arrived in 1862, stood their ground and were able to undermine the heathen resistance. As a result, Ralik and Rabak, the two island chains making up the Marshalls, willingly welcomed the Gospel message around 1870, all the more so in that it was preached most effectively by native converts, to whom the American missionaries were able to pass the torch. In the Marshalls, as in the Carolines, their effort met with success in under twenty years, and this was achieved without their being used as the thin end of the wedge for any kind of imperialist venture by the Washington government (whereas France, England, and Germany frequently used the religious conquest of souls as a first step to territorial conquest). The American Protestant mission in Micronesia was a private undertaking, and went no further than to carry out the dual task it had originally undertaken, which was to save the heathen from hell fire—or at least those of them who did not perish in epidemics or tribal warfare—and to introduce these "barbarous" peoples to modern civilization by teaching them to read and write, tending to their sicknesses, and instilling in them a way of life less ferocious and more respectful of human dignity than the one they had known.

From a cultural standpoint, the Pacific still was a frontier for the Americans when the Civil War broke out. Contacts with the native peoples were few. Those made by traders and military diplomats were no more than sporadic, while those made by the missionaries were spectacularly successful in only one location: Hawaii. Elsewhere, for the handful of islands or atolls which were converted to Christianity, they had to neglect as many more, to the considerable sorrow of Protestants tormented by the urgency of their task. As for China, with its masses of humanity, appealing like nowhere else to those who dreamed of saving millions of lost souls from the clutches of Satan, the surface had barely been skimmed. Such a limited familiarity gave rise to stereotypes, to an incomplete perception of reality. Superficial cultural understanding opened the door to all kinds of interpretations. For Melville, the American who had lived among the cannibals of the Marquesas islands, the Pacific was a paradise inhabited by noble savages—though it did have its disquieting side. For others, the islands were inhabited by devilish heathens, China by callous fiends, and the coast of Latin America by abominable papists. But rare was the American who did not come home from the Pacific convinced of the superiority of his own, modern, Christian civilization. Apart from their picturesque qualities, cultures are not equal. When they come into collision, the superior strength of one invariably upsets the equilibrium of the other, and may bring about its eventual disintegration by totally destroying what binds it together. At the time of the Civil War, the clash between American culture and the other cultures of the Pacific region was still at a stage when the initial impact was being felt. The immense, virgin expanses of the cultural frontier still stretched on forever.

PART II

TOWARD MATURITY
(1868–1941)

The initial stage of the frontier's history came to an end during the Civil War decade. The low-density exploitation of natural resources was now ended, and whaling fell into an irreversible decline. Distance, which until then had made the Pacific marginal to America's center of gravity, became less of an obstacle. With steam navigation came regular communications from one coastline to the other, while the invention of the telegraph meant messages took hours instead of weeks to reach their destination. As a result, economic relations became more intensive, concentrating on the North Pacific, and particularly Japan. Indeed, after exerting its pull for a century, the Chinese magnet had lost its power, and China was no longer the focal point for trade in the Pacific. Little by little, it was left ever further behind by Japan, whose situation, however, was different, since Japan's economic relations with the United States were essentially bilateral. One indication of the frontier's growing maturity was the development of more intensive trade while at the same time the spatial factor was reduced. As a result, American political and strategic interests in the area became much more crucial. More or less by design, the United States became one of the imperialist powers in the Pacific Ocean, carving out its own colonial empire and confronting other expansionist nations, the most threatening of which was the Japanese Empire. As antagonism grew between these two naval powers during the first half of the twentieth century, the early contours of the frontier became more sharply defined. The implicit limits of the various zones of national influence fragmented the region, multiplying the points of contact and, as a consequence, of confrontation or cooperation. Indeed, in relations between nations there is no simple correlation between the creation of closer ties and a guarantee of more cordial feelings. Increased cultural contact does not necessarily lead to more friendly attitudes, and no model of behavior is able to predict whether public opinion will be more sensitive to the qualities or the defects of another people—or, rather to a perception of defects or qualities which generally results from group stereotypes superimposed on the only real differences, which are individual. In this second stage of the frontier's development, as it progressed toward maturity, the shifting nature of many relationships leaves a great deal of room for particular events and circumstances to play an important part. History records the major channels along which human actions flow, but this does not mean they could not have moved in a different direction.

FIVE

More Intensive
Economic Relations

⎯ In spite of the fact that prior to the Civil War American merchant vessels were to be found everywhere in the Pacific, the volume of trade was small. The commercial geography of trade was still that of an age when the globalization of economic relations was in its infancy. Technology did not allow for the conveyance of cumbersome goods over long distances. Because of its enormous area, therefore, the Pacific had been relegated to the margins of international trade. But this all was to change starting in the last third of the nineteenth century. The steamship and the telegraph partially abolished the tyranny of distance, making it possible to link regions which until then had been relatively cut off from one another. The transport revolution provided the United States with opportunities to intensify relations with the Pacific. However, this development took place against a background which at first sight was not very encouraging, for between 1880 and 1940, with the exception of the World War I years, foreign trade tended to play an increasingly minor role in the American GNP, as it shrank from 15 percent to 6 percent.[1]

A New Commercial Geography

Major Trends

Expressed in current dollars, both imports and exports showed strong growth between 1868 and 1941 (cf. Appendices, Table 8). Imports, which remained at under $30 million early in the period, grew to more than a billion dollars by the eve of World War II, while exports did not lag far behind. Annual growth rates of 5.2 percent for imports and 6.5 percent for exports were typical of this expansion, which lasted for three-quarters of a century. In constant dollars, the figures were similar—5.4 percent and 6.3 percent, respectively—since overall price trends remained basically stable, despite occasional fluctuations. These figures indicate a definite acceleration of trade with the Pacific as compared to the preceding period, from 1821 to 1867.[2] The growth is even more marked when seen in terms of volume per head of population, for the growth of the American population was beginning to slow down. If we take this into account, the rate of growth was three times as fast.[3]

This increase in trade relations did not escape repercussions from the major events which were shaking the world at the time. Before 1914, imports and exports each went their own way, except between 1877 and 1896, when they rose and fell almost in lockstep. Imports grew very slowly between 1881 and 1912, while exports benefited from a period of increased economic activity at the end of the nineteenth and early in the twentieth centuries, reaching a peak in 1905, when an explosion occurred in the sales of American manufactured goods. After World War I, on the other hand, imports and exports fluctuated in relative unison, with the prosperous years between 1915 and 1929 contrasting with the Great Depression of the 1930s. The disparity between the two rates of growth must, therefore, be attributed to the evolution which took place before 1914.

Thanks to the stronger growth in exports at that time, the trends observed in the preceding period continued. The United States had almost always incurred a deficit in its trade with the Pacific, but the average ratio of exports to imports (64 percent) was an improvement compared with before the Civil War, and it was improving at an average rate of 0.72 percent per annum.[4] The Pacific region therefore occupied a special place in American foreign trade. But while the United States was building up trade surpluses during this period (during which the overall trade balance went into the red only three times after 1876) its sales never achieved a balance with its purchases, apart from three occasions, in 1905, 1921, and 1938. Clearly, the Pacific countries were important not only as customers but also as suppliers.

This dual function took on even more importance as the Pacific's share grew strongly, in both imports and exports. The average figures for the period 1868–1941 can give a false impression. During this period the United States did indeed make only 20 percent of its purchases and 10 percent of its sales in countries of the Pacific basin—percentages much lower than those involved in trade with Europe and Canada. However, the more meaningful figures are those which indicate the trends. Around 1870, imports from the Pacific region amounted to 9 percent of total imports, but were a third just before World War II. Exports also climbed, from 4 percent to over 20 percent. Increasingly, Americans were coming to rely on goods from the Pacific, while gaining larger and larger markets in the region.

However, the picture for the Pacific as a whole was far from uniform. American trade with some regions developed more rapidly than with others. The major phenomenon was China's decline, as it lost its dominant position to Japan, after 1890 in imports and after 1907 in exports. Very quickly, Japan, which at the beginning of the Meiji era had scarcely been open to trade, now became the United States' primary trading partner in the Pacific. During the first third of the twentieth century Japan developed close ties with the United States, while China, first as an empire and later as a republic, continued to lose ground, soon forfeiting the unrivaled magnetic power it had possessed before the Civil War. From a purely economic point of view, starting from the late-nineteenth century, America had more interests in common with Japan than with China. Japan—which, apart from the period 1932–1940, had large trade surpluses with the United States— looked on the Americans as special customers, indispensable to the stability of their empire, and also increasingly important as suppliers of the fuel needed for economic growth. Japan's role was not as vital for the United States, but they did see it as an excellent customer, who bought an increasing share of its production, while supplying it with raw materials and other commodities which it was not able to produce for itself. Even if the balance of American trade with the Japanese Empire was lopsided, it was much less so than with Southeast Asia, which imported relatively little from the United States—even though the amount was increasing constantly—and which shipped increasing quantities to America, to the extent that they accounted for almost half of its imports by the time of the New Deal. The Dutch East Indies, Malaya, and the Philippines, together with Japan, were the principal source of commodities in the Pacific from the 1890s on. These two areas made inroads, at the expense not only of China, but also of the other subregions of Latin America and Oceania. Japan and Southeast Asia accounted on average for half of America's imports from the Pacific, while half of the exports to Pacific countries were destined for Latin America and Oceania. However, while the relative share of Japan and Southeast Asia continued to increase, the predominance Latin America and Oceania had enjoyed early in the period was being slowly eroded. Nevertheless, even if the economic interest of the latter was declining, they offered the United States appreciable advantages, namely, customers for manufactured goods, more balanced trade relations, and even a positive trade balance where Oceania was concerned.

These divergent trading patterns can be explained by changes in the nature of the commodities traded during the period.

Imports: Silk Takes Over from Tea

While United States trade policy was highly protectionist, except for a brief spell of liberalism under the Wilson presidency, the strong growth of imports from the Pacific was due to the fact that the commodities produced in the region offered little or no competition to American manufacturers and farmers. The tropical zone was a source of commodities and raw materials derived from the vegetable and mineral kingdoms (coffee, tea, cocoa, bananas, spices, copra, rubber, gums, and fibers from the former, nitrates and tin from the latter), which the United States was unable to produce either profitably or at all. The Pacific also satisfied certain needs which could not be completely fulfilled by American producers—for sugar, wool, and copper, for instance. On the other hand, except for some ores which had undergone the initial stages of refining, very few manufactured goods are recorded—silks, matting, porcelain, hats, and fireworks (cf. Appendices, Table 9)—and these originated exclusively in China or Japan.

In trade with the United States, each Pacific country specialized in one or two dominant exports (cf. Appendices, Table 10). Production was little diversified in the region—a typical symptom of underdevelopment.

Even in the case of the most advanced country, which was Japan, at the beginning of the twentieth century 70 percent of its exports to the United States was accounted for by two products, silk and tea. In the circumstances, it is not surprising to see that Chile had little to contribute apart from nitrates, Colombia from coffee, Ecuador from cocoa, Malaya from tin, the Philippines from Manila hemp (abaca), the Dutch East Indies from sugar, and Australia from wool. China was the only exception, for its exports to the United States seem to have been more diverse than before the Civil War, despite a major decline in textiles.[5]

Tea, the commodity which had been at the heart of all American trade in the Pacific during the first half of the nineteenth century, was destined to be toppled from its perch. This was because per capita consumption in the United States, having reached a peak between 1870 and 1895 at around 21 ounces per annum, fell into a rapid decline throughout the twentieth century, dropping to approximately 12 ounces (350 grams) in the 1920s, and falling even farther during the Great Depression. This was in contrast with the rapid rise in coffee consumption, which doubled between 1870 and 1940, in spite of the fact that the price of tea fell continuously after 1874. In addition, tea, which was considered an essential commodity, was exempted from customs duty—a privilege which it lost only between 1898 and 1902, to help finance the war with Spain.[6] The average American consumer, in spite of an improved standard of living, was showing less and less interest in this infusion for whose sake previous generations had been ready to confront the perils of the deep. No doubt this can partly be accounted for by the country's changing ethnic composition, with the influx of coffee-loving Germans, Scandinavians, and Italians. At the end of the nineteenth century, dealers accused their Chinese suppliers (who dominated the market at the time) of shipping them an inferior product, but no such accusation could be leveled at the Japanese, who had been gradually encroaching on the market of their main Asian competitor ever since they had shipped their first cargo of tea in 1863. Owing to Japan's concern for the quality of their product, from 1876 on the volume of Japanese tea sold in the United States was greater than that of either Chinese black or green teas. Thirty years later, the quantity of tea imported from Japan was greater than that of all the Chinese teas put together.[7] When steam navigation came into its own in the North Pacific, Japan was well placed to supply the American Midwest through San Francisco and via the transcontinental railroad—though New York would remain the main market for Chinese tea for a long time to come, thanks to the Suez Canal. But the transport revolution sounded the death knell of the careers of the great ship-owning traders who had formed the merchant aristocracy of the northeastern seaports. The decline in their fortunes went hand in hand with that of tea.

Tea's predominant position was taken over by raw silk, which, however, would never come to dominate trade the way tea had. The success of silk was due to American protectionism, the objective of which was to lessen dependency on silk imports from Europe. Exceptionally high tariffs imposed during the Civil War had allowed a new silk industry to develop, so that Paterson, on the outskirts of New York, found itself in a position to compete with French silks from Lyons. Greater purchasing power in the United States, and the growing prosperity of the middle class, brought about a much larger market for materials previously reserved for the rich. American manufacturers continued to enlarge their range of products to satisfy this enormous clientele, leaving luxury items to the French and the Italians. In 1914, only 10 percent of the silks purchased in the United States was imported. But, since silkworm farming had never caught on, raw materials had to be purchased abroad, some from Italy, but most of it from the Far East, where the major producers were China and Japan. As had happened in the case of tea, but even more quickly, Japan squeezed China out of the market. While the Chinese had a disturbing tendency to mix in a low-grade product, the Japanese government, in its concern to maintain the quality of its exports, initiated quality controls for silk. It was only by cutting prices that the Chinese merchants managed to recover their lead between 1878 and 1883, but this was a shortsighted policy which resulted in leaving the field open to Japan. From 1885, silk accounted for a larger share of imports from Japan than tea. The impressive expansion of the American silk industry, particularly in the 1920s, laid the foundations of a special economic relationship between the two countries. At the time, raw silk accounted for 10 percent of American purchases abroad, well ahead of coffee and sugar, while tea now made up less than 1 percent of the total. Such substantial quantities of silk were imported that the United States' balance of trade was always negative, a situation which did not change until the 1930s, when the silk industry was severely hit by the Depression, resulting in the collapse, not of the quantities imported, but of the unit price, to the considerable detriment of Japanese producers.[8]

Sugar was the second largest import, after tea. Indeed, per capita sugar consumption increased in America, with demand being positively impacted by improvements in income and negatively by price increases. Up to 1900 domestic production was able to supply only 10 percent to 15 percent of the demand, the rest being imported. Even with the annexation of Hawaii and the extensive cultivation of sugarcane in Louisiana, and of sugar beet on the Great Plains, America still bought more than half its sugar abroad, at least until the Depression in the 1930s, when domestic production took over. The main suppliers were Cuba and Europe. The Pacific played a relatively marginal part in this particular trade, for its great distance from the major centers of population in the United States put it at a disadvantage, despite production low costs. In the last quarter of the nineteenth century, the kingdom of Hawaii, protected by the reciprocity treaty of 1876, was exceptionally well placed to supply the West Coast, where California was booming. Hawaii had become so well integrated into the United States economy that the new customs policy of 1890, allowing sugar to be imported duty-free, deprived it of its privileged treatment and dealt a severe blow to the little universe of the American-managed sugar plantations, which had prospered under the provisions of the treaty. This new development made annexation much more attractive than independence, with the result that one of the ingredients of the republican revolution of 1893, foreshadowing the islands' absorption by the United States, was undoubtedly sugar. After 1898 the Philippines, along with Hawaii, became one of the sources which allowed the United States to become less dependent on foreign suppliers. The other major sugar producer in the Pacific, Java, played a substantial role only whenever the Cuban harvest was affected by bad weather, or when it was destroyed during the struggle for freedom from the Spanish colonizer. Like silk or rubber (which would benefit from the demand for automobile tires after 1910), sugar was one of the products—a new one compared with the preceding period—which was shipped to the United States from the Pacific, providing an example of the closer ties and the intensification of trade that are typical of a frontier region beginning to shrink.[9]

Exports from a Diversified Economy

The range of products listed in customs records provides a good indication of the level of development of the different nations involved in mutual trade. The general brevity of the list where imports from the Pacific region are concerned is in contrast with the long and varied list of exports to it. The United States was in a position to offer its trading partners a much greater range of products than they were able to offer in return (cf. Appendices, Table 10). As a result, it was unusual for a commodity to dominate exports to a given country in the same way as a single commodity or raw material would dominate imports from its underdeveloped supplier. Two notable exceptions, at the beginning of the twentieth century, were the export of cottons to China and of petroleum to the Dutch East Indies. Another asymmetrical factor was the relatively small part raw materials played in American exports (cf. Appendices, Table 9), the exception being cotton, which in any case was really shipped to only one market in the Pacific, namely, Japan. This was because from the end of the nineteenth century Japan had developed a dynamic textile industry of its own, using raw material mostly imported from India until World War I.[10] Around 1900, the southern states still sold scarcely more than 2 percent of their cotton to Japan, but this would change rapidly with the outbreak of war in Europe, which left the field clear for Japan to carve out a secure niche in cotton manufacturing. To do so, it needed to import increasing quantities of raw cotton, amounting, after 1920, to between a quarter and a third of its total imports. In some years, almost half of this raw material, which had become so vital to the Japanese empire's stability, came from the United States. In this way, a kind of symmetry came into being after World War I, with cotton taking a role equivalent to that of silk. Southern cotton growers had as much of a stake as the Japanese silk farmers in the maintenance of good relations between their two countries. Had it not been for American protectionism, trade between them might have expanded even further, and become more diversified, instead of being dominated by nonmanufactured goods.

Apart from cotton, United States exports were essentially made up more or less of finished products. Cottons lost the predominant position they had held prior to the Civil War. This was not only as a result of a shortage of raw material during the hostilities, since, even when the South reattained and then surpassed its earlier level of production, cotton cloth amounted to less than 20 percent of American exports to the Pacific, except for two years, 1881 and 1906. Nevertheless, the countries around the Pacific were still the principal markets for factories in New England and in the South. Between 1881 and 1914 they often accounted for almost half the sales of American cottons abroad, and for an even larger share between 1902 and 1906, particularly during the Russo-Japanese War, when a speculative market grew up in China. If they had wanted, American industrialists could have cornered the market, but until the crisis at the end of the nineteenth century they took little interest in exports, displaying an obvious preference for the domestic market. But economic difficulties, together with the technological advances which made American factories more competitive, encouraged mill owners in the South, and particularly in South Carolina, to specialize in the export of unbleached cloth to China. It was all the easier for the United States to make this its own preserve since it encountered no competition from the British, giving it a clear field for coarse cloth and for the cotton drill so much in demand in northern China and Manchuria, where clothes had to be quilted for protection against the frigid winters. It left to the British the market for the lighter, better finished, and less standardized materials, which were in greater demand among the southern Chinese, whose climate is less extreme. But the success of the American cotton industry was not to last, and it was to be no more than a memory after World War I. Consular reports provide a good insight into the reasons for its failure, in both Asia and Latin America. They blame the inadequacy of the shipping links and of the credit institutions compared to those of the Europeans, who were well provided for in these respects. But perhaps this is to confuse cause and effect, for if the quantities exported had been constant and substantial, the necessary services would have been put in place. However, this was not the case. Exports were just not a primary concern for American manufacturers. Their merchandise was poorly packed, and often suffered damage during transshipment. There was no real marketing policy, little concern about delays in delivery, and a readiness to cancel sales whenever a more lucrative opportunity arose in the home market. Little thought was given to catering to the tastes of the Chinese clientele, and low prices—requiring a standardized product to make them possible, even if the quality was good—were not enough for them to hold onto their share of the market for a commodity for which the demand was very diverse and variable. As the U.S. Census of Manufacturers for 1905 noted, the export of cottons was "an accident rather than a business. Up to a certain point it grew rather than it was created"—a remark which suggests a certain lack of initiative on the sellers' part.[11]

Petroleum, on the other hand, which—since the earliest wells in Pennsylvania date only from 1859—did not appear in customs statistics before the Civil War, quickly gained an important place in United States exports, being actively marketed by those in control of the industry. It is true that the Pacific countries were slow to accept kerosene as a lighting fuel. Prices, until they were stabilized by Rockefeller's Standard Oil, stood in the way of the popularity of a commodity whose price in certain years was relatively high for the poorest members of society—particularly when shipping and handling costs were added to the retail price. As a lighting fuel, the Chinese preferred the cheaper vegetable oils. Around 1880, however, the situation changed completely, as kerosene began to cost three or four times less, and lamps were being manufactured which were within reach of almost any pocket. American petroleum exports to the Pacific began to boom as the region absorbed more than a fifth of the foreign sales of the Pennsylvania oil wells, with the Texan and Californian wells joining in at the turn of the century. By around 1890, refined petroleum—the Pacific countries did not import any crude—accounted for more than a quarter of the total exports to the area.

For a while, things went very well indeed for Standard Oil. The company enjoyed an almost total monopoly. In Latin America it sold to both small- and large-scale markets through brokers, before employing its own sales force. In the Far East, kerosene was shipped to local merchants in cans. But these monopolies were short-lived. At the end of the nineteenth century, competition arose from two sources. First, there was the Russian oil from Baku, controlled by the Rothschilds, which, in 1892, was authorized to be shipped in bulk through the Suez Canal in steam-driven tankers—a much more economical form of transport than in cans piled up in the holds of sailing ships. Secondly, there was oil from Sumatra, which Royal Dutch began pumping on a large scale in 1896. The merger of the Dutch company with the British company Shell in 1907 created a formidable rival throughout the western Pacific. In 1912, it even had the gall to send a tanker to Seattle with a cargo of gasoline refined in Sumatra. Suffering a temporary setback, Standard Oil defended its market with a substantial marketing effort: small kerosene lamps were given away free; a lower grade of kerosene, made from Texas, Kansas, and California crude, was introduced; warehouses were constructed; tankers were used for transportation; and sometimes, as happened in 1898, it even bought cheaper Russian oil so that it could undersell Royal Dutch. During World War I, the American and British governments called for an end to this vicious competition, in the name of allied solidarity. The agreement of July 1918 divided the region into zones, with Standard Oil of New York being awarded 63.8 percent of the Chinese and Japanese markets. But it was a precarious agreement. As a punishment for refusing to boycott Soviet oil, on Royal Dutch's insistence, the New York company was plunged into a price war in 1927 from which it only escaped in the following fall thanks to the celebrated Achnacarry agreement, which split up the markets outside the United States on the basis of the market shares held by each of the signatories at that particular moment. In China, in 1934, a similar type of agreement put an end to the increasingly bitter competition by guaranteeing 30 percent of the market to Royal Dutch Shell, 25 percent to Standard Oil and the Chinese company representing Soviet oil, and 25 percent to Texaco. The demand, on the other hand, was becoming more diversified. The age of the oil lamp was drawing to a close, and between the wars Pacific consumers were buying more and more automobile and aviation fuel, and, in the case of Japan, industrial lubricants as well as crude, in order to free itself of its strategically excessive dependency on the United States and the Dutch East Indies—though it was never able to do so sufficiently to allow it to meekly accept an embargo on high-octane fuel, which was one of the immediate causes of the attack on Pearl Harbor in 1941.[12]

However substantial the exports of cottons, oil, flour, pork and beef at the beginning of the twentieth century, they were overshadowed by the range of articles produced by the metallurgical and mechanical engineering industries. Prior to 1896, the United States had met with only very modest success in this field.[13] But with the end of the crisis came productivity gains which suddenly made American products extremely competitive. American manufacturers were now well placed to make life difficult for their European competitors. At the turn of the century, the novelty and extent of the vigorous surge in exports was so unexpected that some even went so far as to speak of American industry taking over the entire world. The range of goods was very large, extending from partially finished products, like structural steel and railroad tracks, to the most highly finished goods, such as scientific instruments and electric- or steam-powered machinery. Japan and Australia, the most highly developed nations, were the best customers, for they possessed the know-how required to use these American-made producer goods. During World War I, industrialization opened up new markets. Thus, between 1916 and 1922, Saco-Lowell, the makers of textile-manufacturing machinery, won major Chinese and Japanese orders. However, this success was short-lived, because the costly American looms had been devised to reduce labor costs, whereas mill owners in the Far East had little difficulty finding employees willing to work for low pay, and, lacking capital, preferred to buy less expensive European machines, or to make copies of machinery imported from the United States.[14] The prosperity between the two wars was an opportunity for American automobiles to become popular in Australia and New Zealand, the Pacific countries which enjoyed the highest standard of living, while elsewhere they were within reach of only a wealthy minority. The worldwide depression of the 1930s put paid to this market. Japan reverted to a policy of self-sufficiency (it was symptomatic that the principal metal commodity it purchased from United States at the time was scrap metal, the raw material required for its armaments industry). On the eve of the Second World War, the dreams of an industrial takeover of the world which had been felt early in the century were far from being realized. Trade in unfinished or partly finished goods was far ahead of trade in finished articles. This situation, born of the crisis and of the nationalistic reactions it had engendered, was unsatisfactory in the eyes of decision makers in Washington, as proponents of a new liberal order. It also was a cause of the deadlocks which led to aggressive behavior and paved the way for the Pacific conflagration.

What blame must be borne by American trade policy for this state of affairs? A review of customs tariffs in force between 1922 and 1930 can give a misleading impression. It seems to imply that a large number of commodities from the Pacific were imported duty-free. This was indeed the case for raw silk, tea, coffee, cocoa, rubber, and tin, because the United States produced none of these commodities, and exempting them from customs duty lowered the cost for consumers and industrialists. But as soon as a sector of the United States economy felt threatened, tariffs were imposed, sometimes at prohibitive levels, and certainly much higher in 1930 than in 1922. Wool, for example, which might have provided a profitable trade opportunity with Australia and New Zealand, was subject to an ad valorem duty of 53 percent in 1922, rising to as high as 89 percent in 1930. The textiles Japan might have sold in the United States, allowing it to buy a greater number of manufactured articles in return, were subject to heavy tariffs of 61 percent on silks, rising to 63 percent, and 53 percent on cottons, rising to an average of 60 percent.[15] In the opposite direction, China was not allowed to impose a duty of more than 5 percent ad valorem until October 1926, and even then was only allowed to raise the maximum to 7.5 percent. Japan, which at the end of the nineteenth century had regained its autonomy in setting tariffs, opted for protectionism rather than for free trade, generally relying on specific import tariffs which were increased when prices fell, but these seem moderate when compared to those imposed by the United States: in 1910, 1926, and 1930, with machinery subject to a 20 percent duty, automobiles to 50 percent (later 35 percent), and railroad cars to 30 percent.[16] The only possible conclusion is that the Americans, while remaining attached to protectionist dogma and to the virtues it was supposed to engender, did little to encourage the emergence of regional prosperity in the Pacific, something their position as the leading world economic power gave them the ability, if not the responsibility, to do. Saying this, however, in no way exonerates Japan for the heavy responsibility it must bear for the coming of war, because of its aggressive imperialist behavior in China.

WAYS AND MEANS OF COMMUNICATION

Before the Civil War, American ships, both merchant vessels and whaling ships, were to be seen everywhere in the Pacific, in all latitudes, whatever the prevailing wind. However, there were still no shipping lines with set arrival and departure dates, and information was slow to move from one shore to the other. Several months would pass before information about events in Europe or in the United States arrived. This made for some splendid opportunities, but it also meant formidable risks for speculators and adventurers. This all changed around 1870, when steam navigation and the telegraph brought about a revolution in communications and in the transmission of news—a revolution continued by the first commercial flights across the Pacific in the 1930s.

The Decline of the Merchant Fleet

The Civil War coincided with a decline in the American merchant fleet. This was not so much because of the losses incurred during the first four years of hostility, or of the activities of Confederate marauders

and the high insurance premiums required of any shipowners still rash
enough to send ships to sea. It occurred, rather, because the Americans—
and this had already become evident shortly before the attack on Fort
Sumter—had lagged behind in the technical revolution in shipbuilding, in which iron was replacing wood and steam was more and more
taking over from sail. The reasons for this backwardness are well known.
Shipowners wanted to acquire vessels inexpensively to lower their
operating costs, to allow them to compete with their British, Norwegian, or German rivals. The production costs of American shipyards
and the steel industry were too high, but they protected their market
jealously. Under pressure from so many different directions, the federal
government leaned toward protectionism, making inevitable the
decline of an American merchant fleet, which for a brief time had
seemed capable of threatening the British supremacy. The decline of
whaling, which the competition from petroleum had made inevitable, added its effects to Washington's political choices. American
merchant ships were to become a much less frequent sight throughout
the Pacific.[17]

Yet it was in the Pacific that they met with the most success. They
were eliminated almost entirely from the Atlantic by competition of
the great British, German, and French shipping lines. The coming of
steam was decisive in the downfall of a fleet which failed to keep up
with modern developments. In the Pacific, on the other hand, because
of the enormous distances and of the paucity, in both directions, of
any activity comparable in scale to the conveyance of cereals, cotton,
and emigrants across the North Atlantic, sail was able to hold its own
for a long time, except in coastal shipping to British Columbia and the
isthmus of Panama. In 1880 sail was still paramount in shipping to
South America and Southeast Asia as well as to the Pacific archipelagoes. However, steam finally triumphed in the last decade of the nineteenth century, with only the routes to Oceania remaining loyal to sail.
Starting in 1893, the steamships of the Grace Line took over most of
the traffic between New York and Chile and Peru. The long and uncertain voyages on the three-masters of yesteryear were no longer acceptable for shipping sugar from the Dutch East Indies, so that it
followed the example of Chinese tea, which since 1876 had used the
fast freight-carrying steamships taking the Suez Canal to compete with
Japanese tea coming in through San Francisco and traveling onward on
the transcontinental railroad.[18]

Steam-powered ships were indeed what the American merchant fleet lacked. The result was that it could only give way to foreign competition. New York merchants, for instance, would hire British vessels in order to take advantage of their better freight service, while the disappearance of the merchant-shipowners had put paid to any will to resist inspired by patriotism or chauvinism. In 1900 not a single American steamship sailed to the Pacific from the Atlantic coast ports. Grace equipped his line with vessels registered in England, and the cargo ships bringing tea from Shanghai all flew the British flag. Robert Dollar adopted an identical policy at the beginning of the twentieth century. If the American merchant fleet survived at all, it was thanks to the special relationship with Hawaii, particularly after it was annexed in 1898, and to the dynamism of a small number of companies, such as the Pacific Mail Steamship company, which was subsidized by the postal service. This company originally prospered thanks to what was in effect its monopoly over links with Panama and San Francisco, but this rich vein ran out when the transcontinental railroad came into service in 1869. Fortunately for the company, two years earlier it had established a line between California and Japan which throve on the growing volume of imported tea and silk. It managed to withstand competition by making compacts with competitors, or taking them over, until it eventually went out of business in 1915, after chaotic spells of tariff wars in the North Pacific. What the United States lacked was a sufficient number of enterprising shipowners eager to compete with foreign companies in the way the Russell Company had been able to do with the British in Chinese waters between 1864 and 1877.[19] But these few successes could not prevent the decline which set in at the beginning of the twentieth century. In 1900 only a third of the shipping between the United States and the Pacific was carried on American ships, instead of at least half twenty years previously, and American vessels had a significantly smaller share of long-distance traffic to the Far East, Southeast Asia, and South America. In terms of tonnage, only foreign fleets were able to benefit from the intensification of trading relations.

In links with the Pacific, it was the ports situated on the Pacific coast itself which accounted for the major share, but if we discount the shipping to and from British Columbia, which was in fact a kind of cabotage, their advantage was much less marked (cf. Appendices, Table 11). San Francisco, which was increasingly losing out to competition from Portland and Seattle, surrendered its first place to New York in trade with South America, Southeast Asia, and even China, though it should be said that the port on the Hudson had by now become the focus of almost all the shipping between the Atlantic coast and the Pacific. It is worth noting that Boston, whose ships had played a pioneering role before the Civil War, would have been wiped out almost entirely if it had not been used as a port to supply the sugar refineries of Philadelphia and Baltimore, as well as its own, with a raw sugar imported from Java or the Philippines. Likewise, the South only counted at all because of its exports of cotton to Japan. This success by New York was due to the fact that in the vast consumer market of the Northeast it had the richest hinterland in all the United States, while San Francisco, surrounded by a still sparsely populated California, was handicapped by transcontinental transportation costs, which were high enough to cancel out any advantage it gained by being closer to the China Sea.

In 1914, two events were to transform the maritime geography of the Pacific, one being the World War, which diverted the attention of the European rivals, and the other the opening of the Panama Canal. These afforded a unique opportunity for American merchant shipping to regain the position it had begun to lose several decades earlier.

The Construction of the Panama Canal

Strategically, the United States had a great deal at stake in the isthmus of Central America. It could not permit it to be controlled by a foreign power, and least of all by Britain. Between 1848 and 1869 it was the only short route between the United States' two ocean coastlines, despite the two transshipments involved. Later, the transcontinental railroads would make it less useful, and they used their powerful lobby to block any proposals for a canal linking the two oceans. Interminable discussions went on about which was the best route for such a canal. The narrowest point was in Panama, but the unhealthy climate was a severe disadvantage. The advantage of a route through Nicaragua was that a natural lake would provide a considerable part of the crossing, but even so the many miles of excavation required, together with the need for a system of locks, were enough to deter even the most foolhardy.[20]

Although it was the Americans who stood to gain the most from a canal, it was not they who first attempted its construction, despite President Grant's efforts, but the Frenchman, Ferdinand de Lesseps, who was hoping to repeat the prodigious achievement of digging the Suez Canal to which he owed his fame. Since de Lesseps was entirely reliant on private capital the French state was not involved, meaning that the United States could not invoke the Clayton-Bulwer treaty of 1850 to prevent this incursion into what it had already come to consider its backyard. Anyone could invest in the Universal Interoceanic Panama Canal Company, irrespective of his nationality. The company's objective was to carry out the project approved in May 1879 by an international study conference, which had been made to measure by de Lesseps himself. The project failed, as it was bound to do. Even if de Lesseps, half confidence trickster and half visionary, had managed the matter with all the requisite intellectual and financial integrity, in the 1880s conditions were not yet right for the success of such an enormous undertaking. Even if, instead of pushing on with the construction of a canal at sea level, on the Suez model, he had from the beginning supported the plan of a canal with locks presented by Godin de Lépinay, he could never have completed a shipping canal between the two oceans at Panama. This was because there was nothing to stop the climate and the diseases from decimating the ranks of the engineers and laborers, for the requisite scientific knowledge was still lacking. In addition, the earth-moving machinery was still not powerful enough to remove the thousands of cubic yards of earth from the Culebra Cut, and the industrial applications of electricity were still in their infancy. The French had enough overall financial resources to finance the project, but the length of time required was a disadvantage for people needing income from their investment. When we add to these insuperable handicaps the excessive expenditures—including the sums of money spent on generous bribes to the press and to politicians—it is easy enough to understand why the Universal Company declared bankruptcy in February 1889.

Such an outcome was naturally much to the delight of the Americans, who from the outset had been hostile to the French project. Yet, at the very same time, their companies were unable to do any better in the isthmus of Tehuantepec and in Nicaragua. In this small Central American republic, the Nicaragua Canal Construction Company began work in October 1889. It required federal aid to raise the financing, and Congress refused several times to provide a guarantee, having been convinced by the railroad lobby that this would be a violation of the 1850 treaty. The crisis of 1893 drove the American enterprise also into bankruptcy. However, the Nicaragua Canal project still had its fierce

supporters, like Senator Morgan from Alabama. In 1898, its time
seemed to have come at last. The annexation of Hawaii and the Philippines made a shipping canal through the isthmus an absolute strategic
necessity. There was substantial agreement on this point, but the various parties disagreed as soon as they had to move beyond this basic
principle. Which was the better route? Nicaragua or Panama? Who
should be responsible for the project? Should it rely on private capital
or be funded by the federal government? The committees formed to
study these questions provided contradictory answers. In November
1901 the Isthmian Canal Commission recommended the Nicaragua
route. Two months later, it changed its mind, preferring Panama. The
reason for this change of heart was that, in the meantime, in January
1902, the New Panama Canal Company, which had taken over the
Universal Company's concession, and which had been created to sell its
shares to Americans, rather than out of any enthusiasm for recom-
mencing the excavation, lowered its estimate considerably, from $109
to $40 million. It was a victory for the "Panama lobby" headed by
Frenchman Philippe Bunau-Varilla, business lawyer William Cromwell,
Senator Hanna, and engineer George S. Morison. The new American
president, Theodore Roosevelt, now convinced of the superiority of
the Panama route, threw all his influence and energy behind the project. Congress, also influenced by the eruptions of Mount Pelée in Mar-
tinique and Mount Momotombo in Nicaragua in May 1902, voted the
following month for a canal across the isthmus of Panama, to be financed by the federal government. There was no longer any British opposition, for the United Kingdom was preoccupied at the time by the
Boer War. After a lengthy hesitation to overcome which required the
negotiation of two treaties, Britain finally gave the United States a
green light with the Hay-Pauncefote Treaty (November 18, 1901),
which allowed the latter the exclusive right to draft and to administer
regulations. The canal would, of course, be neutral and open to ships
of all nations without any discrimination, but the Washington government was allowed to maintain a security force, and there was nothing
to prevent it from constructing fortifications.

Once the question of the route had been finally decided, it remained for Colombia, of which Panama was still a province at the
time, to allow the new company to transfer its assets to the government
of the United States. Negotiations dragged on interminably, complicated, in addition to Bogota's remoteness, by the civil war which was

being waged in the Andean republic. By exerting pressure on the Colombian diplomat Tomás Herran, Secretary of State John Hay managed to obtain his signature on a treaty, which the United States Senate ratified on March 17, 1903. However, the Colombian Congress was not satisfied with the sum of $10 million as payment for a six-mile zone, even if the country's sovereignty was respected. Colombia considered it had been less generously dealt with than the new company, which had been awarded four times that sum. It therefore decided to ask for better terms. The United States might have short-circuited these blackmailing tactics by reverting to the Nicaragua proposal, or by waiting for emotions to calm down in Colombia. Theodore Roosevelt, however, was not prepared to tolerate what he considered extortion, and swore at the "bandits" in Bogota. He had another reason to persevere with the Panama project, since that province, which had always been cut off from the rest of Colombia, periodically attempted to increase its autonomy, and even to achieve independence. When revolution broke out in Colón and Panama City in early November 1903, the American fleet prevented Colombian military reinforcements from intervening, thus ensuring the victory of the party favoring independence. On November 6, Washington recognized the new state, and on November 18 eagerly signed a treaty with the special envoy and Minister Plenipotentiary P. Bunau-Varilla. The provisions of this treaty were more favorable to the United States than those previously negotiated with Herran. The United States was granted in perpetuity the concession of a zone 10 miles wide for the construction of the canal in return for an immediate payment of $10 million and annual payments of $250,000. Within this zone they would enjoy the same rights, powers, and authority as if it were their sovereign territory. Article XXIII authorized them to use their police and armed forces at any time, at their own discretion, for the protection of the canal, and to construct defenses. The United States was now the master of the isthmus, in return for the relatively modest outlay of $50 million.

Now the Americans had to continue the task begun by the French—and this would require ten years of work. But they would succeed where their predecessors had failed because twenty years had passed, enabling them to take advantage of certain advances which had been made in the meantime. These were in the area of health first of all. In 1897–1898, the Englishman Ronald Ross and the American Walter Reed had discovered that the mosquito was a carrier for the parasites responsible for malaria and yellow fever. By systematically combating this pest, Colonel Gorgas, a specialist in tropical diseases, was able to prevent deaths in the numbers which had worn down the French—though he was unable to prevent pneumonia from wreaking havoc among manual laborers from the West Indies. Second, the Americans were able to use much more powerful equipment—massive earth-moving machinery. They used concrete, special steel, and electricity in the construction of the locks. Finally, they were able to rely on an annual source of funding, namely, the budget approved by Congress. This meant they had no need to depend on annual general meetings with impatient shareholders, or to concern themselves with short-term profits. In total, between 1904 and 1914, they spent $352 million, including the $50 million in indemnities. For the sake of comparison, let us remember that between 1881 and 1903 the French companies had spent $287 million with very little to show, and had left four times the number of dead behind them.

But the ultimate success of the venture should not obscure the difficulties encountered. Initially, the federal bureaucracy was an obstacle. Fortunately, Roosevelt, to whom the project was to owe its success, took things in hand. In February 1907, after the failure of two civilians, John Wallace and John Stevens, he gave complete authority to Major George W. Goethals, an officer seconded from the army, a cool, competent, and self-assured individual. If there was a delay in getting work underway, this was caused by the lengthy hesitation about the type of canal it should be. The ideal of a sea-level canal still had its defenders, if only out of a reluctance to seem to take second place to the French. In November 1905 an international commission still came out in favor of this more costly and more lengthy solution, as did a Senate committee in May 1906. Finally, however, John Stevens convinced Roosevelt that a canal with locks was less risky, would have lower maintenance costs, and could be completed in eight years. On June 21, 1906, the Senate gave its final approval to the project by a small minority, with 36 for, and 31 against. It only remained, as the president put it, to "make the mud fly." But this was no small matter. The amount of earth that had to be removed kept on growing, for, to prevent the sometimes enormous landslides from recurring, it was necessary to slope the sides more gradually, and the minimum width of the channel at the Culebra Cut was increased to 90 meters. In eight years, 220 million cubic yards were excavated, in comparison with the 76 million removed by the French. The engineers found a solution to the problem of soil disposal by perfectly coordinating the excavation and the removal of earth by rail. The six pairs of locks constructed at each end to raise and lower ships between sea level and the artificial lake of Gatun, were wide enough, at 36 yards, to allow the largest battleships of the day to pass through. When the canal was finally opened to shipping, on August 15, 1914, twelve days after the Great War broke out, the United States controlled the sea route between the Atlantic and the Pacific Oceans. It could easily move its fleet from one ocean to the other, instead of having to make the lengthy voyages via Cape Horn or the Cape of Good Hope. At the same time, its commercial shipping enjoyed a considerable reduction in the time required for certain voyages, undeniably giving it a new edge in competition with other nations.

Transpacific Trade between the Two Wars

But, despite this, the American merchant fleet was still unable to regain the supremacy it had enjoyed prior to the Civil War. The government in Washington did provide support, though it never put into effect the grandiose plan that President Wilson had hoped would bring about a new international economic order. Several laws, in 1916, 1920, 1928, and 1936, forbade foreign shipowners to practice any discrimination whatsoever against American exporters, encouraged the equipment of the fleet with modern ships which could be used as support vessels in wartime, raised the possibility of establishing publicly owned shipping lines, awarded generous subsidies, and put it all under the control of a federal agency, the United States Shipping Board (replaced in 1936 by the United States Maritime Commission). In spite of these efforts, the merchant fleet was unable to consolidate the advantages it had gained during World War I. The American International Corporation (AIC), created in 1915 by the National City Bank of New York had reached an agreement with the firm of W. R. Grace and Company to buy back the Pacific Mail Steamship Company and take over market shares of the market in shipping to various Pacific countries, under the American flag. The enormous tonnage accumulated by the federal government during the war, which had made it the foremost shipowner in the world in the 1920s, could have provided a basis for supremacy in shipping if some ill-advised choices had not been made in putting this policy into effect. A number of ships built during the war were soon technologically out of date and were no equal for foreign competitors' more modern vessels. A good number of them should have been sent to the scrap heap, but this was a difficult decision for a bureaucracy which was often static and unable to react quickly to changing conditions. American law systematically favored the creation of established shipping lines in preference to tramp steamers traveling from port to port in search of cargoes. A regular line brought more prestige, but suffered from the handicap of fixed costs, and was ill-suited to carry the cumbersome commodities which, as we have seen, were a fundamental part of trade between the United States and the Pacific. In the 1930s there were virtually no American tramp steamers in the Pacific, leaving the field open to the Japanese and British. The law also forbade American companies from entering into compacts with foreign shipping lines, thus excluding them from taking advantage of secret agreements between the members of such cartels to share the business, fix the rates, and pay rebates to loyal clients.

In competition with an increasingly weaker American merchant fleet, the British, and particularly the Japanese, took advantage of the situation. At the end of World War I, Japan possessed the third largest merchant fleet in the world. The imperial government supported three companies with substantial subsidies: Nippon Ysen Kaisha (NYK), Osaka Shosen Kaisha (OSK), and Toyo Kisen Kaisha (TKK), the first merging with the last in 1926. It thus enabled the first shipping lines between Japan and Seattle and San Francisco to be established at the end of the nineteenth century. Other shipping companies, with no government assistance, invested in a fleet of tramp steamers, or, like Mitsui, chartered steamers to enable them to carry on business on a worldwide scale. Between the two wars, the Japanese fleet had the advantages of very low operating costs, flexible management, and often close links between the owners and the sogoshosha (general trading companies)—for instance, NYK with Mitsubishi and OSK with Sumitomo.[21] Under these conditions, between the wars a growing proportion of trade between the United States and Japan was carried on under the Japanese flag, with a very marked decline in the American percentage during the Great Depression and the New Deal. In 1936, for example, Japanese steamship lines accounted for 43 percent (by volume) of exports to the Far East from Northwestern ports (Seattle, Portland, Gray's Harbor, etc.), tramp steamers (mostly of the same nationality) for 33 percent, and American ships for only 16 percent (although this was merchandise with a higher value, such as automobiles and machinery), with the remaining share going to British, Dutch, Danish, and Norwegian lines.[22] In the same year, only 8 percent of exports from the Atlantic ports to

Japan, China, and the Philippines (scrap metal, steel products, and fertilizer) was carried by American vessels, compared with 36 percent by Japanese shipping lines and 23 percent by Japanese tramp steamers. There was a similar picture for imports from the same region, calculated by weight (a method which, as opposed to calculating the value, reduces the share of silk, compared with sugar, vegetable oils, or rubber) and passing through the Panama Canal. This puts the share of American ships at only 10 percent (despite the almost colonial relations with the Philippines), 58 percent (and practically all the silk) going to the Japanese lines and 16 percent to tramp steamers.[23] In the 1920s, the Japanese silk used by the Paterson factories was generally unloaded in San Francisco and conveyed to the New York suburb by special trains, thus reducing the insurance costs on such a valuable commodity. The collapse of the price of silk during the Depression led to a search for even greater savings in the area of freight costs. In 1930, OSK started up an express line between Yokohama and New York with new diesel-powered vessels able to complete the voyage via Panama in twenty-eight days instead of the forty-six required for the same trip by older vessels. The difference between the sea route and the combined sea and overland route was reduced from thirty days to twelve, with the result that, where the interest on the capital tied up and the insurance premium were concerned, the handicap of the former fell from 10.48 to 2.47 yen per picul, while the freight differential in favor of the latter rose by only 2 yen. In August 1929, when a picul of silk was worth 1,330 yen, the extra cost of 7 yen involved in shipping it by San Francisco was acceptable because of the time saved, but by February 1931, when the OSK express line came into service, the price had fallen to 685 yen, making it preferable to ship it via Panama in order to take advantage of the 13 yen savings on freight costs—a much more substantial economy relative to the value.[24] The Americans viewed the successes of the Japanese merchant fleet as a consequence of unfair competition, though they themselves had failed to ensure that the playing field was level—showing that there is nothing particularly new about the kind of complaint we often hear today.

Dependence on the Telegraph

Its merchant fleet was not the only weakness of the United States in the Pacific. It also depended heavily on foreigners for the transmission of information. The Collins project which Western Union supported might have given it a strong position if it had not been preempted, as we have seen, by the laying of the transatlantic cable in 1866. Instead of crossing the North Pacific, news traveling between America and the Far East would come via Suez and the Indian Ocean, an objective achieved in June 1871 when the Singapore-Hong Kong link was completed, soon to be extended to Japan, while a cable laid across Siberia opened up a second route. Two companies, one of them English (Eastern Extension) and the other Danish (the Great Northern Exchange Company), were the driving forces behind this network linking the Far East to Western Europe and, consequently, to North America. Instead of competing and reducing rates, they combined forces to share the traffic on a geographical basis. This was not at all to the liking of the Americans, who balked at Danish attempts to create a monopoly in China, but in 1883 the two Western companies consolidated their relationship, clinging strongly to their advantage until the end of the century. Several draft bills were put before Congress to encourage the construction of a transpacific telegraph cable under American control. The desire to be independent of the European monopolies and the necessity of protecting national defense secrets were invoked. But for a long time the obstacle remained the eternal problem of who should undertake responsibility for the project: should it be a public or a private undertaking? Western Union was quite prepared to challenge Eastern Extension in the Far East because of the unfair way in which it competed along the Pacific coast of South America. However, subsidies from Congress would be needed to lay a cable across the Pacific. In 1898 the premises changed with the conquest of the Philippines, which made a cable necessary to link the United States with its new colony, rather than Japan with North America, which had been the basis of earlier projects. Finally, in 1901, businessmen John W. Mackay and James Gordon Bennett founded the Commercial Pacific Cable Company to lay a cable from California to Manila via Hawaii, Midway, and Guam, without any reliance on any subsidies or exclusive rights. Private enterprise had carried the day. In July 1903, when work was completed, Americans finally felt they no longer had to depend on the Anglo-Danish monopoly. But this was an illusion, for the supposedly American company was in fact secretly 75 percent owned by these very same monopolies. It, too, played its part in controlling competition in the Pacific, and like the German-Dutch company connected to Guam, it participated in all the agreements, particularly in those of 1904 and 1913, which regulated the sending of information across the Pacific, using the rate structure to discourage the routing of messages via the United States. As far as the telegraph was concerned, the Americans had come on the scene too late to pick up anything but the crumbs.[25]

The Beginnings of Aviation

The first American airplanes made their appearance in Peru in 1927, where they were used to spread fertilizer on the cotton fields, taking advantage of the fact that the seasons were the reverse to those in the northern hemisphere. This inspired the idea of establishing a mail service. C. E. Woolman, the future founder of Delta Airlines, was granted a concession by the Peruvian government to carry passengers, freight, and mail in a country where road traffic was made difficult by the mountainous terrain of the Andes. It was not long before he transferred his concession to a new company, Panagra, which merged Pan Am, controlled by Juan Terry Trippe, with W. R. Grace and Company, the business firm. The service began in September 1928, linking Lima and Talara, the headquarters of the International Petroleum Company. In the following March, Panagra signed a contract with the American Postal Service, establishing an air link between Panama and Chile and, later, Argentina. Beginning in October, with support from Washington, a weekly flight was introduced, making the New York to Buenos Aires trip in nine days, passing throughout the countries of the Andes. Even the crisis of the following decade was not enough to bring the growth of commercial aviation along the Pacific coast to a halt.[26]

Too many difficulties prevented a transpacific service in a westerly direction from being introduced right away. There was, for instance, the disappearance of Amelia Earhart, who had left Papua on July 2, 1937, in an attempt to set a new around-the-world record, and who may have died of thirst on the uninhabited atoll of Nikumaroro in the Gilbert Islands, as a result of a navigation error.[27] Nevertheless, the United States and Britain vied actively for possession of previously uninhabited islands as sites for relay stations to direct airplanes in distress and to serve as refueling stops. Taking advantage of the 1856 guano law, President Franklin D. Roosevelt extended American rule in the Central Pacific to Kingman Reef and Howland, Baker, Jarvis, and Johnston Islands, installing a few settlers and troops to provide support for the Pan Am seaplanes flying between Hawaii and New Zealand via Samoa. Some of these atolls would later become important links in the supply chain between the United States and Australia during World War II. In peacetime they would be useless, for after 1945 planes had enough range to fly the entire width of the Pacific.

AMERICAN INVESTMENT IN THE PACIFIC

Before the Civil War the Americans had very few investments in the Pacific area. At that time they were themselves net importers of capital to such an extent that they had no surplus available. When peace returned, in spite of the export boom, the United States remained a debtor until the end of the nineteenth century. Then, suddenly, between 1897 and 1905, it found itself in a surplus position and recognized the possibility of increasing its direct investments abroad— although this situation was not to last. Only World War I, which radically transformed it from a debtor into a creditor nation, created the opportunity for massive investments abroad.[28] There were two fundamental reasons for such movements of capital abroad. The first was the quest for new markets, for this provided a way to dispose of surplus production, to get around protectionist customs laws, or, in labor-intensive industries, to take advantage of a less costly workforce than in the United States in an attempt to win a share of the foreign market. The second was the desire to take advantage of natural resources, either fearing that America's reserves, which had been partly responsible for the country's exceptional prosperity, would soon be exhausted,[29] or because a better return on investment could be found abroad. In the first case, demand was the deciding factor, while in the second it was supply. In both respects the end of the nineteenth century represented a break with the preceding period. In each period the massive exports of capital coincided with a phase of imperialism, without there necessarily being any link between the two phenomena and even less so any relationship of cause and effect.[30]

A Twentieth-Century Phenomenon

As Table 12 in the Appendices shows, according to the estimates by Cleona Lewis and Mira Wilkins, American direct investment in the Pacific region in 1897 came to only $45 million, or just 7 percent of all capital investments outside the country. This was not a large amount, but all the conditions required for a spectacular expansion were in place. The "hard times" of the 1893–1897 crisis led investors to turn abroad, fearing that permanent surpluses might lead to a social explosion. The wave of mergers beginning in the last decade of the century led to the formation of powerful companies with enough financial resources to reach beyond the nation's borders. With no choice but to grow, large companies could not absorb all their rivals in the United States without running afoul of the antitrust laws, so it became attractive for them to seek growth outside the country. The decision to invest abroad was an offshoot of the individual strategy of each individual firm, without any overt encouragement from the federal government, which, however, did occasionally offer support, as in China, where it threw its weight behind the quest for railroad concessions and membership of banking consortiums.

The period between 1914 and 1929 was the golden age of American investment in the Pacific. European companies were suffering from the decline of their countries of origin, whereas the United States, who would be a net creditor for the next sixty years, had surplus funds available. Furthermore, the squandering of American natural resources was causing some fears for long-term national self-sufficiency. Concerns for the conservation of resources encouraged the exploitation of countries' natural riches, using them not just to supply the United States' needs but also those of other developed nations in need of raw materials. The general prosperity lasted only fifteen years, giving way to a lengthy and intense depression, which created a climate unfavorable to investment abroad. In the United States, local producers and unions made no secret of their hostility to what they considered unfair competition from countries with lower wages. Any relocation of activities and the mobility of capital were extremely unpopular. In host countries suffering from a particularly severe collapse in the price of raw materials, foreign investment was criticized for often failing to serve the national interest, for an excessive reliance on decisions imposed from New York, and for channeling the national economy into a kind of overspecialization, which was very risky in times of crisis. Furthermore, the colonial empires, which had opened their doors by only a crack even during prosperous times, tended to close them again, a case in point being the British Commonwealth with its preferential tariffs. In the Pacific, American investments fell by an average of 1.8 percent per annum between 1929 and 1940, compared with a growth of 11.5 percent per annum since 1897, and as much as 14.6 percent between that year and 1914.

Because it grew faster than the national average up to the time of the 1929 crisis, the Pacific's share increased substantially, particularly before World War I, reaching a peak of close to 20 percent. American investors, who until then had demonstrated their preference for Europe and the neighboring countries of Canada and Mexico, were no longer deterred by distance, and willingly accepted a greater amount of globalization. This gave rise to a parallel trend in the movement of both capital and goods, without any obvious single cause seeming to underlie both phenomena. When we look at the distribution of investments, we find significant variations. In some cases there was concentration on a single industry, such as on plantations in Central America or on mining in South America. In other cases, as in Asia and even in Oceania, it was distributed quite evenly among the various sectors (cf. Appendices, Table 13). In 1929, the supply factor seemed to be the most successful in attracting investments, accounting for 57 percent of the total, thanks to the importance of mining, oil fields, and agriculture. Sectors catering to a demand, such as public utilities, petroleum distribution, and to a much lesser extent, industry and trade) were subject to the greater attraction of countries on the eastern shores of the ocean. Even though direct investments in the Pacific increased by a factor of 33 between 1897 and 1929, American capital still gave the impression it relied on its continental base and was showing some reluctance to cross the ocean.

Railroad Misadventures

It might have been expected that the Americans, as possessors of the longest railroad network in the world, would play a role in railroad construction in the Pacific Rim countries. Actually, their contribution was relatively insignificant. Henry Meiggs, the "Yankee Pizarro," did not leave an entirely positive memory behind him in Latin America.[31] Fleeing California in 1854, where he was wanted for fraud, he headed for Chile, where he made himself a reputation in railroad construction by completing the Santiago-Valparaiso line in only two years, from 1861 to 1863—a feat none of his forerunners had been able to bring off. Thanks to his fame, he was summoned to Peru in 1868, where President Balta, wishing to forestall any negative consequences arising when the guano deposits were exhausted, developed a plan to mine the natural resources of the Andes. For this to be feasible, a railroad through the mountains would have to be built. Meiggs, with the help of kickbacks, signed several contracts with the Peruvian government, which assumed responsibility for raising the requisite capital, thus assuming a role as service provider rather than as investor. Meiggs succeeded in some considerable technical achievements, such as the Oroya railroad, climbing from sea level to 14,500 feet, crossing sixty-one bridges, and running through sixty-five tunnels. But financial difficulties soon began to accumulate. Peru had been counting too heavily on the project's short-term profitability, and international creditors began to have misgivings as early as 1872. Three years later, Meiggs was obliged to halt construction. In May 1876 he proposed work be resumed in return for the concession of the Cerro de Pasco copper and silver mines, accepting personal responsibility for the financing, with Peru guaranteeing the interest—a contract he was never able to fulfill, since he died the following year and in any case no longer had any credibility on the financial markets. Peru was ruined, and things only got worse when it lost its nitrate mines to Chile in the war of the Pacific.

It was at this point that another American, a recently naturalized British subject, Michael Grace, the brother of William (who around this time was twice elected mayor of New York), came on the scene. In 1885 Grace purchased all the shares of the Oroya railroad and of the Cerro de Pasco mines, promising to finish construction of the railroad and raise capital in the United States to develop the mines. Most important, acting as mediator between the Peruvian government and its British creditors, he resolved the foreign debt problem for once and for all. After several reverses arising from the domestic turmoil in Peru and a lack of cooperation by Chile, by 1892 he had managed to satisfy the two adversaries, at the same time pocketing a commission of between $500,000 and $1 million. The Grace contract has been viewed differently by Peruvian historians, depending on their ideology. For the liberals, it provided an opportunity for the national economy to get started again with the help of the international capital market. For the Marxists, it condemned the country to a dependency which left it no real freedom of choice.[32]

American investors had a few other successes in railroad construction, in Costa Rica, Bolivia, and Ecuador. In the republic of Ecuador, Liberal President Eloy Alfaro entrusted Archer Harman with the construction of the Guayaquil-Quito line. Harmon undertook to raise the necessary financing in London and New York and completed the project in 1908, but in Ecuador's case also the government was bank-rupted.[33]

On the opposite shore of the Pacific, in China, however, every ini-tiative met with failure. In 1896, the imperial government, attempting to play the "barbarians" one against another, was persuaded, with at most the lukewarm support of the American government, to grant the concession for the Peking to Hankou line to the American-owned China Development Company, but the latter found the terms unac-ceptable and lost the contract to a Belgian syndicate.[34] Three years later the same company had a similar experience with the Hankou-Canton line. In Manchuria, American Consul Willard Straight supported the project put forward by Edward Henry Harriman, the railroad magnate, to build a line to compete with the Japanese company which had been solidly in place since Japan's victory over the Russians in 1905. It was no good, however, for President Roosevelt preferred to avoid any con-flict with the new imperialist power of Japan. His successor, Taft, was less inhibited. Under "dollar diplomacy," Secretary of State Philander C. Knox provided solid support for American capital investment, which he saw as being crucial for the protection of trade. In 1909, to chal-lenge what was in effect the monopoly established in Manchuria by the Russians and the Japanese after their reconciliation, he put forward a proposal to neutralize the railroads. The suggestion was that all inter-ested foreign powers would combine to purchase the railroads in the region, and then resell them to the Chinese government, while keep-ing control until the loan granted for the purchase had been paid off. The United States was the only nation to support this idea, which was therefore easy for the coalition to reject.[35] Similarly, Knox insisted that American banks should be allowed to participate in the Anglo-French-German project to construct the 560-mile-long Peking-Canton rail-road, and did succeed in obtaining a minor role for them, just before the empire fell and chaos set in.[36] But there was very little to show for all the effort expended.

From Natural Resources to Consumer Goods

More than railroads, it was mining that attracted American capital after 1900. This is why the Pacific coast of South America—and Chile in particular, with its wealth of copper and nitrates—was able to recover its previous importance. In 1904, William C. Braden, a former senior executive of the major corporation dealing in nonferrous metals, the

American Smelting and Refining Company, purchased the El Teniente
mine in the Andes, and reached an agreement to operate it with the
Guggenheims, who had made their fortune in the mining and process-
ing of ores from the Rockies. The low grade of the Chilean copper deposits had made potential investors keep their distance. However, in the
United States, new, so-called nonselective technologies had recently
been developed, allowing open pit mines to be dug using powerful machinery, and making it economical to mine less high-grade ore deposits.
In addition, the flotation process meant that greater quantities could be
extracted. Before the First World War, Chile had been well placed to recover its place as a major copper producer, on condition that the required capital could be found and modern technology be employed.
Only American investors could provide both money and know-how,
and were all the more eager to do so because they found Chile's political stability reassuring at a time when anti-American feeling in Mexico,
which was caught up in the maelstrom of revolution, was causing them
some uneasiness. In 1910, the Guggenheims bought the Chuquicamata
mine. This gave them a dominant position in Chilean copper production, and they turned a very handsome profit during World War I. In
1923 Chuquicamata was sold to another American firm, Anaconda, the
proceeds of which allowed the Guggenheims to pursue their activities
as part of the Kennecott Company and, in particular, to develop their
stake in nitrates. They had been involved since 1916 in this sector, indispensable for the manufacture of fertilizers and explosives, which until
then had been dominated by the British. Thanks to their method of refining using a refrigeration process, they were able to process low-grade
caliche in large quantities, lowering their production costs by between
20 percent and 25 percent. This seemed enough to ensure the survival
of a branch of mining that was being threatened by the development
of less costly synthetic nitrates. This time, however, the Guggenheims
were overoptimistic, for to acquire a near monopoly over nitrate production they had had to incur heavy debts on Wall Street, and these
became a millstone around their neck when the Great Depression set
in. The only way out, as the future President Hoover had already suggested in December 1928, was to share the burden with the Chilean
government by creating, in March 1931, a joint company which they
would manage. But the solution was short-lived, for prices continued
to fall, and two years later the new company went bankrupt, showing
that American investment in the Pacific was not always crowned with
success.[37]

There were more impressive results in oil, mainly in the Dutch East Indies. Long held in check by the Dutch colonial administration, Standard Oil of New Jersey was finally granted concessions, thanks to support from Washington, who threatened reprisals against Royal Dutch Shell in America. In 1926 the first barrel left the Palembang refinery. Seven years later, Standard Oil merged with Socony-Vacuum, the ancestor of Mobil Oil, to form Stanvac (Standard Vacuum Oil Company), combining its production and refining operations in Sumatra with the marketing networks of Socony-Vacuum in Asia and Oceania. A third of the investment in Indonesian oil was now American—a weapon Roosevelt would use in the struggle against Japan in 1941. Warehouses, storage tanks, service stations, and a fleet of tankers all showed the importance of the major American oil companies in the western Pacific, particularly in China, Japan, and the Philippines.[38] The expansion of the automobile industry also encouraged the tire producers of Akron, U.S. Rubber and Goodyear, to guarantee their natural rubber supplies by investing in the plantations of Sumatra and Malaya, although their involvement would never equal that of the British.

To satisfy local demand, American companies constructed electric power stations and automobile assembly factories in Japan and Australia. On the urging of its president, Sosthenes Behn, in 1925 the International Telegraph and Telephone Company (ITT) took over the foreign markets of ATT, and gained a solid footing all around the Pacific. The Chinese market, with its millions of potential consumers, was a source of many disappointments, except for cigarette manufacturers, whom it provided with one of their principal markets. In the attempt to make inroads into the Chinese market, James B. Duke, the tobacco tycoon who in 1902 had founded a powerful Anglo-American multinational company with its head office in London, the British-American Tobacco Company (BAT), decided to construct factories in Shanghai, Hankou, and Manchuria, where labor was plentiful, skilled, and hard-working, and taxes were low. Initially, the raw leaf came from the United States until Duke encouraged local production. Similarly, in marketing, he depended, to begin with, on young bachelors from the southern United States, hired at $1,200 a year to travel the roads of China as salesmen. But soon this task was mostly taken over by salaried compradors or independent agents like H. H. Kung, the future magnate, financier of the Guomindang, and brother-in-law of Chiang Kai-shek, who acted as intermediaries in setting up a tight-knit sales network with the local merchants. It was not long before the BAT came to dominate the market, arousing the resentment of its Chinese competitors. As a result, it became one of the principal targets of the anti-American boycott in 1905–1906, from which it emerged victorious, establishing what was to all intents and purposes a monopoly. In 1915, World War I, which had stirred up Chinese xenophobia, encouraged the Chien brothers, founders of the Nanyang Brothers Tobacco Company, to try to compete. Starting in Canton, the competition rapidly spread to the Yangtze valley and to the north, in a war without any quarter. Fortunately for the two rival companies, the rapidly expanding market allowed both of them to garner magnificent profits. This golden age came to an end in 1923, when the Americans lost control of BAT to the British. They were obliged to give up the direction of the Chinese branch, which they had managed so successfully until then, and Duke had to resign.[39]

Portfolio Investments

If the growth in direct investment was proof of the increasing strength of the nation's industry, the amount of short- or long-term loans made abroad was an indication of the extent of the banking system's financial weight. Before 1914, the state of the United States current account balance barely allowed Wall Street financiers to look for potential foreign borrowers, or to offer terms comparable to those available in London, Paris, or Berlin. According to estimates by Cleona Lewis, the amount of loans made to the Pacific area by the United States between 1897 and 1914 totaled $300 million, or 30 percent of all overseas loans. The major portion (79 percent) seems to have been made by New York banks to Japan to help it finance the war against Russia in 1904–1905. China, on the other hand, accounted for only the insignificant amount of $9.5 million, hardly more than Bolivia, despite the efforts of the Taft administration to gain admittance to the European-dominated banking consortiums—a policy abandoned by his successor, Wilson, as soon as he took office.[40]

The situation changed completely with World War I, when the surplus in the current account balance allowed Americans to become net creditors. Between 1914 and 1935, they issued $10 billion in long-term and $4 billion in short-term loans. Because of European requirements, the Pacific's share fell markedly, compared to the earlier period, making up 17 percent of long-term and 4 percent of short-term portfolio investments. In total, however, more than $1,800 million was borrowed, mostly between 1924 and 1929, when the bankers, throwing caution to the winds (as they seem wont to do from time to time), sought out foreign clients without any concern for their subsequent solvency.[41] If they steered clear of China, where there was obviously too much turmoil, and if they were not running much of a risk in Japan (still the major borrower, though it accounted for only a quarter of the total), in Australia, or in the Dutch East Indies, they were laying up serious future problems in store for themselves in the Latin American nations that went bankrupt one by one from 1930 on.[42] Was the world economic crisis to blame, or was it just a lack of caution, occasionally combined with dishonesty and the misappropriation of funds? The debate was brought into the open with Congress's 1932–1934 inquiry into Wall Street financial practices, which came down on the side of the second interpretation. So it was that this initial massive intrusion of American capital into the Pacific zone ended with a whimper rather than a bang.

Between the Civil War and World War II, the economic frontier progressively lost its initial characteristics. The obstacle of distance, when measured by the time required, was becoming less and less serious. In 1914, the new shortcut through the isthmus of Panama was a decisive step forward, for the two seaboards of the United States were no longer pulled in opposite directions or cut off from one another. If the American Fleet no longer had the kind of supremacy over the entire Pacific it had enjoyed before 1860, economic ties had been strengthened, trade was increasing, and investments which had previously been nonexistent gave an indication of the developing power of American capital. Nevertheless, all these contacts still had something of the frontier about them. The range of products involved, which was still relatively limited, underlined the lack of balance between the center—for since the 1880s the United States had become the major world economic power—and the marginal zone which the Pacific still was. Only Japan now emerged as a worthy partner and rival. This new geography of wealth and power was accompanied by certain strategic, diplomatic, and cultural implications.

SIX

Closer Cultural Contacts

— When the Civil War broke out, cultural contacts between the
United States and the Pacific countries were still minimal. The Protestant missions could boast of only one real success, in Hawaii, while
China, which they had hoped to convert to Christianity, had opened
its door by only a crack. Americans and the other inhabitants of the
Pacific coastal countries had at most a superficial knowledge of one another. It was difficult to find any signs of Americanization anywhere
other than in the Polynesian kingdom of Hawaii. Furthermore, there
was a profound disparity in these contacts, for while American ships
put in to all the Pacific ports, Asians were just beginning to arrive in
California, along with Chileans and Peruvians, to prospect for gold in
the Sierra Nevada. The immense western ocean seemed rich in promise.
With an increase in trade, the cultural frontier would also begin to disappear as contacts became more numerous, but as this reality became
more imminent, it conjured up the specter of danger, just as much as it
did the vision of a more modern world.

THE AMERICANS, AGENTS OF MODERNIZATION

At the end of the nineteenth and at the beginning of the twentieth centuries, Americans thought that, in introducing their ideology of progress
to the backward peoples of the Pacific region, they were helping to

bring them into the modern world. Americans had the technology, the know-how, and also the ideals needed to help these countries emerge from underdevelopment, if only they could be convinced that the solutions they had to offer could work for all. There was a certain over-confidence in this attitude, and it would lead to disappointments later on, so that the final picture is one of both failure and success, whether the initiative was taken by the private sector, as was generally the case, or by the state, as an instrument of colonization.

Private Development Aid

To the extent that development is mostly a matter of people rather than of capital, the United States contributed to the modernization of the Pacific by educating the elite and training skilled workers, and by instilling in them values which would contribute to their spectacular advance. Increasing the efficiency of human capital could be done either locally or in educational establishments in America. Before the 1870s, colleges and universities offering a curriculum based on the classics and on mathematics, along the traditional lines inherited from Britain, had little to offer young Asians eager to revolutionize their countries, apart from introducing them to Christianity when they were run by a religious denomination. It was not until modern universities were introduced, partly under German influence, and until professional schools grew up at the end of the nineteenth century, that America began to have an appreciable impact. But by 1873, only five years after the Meiji revolution, between 200 and 300 Japanese students were enrolled in American schools, soon to be followed by other groups which, like the members of the Iwakura mission in 1871–1872, numbering the future modernizers Okubo and Ito among them, would be seduced by the marvels of science and technology, by everything from railroads to house interiors. Chinese students also sought, less effectively, to learn the secrets of power so that they could arouse imperial, and later Republican, China from its somnolence, and equip it to resist Europeans and Japanese attempts to carve it up between them.[1]

At the time, it seemed more expedient to modernize the system of education in the countries concerned rather than send the young people to a far-off foreign country, where often they would be far from welcomed with open arms. For the Protestant denominations it became increasingly important to open schools on various levels in order to transmit knowledge and educate the future elite who, it was hoped, would eventually assist the spread of Christianity. So, until just before the Second World War, it was in China more than anywhere else that university institutions inspired by the American model, such as Yale-in-China, proliferated.[2]

Early in the Meiji era, the Japanese government, carried away in a frenzy of modernization, attracted Western technicians and scientists by offering them handsome salaries. These *yatoi*, among whom there were more Americans than any other nationality, failed, it is true, to play a decisive role in Japan's development, if only because the Japanese governing class was determined to cling to its decision-making power, although in certain areas the *yatoi* did make a substantial contribution.[3] For instance, to assist development in the large northern island of Hokkaido, where the climatic conditions are not unlike those of the United States frontier, the Japanese modernizers turned to the federal Agriculture Commissioner Horace Capron, who was unable to resist the salary of $10,000 per annum he was offered—three times what he earned in Washington. During the four years he spent in Japan, between 1871 and 1875, he introduced new crops and agricultural implements, convinced the people to replace rice with wheat, which was better adapted to the conditions, and improved the breeding stock. However, his memories of his stay were not all positive. "How it is," he reflects in his memoirs, "that a people naturally so intelligent, ingenious, appreciative, and so capable of imitating everything they see, should remain so long in a state of semi-barbarism, is perfectly incomprehensible."[4] On Hokkaido, where he spent only eight and a half months, in 1876 and 1877, William Smith Clark nevertheless managed to found the Sapporo Agricultural College, along the lines of the college of which he had been president in Amherst, Massachusetts. He introduced an up-to-date program of study, established an experimental farm, and, together with Edwin Dun, the son of an Ohio farmer, developed dairy farming practices adapted to colder climates. As he was about to leave, on horseback, he turned to his Japanese students and exhorted them in words as brief as they are famous: "Boys, be ambitious!"—a typically American sentiment, but one which showed little understanding of the Japanese Empire's hierarchical and deeply anti-individualist culture.[5]

If it is hardly surprising to find few Americans among the military or legal advisers[6] (the long career of Henry W. Denison, who served in Japan from 1880 to 1914 is a notable exception), Americans did make an impact in the area of education, following the trail blazed by Guido Verbeck, a missionary of Dutch origin. Among the better-known personalities are David Murray, a professor at Rutgers University, who was hired to create a modern education system. If the centralized administrative structure was modeled on that of France, the educational objectives were largely American-inspired. The aim was to turn out independent, highly moral, and patriotic individuals. There was no discrimination in admissions: anyone—even girls—could enroll. Subjects useful for the creation of a modern society were emphasized. The new University of Tokyo conferred degrees patterned on the American system. By the time Murray returned to the United States in 1879, Japan was in possession of the resources to educate its youth, although some stresses and strains remained, for, after a short period of decentralization, central control was reestablished in 1880. This was a departure from the American system, which in 1886 was replaced by one along Prussian lines that was better adapted to the needs of the empire, which was increasingly expecting its subjects to display the unwavering loyalty called for by Confucianism. Similarly, John Dewey's progressive ideas, which had made an impression in the 1920s, were swamped by the frenzied nationalism of the next decade. Other Americans made their mark through teaching. One example is the zoologist Edward S. Morse, who introduced his subject to the University of Tokyo in 1877–1879. Morse established the Imperial Museum, popularized Darwinism, and launched the first archaeological dig near the capital. Ernest Fenollosa, recommended by Morse, was professor of philosophy from 1878 to 1886, and from 1897 to 1900, but is best known for his research on Far Eastern history and art. The most famous of all the specialists on Japan was Lafcadio Hearn, born in Greece of a British father. He had emigrated to the United States in 1869, where he made his name as a journalist. Twenty years later he moved to Japan, where he took up a position as professor of English. He married the daughter of a samurai, fell in love with his adopted country, and took out Japanese citizenship under the name of Koizumi Yakumo. Between 1894 and 1903, when he occupied a chair at the University of Tokyo, he introduced his Japanese students to English literature, while devoting his spare hours to writing books (widely read in the West) eulogizing the traditional Japan, which in his eyes was unfortunately dying out. This ultimately led to his disillusionment and to his decision to return to America, which was prevented only by his death in 1904.[7]

Apart from the Philippines, the only country of the Pacific region where the Americans made a substantial contribution to development was China. Unlike in Japan, this contribution was not a substantial one initially, for the Manchu regime had no desire to modernize an empire where revolt was constantly simmering. Its hostility to the Christian missions, which could have provided the required assistance, delayed the creation of Western educational institutions. After the failure of the Boxer insurrection, the Manchus' attitude became more favorable. So, too, was that of the Republic in its early days. As a result, the twentieth century saw a rapid growth in the number of schools and students enrolled.

Since the Washington government showed little inclination to develop cultural ties, this task fell to private organizations, and particularly to certain large foundations motivated by philanthropic concern and a desire—in the name of what some historians have termed "sentimental imperialism"[8]—to improve the lot of Asians. Beginning in 1914, the Rockefeller Foundation made the modernization of China one of its principal objectives. It thought the country was ready for it, if only it could be inspired with a genuine scientific spirit. To bring about such a cultural metamorphosis, Western medicine, which the Chinese accepted eagerly, seemed the best starting point. The foundation created a Chinese Medical Bureau and established a medical college in Peking which it expected to provide the best Western civilization had to offer: "not only in medical science but in mental development and spiritual culture."[9] So, from the outset, science was seen as a means to transform traditional Chinese society. The emphasis was placed on pure research rather than on expanding medical care. However, such lofty ambitions could not succeed overnight. The foundation was obliged to abandon this policy, not just because the medical college was proving too costly, but also because of the need to reckon with the growing chauvinism of the Chinese, increasingly unwilling to take directions from New York. In 1929 the foundation decided to make applied research its priority, since this was more likely to introduce the leaven of modernization into the Chinese countryside. Social scientists were enlisted to give some direction to its "rural reconstruction" program, an experiment in social planning and global reform with liberal and democratic objectives. This might have turned China into a "laboratory" had it not come up against the inertia of the Guomindang administration, and eventually been brought to an end by the Japanese aggression of 1937.[10]

In Latin America, American influence made its impact in the form of expert advice rather than through education. The most celebrated of the experts, Professor Edwin W. Kemmerer of Princeton, nicknamed "the money doctor," undertook several missions to the Andean countries between 1923 and 1931, as well as to China in 1928, acting in a private capacity, though he went with the blessing of the Department of State. Those in power made use of his advice and recommendations, not only to make themselves look better to American bond buyers, but also to modernize their societies. In Chile, for example, where Kemmerer arrived in 1925, the inflation resulting from the unrestricted printing of paper money, while it pleased the large rural landowners, caused much dissatisfaction among wage earners, whether workers, soldiers, industrialists, or importers. On his arrival, Kemmerer was given an enthusiastic welcome befitting a "financial Messiah." As a scientific expert immune to partisan sentiment, he was immediately accorded total trust. The "money doctor's" prescriptions aimed at adapting the monetary, financial, and budgetary policies underlying United States prosperity at the time to Chilean conditions. A central bank was created to stabilize the peso, which was put on the gold standard; a banking system was established subject to the surveillance of an independent watchdog, along the lines of New York State's, to keep a close watch on the budget; and an income tax was introduced. All these reforms were easy to put in place, since they were wanted by the military dictatorship which was in power at that moment, and because they favored the urban classes. With Kemmerer's help, the Chilean government had no further difficulty obtaining loans in New York, allowing it to embark on a policy of debt-financed growth—though this would soon be cut short by the Great Depression. However, all the American expert's good work was not brought to nothing by this contretemps, for the institutions he had created did manage to survive, though under state control.[11]

Cells of Modernization

American optimism tended to underestimate the extent to which the traditional societies of the Pacific could resist a culture so imbued with scientific rationalism. If a rapid transformation of mentalities and social structures was to be brought about by osmosis, it was not enough to create institutions characterized by a typically Western mentality and dynamism. This was even more so when the objectives of American companies were somewhat less noble than those of the Rockefeller Foundation. Such was the case in Latin America, where mining and agriculture were concerned, the only important consideration being the pursuit of profit. The intrusion of American technology into Andean societies did have an effect on the traditional way of life, but its influence was indirect. This was because the mine or the plantation remained a separate world, isolated from the rest of the country. The native workers lived in their own compounds, and did their shopping at the company store. The expatriate American executives resided in their own separate little universe, where they could enjoy all the amenities of life in their own country. Their compounds were really foreign growths grafted from North America. This meant that in the world of work two cultures coexisted which remained strangers to one another, and which were also separate from the indigenous society. But this is not to say that no modernization resulted, for American companies paid better wages, encouraged a lifestyle and hygienic practices which certainly represented progress, and also established schools. Because of their economic influence, they provided models for local modernizers. The creation of an assembled workforce brought about favorable conditions for the creation of a homegrown union movement, able to make demands, and at times willing to challenge not only the policies of the ownership but also the very foundations of a social order controlled by large landowners. Obviously, these consequences were not viewed positively by the multinational companies, who preferred to go on existing as self-contained units, concerning themselves with local issues only when their own interests seemed at risk. The foreign modernizers, even when their intentions were good, were not necessarily welcome in the host country. If they cut themselves off from the society at large they were accused of exploiting the country's resources and workforce, and of displaying an arrogant indifference to the traditional culture. If, on the contrary, they attempted to change the culture, they were accused of arrogantly attempting to impose their own view of the world, in the name of universal human values, with little or no concern for local uniqueness. So, as the urge to preserve special interests combined with a dislike of the American model to produce an instinctive rejection, anti-American feeling developed, on both the political right and left—but then so did envy of a way of life which would remain out of reach for many years to come.

An Ambivalent Colonial Legacy: The Case of the Philippines

Theoretically, the Americans had a much freer hand in imposing their ideal of modernization in their own colonies than in the independent countries of Latin America and Asia. The Philippines, taken from Spain in 1898, and subsequently from the indigenous nationalists, became an experimental zone, providing its new American masters with the opportunity to differentiate themselves from the Europeans by their "benevolent colonialism." But there, too, there was a sizable gap between the ideal and the practice. As soon as the decision was taken to annex the archipelago, Washington set out to turn it into a modern, democratic, and developed nation—a radical departure from the authoritarian and economically exploitative policies which had characterized Spanish rule. This meant destroying the old order, nationalizing property held by the monastic orders so as to eradicate what was considered a reactionary priesthood, and constructing a large number of schools, particularly at the primary level, to prepare future citizens to assume their responsibilities, and also technical schools to help diversify an economy in which agriculture had hitherto played an excessive part. It called for the administrative and political functions to be put in Filipino hands as soon as possible, initially at the local and regional levels, and later, nationally. The Philippines, if they were administered in conformity with the ideal of self-government and their people educated along American lines, would become a showpiece on the very threshold of Asia for the United States' modernizing influence, and bear witness to the superiority of American civilization over that of the European colonial powers.

The American governor generals who succeeded one another in Manila, from Taft to Murphy, put this generous program (which Kipling might well have seen as an instance of the "white man's burden") into effect with varying degrees of enthusiasm, depending on their view of the people they had been sent to govern. The new education system encouraged knowledge of the English language, although on the eve of the Second World War only one-quarter of the Filipinos were able to express themselves in English to any extent. This was because the budget allocated to primary schools grew only marginally, and sometimes even shrank, not only because the Americans had no interest in paying taxes to help modernize their colony, but also because the *ilustrados*, the indigenous elite—members of the liberal professions and large landowners, with the help of their influential connections—

pushed for the expansion of secondary education, which was of greater benefit to them than to the other social classes. American colonial policy thus found itself on the horns of a dilemma. On the one hand it hoped to create a democratic nation in which every citizen would be entitled to participate actively, requiring the overthrow of the traditional oligarchy, the introduction of agrarian reform, and a change of mentalities. On the other hand, in order to keep the cost of administering the archipelago to a minimum, the Americans were led, in conjunction with the members of this same local oligarchy, to develop a policy which historians, depending on their sympathies, call "co-optation" or "collaboration." Relying on their connections, the Philippine *caciques* guaranteed relative peace on condition that there was no interference with their privileges or with their role as intermediaries between the colonizing power and the ordinary people, whose guides and interpreters they purported to be. In the circumstances, the attempt to modernize the Philippines sometimes left a bitter taste in the mouths of its advocates. Some felt that the indigenous people were not yet ready for independence, and that the United States should not abdicate its responsibility by granting it prematurely, for this would only benefit the local oligarchy at the expense of an exploited populace. For others, more inclined to compromise, the American role was not to change a society so deeply impregnated with Spanish Catholicism and marked by centuries of underdevelopment— particularly since the colony was competition for certain domestic producers. It was much better to be rid of it and forget about the white man's burden! And so in 1941 the Philippines were still a case of unfinished modernization of a developing nation which had not yet come close to the American model, with an economy still based on four export commodities—sugar, abaca, tobacco, and copra oil—and an undemocratic but resilient social structure.[12] Despite the promise that independence would shortly be granted under a constitution partly inspired by that of the United States, the traditional social structures of the Philippines remained firmly in place, suggesting that the American colonial presence was no more than an interlude.

EVANGELIZING THE PACIFIC IN A SINGLE GENERATION

Of all the American social groups that participated in cultural interchange with the Pacific Rim countries, none was more important than the Protestant missionaries. That is not to say that the influence of the

Catholics—the Maryknolls in China after 1918, for example—or of the Mormons was completely negligible, but they had neither the numbers, nor the wealth, nor the head start in the enterprise of conversion enjoyed by the more traditional Protestant denominations.[13]

The Heyday of the Missionary Movement (1886–1925)

Preoccupied by many tasks at home during Reconstruction, and suffering from an undeniable decline in public support, American Protestantism tended to lose interest in the missionary effort in the years immediately following the Civil War. And yet new missionary fields were opening up, beginning with China in 1860, and a little later in Japan, with the removal of the ban on Christianity imposed during the seventeenth century. In fact, new missions were created in northern China, and institutions were founded in Japan, but what was lacking was the enthusiasm required to convert the heathen living around the Pacific, for this was an age of materialism and cynicism rather than of idealism or enthusiasm for spreading the gospel message. As had happened at the beginning of the nineteenth century, this reversal of fortunes finally brought about a powerful revival. In 1884 the Reverend Arthur T. Pierson published a work on the theme of "the crisis of missions," at the very moment when the evangelist Dwight L. Moody was convincing young Protestant students to read and reflect on the Bible once more, at his summer schools at Northrop (Mass.). These two influential figures held out a noble and stirring purpose to these idealistic young people—the "evangelization of the world within this generation." This slogan could be interpreted in various ways. Sometimes the emphasis was on the urgency of converting the heathen before Christ's second coming, in accordance with millennialist beliefs, and at other times it was considered sufficient to make the Christian message available to them without expecting such rapid results, but in the hope that, as a result, the missionaries' own lives would be radically changed.

For four decades the American Protestant missionary movement prospered. Neither vocations nor funds had ever been so plentiful. This explains the widespread optimism, for the objective of evangelization within a single generation did not seem so unattainable. But if it was to be achieved, its proponents had to adapt to the spirit of an age in which large bureaucratic structures were paramount, and this compromise sowed the seeds of ultimate failure. At the end of the nineteenth and at the beginning of the twentieth centuries American society worshiped at the altar of efficiency and believed in rational organization. If the Pacific was to be evangelized, it could not wait for the Holy Spirit to descend upon the unbelievers or for the truth patiently to reveal itself to them. The hope was that if the same techniques which had been so successful in the production and marketing of consumer goods were applied to the missionary effort, they would give it a new strength and efficiency. The principal Protestant denominations—the Methodists, Presbyterians, Congregationalists, and Baptists—were infused with the forms of action developed by the "visible hand" of capitalism.[14]

They were able to benefit from the support of specialized institutions independent of any particular church, such as the Student Volunteer Movement (SVM) for foreign missions and the Lay Missionary Movement (LMM). The SVM, founded in 1888 on the instigation of its general secretary, John R. Mott, set out to encourage the greatest possible number of students to volunteer for service as missionaries abroad. Prayer meetings and ceaseless publicity stoked enthusiasm on university campuses in the Midwest and, to a lesser degree, in the East. Several thousand young students decided to devote some years of their lives to converting the Chinese. This form of recruitment was facilitated by a shift in Protestant theology, while it itself led to a more liberal interpretation of the Bible message. Instead of emphasizing the wrath of God and the damnation of unrepentant souls, as had been done in the past, the new gospel had a social orientation, for, in tune with one of the trends dominant in progressive America at the turn of the century, it was at least as concerned with social progress as with personal salvation. It enlisted the assistance of experts in education and medicine, so that the missionary effort became professionalized in a way which profoundly altered its character, but at the same time responded to the aspirations of the minority of educated young Americans willing to travel to the Pacific countries.

The funding of such an undertaking required a substantial financial contribution from the members of the major denominations, and enthusiasm had to be kept alive if cash was to flow in on a regular basis. So, missionaries on furlough spent the time on lecture tours. Special campaigns supported the victory of Christianity. More and more lay missionaries worked alongside the clergy. The membership of their movement, the LMM, founded in 1907, was largely composed of businessmen. These were somewhat conservative in social matters but were

prepared to introduce the principles to which they owed their economic success—rational management practices and the fundamental
importance of organization—into the infrastructure of the missionary
effort. The missionary movement borrowed many of its methods from
big business and apparently did so with some success, since both the
funding raised and the number of vocations reached their peaks immediately after the First World War, in 1919–1921.[15]

High Hopes and Bitter Reality

The missionary movement, run like a large business enterprise, collected statistics (of varying reliability, it is true) which show how well it
prospered, though it was still not able to achieve the evangelization of
the world quite as quickly as within a single generation.[16] All around
the Pacific, potential mission fields lay fallow. In the Catholic countries
of Latin America the missions established a few churches and schools in
the principal seaports of Chile, Peru, and Ecuador, to minister to
sailors and other Anglo-Saxons in residence or passing through, rather
than for the native populations whose "popery" seemed beyond re-
demption.[17] Elsewhere, the British and the Americans worked together
on relatively good terms, sometimes dividing up the territory between
them so as to avoid any harmful competition. The major Protestant denominations on both sides of the Atlantic, as well as the Anglican
Church, developed a deep sense of solidarity as they confronted their
Catholic or heathen adversaries. In Micronesia, the Congregationalists,
closely linked to the Hawaiian gospel mission, decided to leave the
Gilbert Islands to the London Missionary Society, while in the Caroline Islands they cut back their activity when Germany purchased the
archipelago in 1898.[18] The American Protestants, who supplied half the
ordained clergy working in the Pacific zone, focused their main efforts
on two countries: China and Japan.[19]

In Japan, after a lackluster beginning, the work of evangelization
appeared to make stunning advances in the 1880s.[20] The Japanese,
wishing to rid themselves as quickly as possible of any threat of European colonization, were willing to adopt western ways with all speed.
They could see that the developed world at the time was Christian, and
hoped that conversion would endow them with the kind of mentality
which was the basis for industrialization, for modernization, and, consequently, for power. The gospel message was well received by the edu-

cated, wealthy elite, the descendants of the ancient samurai class. In Japan, Protestantism did not attempt to address the common people, its strategy being to win over the upper classes first of all, and to transform society from the top down. The American missions, which were by far the most influential, made the connection between Westernization and religious conversion very early on.[21] They founded schools, particularly kindergartens, or, as in Kobe, institutions for young girls, who were often excluded from the public education system. They also helped to open universities, the most famous of these being Doshisha University in Kyoto, where the teaching of languages and of management was emphasized.[22] Yet Japan took second place to China in both the number of missionaries and in funding. This was because the Japanese converts, by virtue of the fact that they belonged to the elite, very soon took control. They founded their own independent churches, entirely financed by themselves, and set out to convert the populace. They took an active role in developing education—Doshisha University was founded jointly in 1875 by the Reverend Jerome Dean Davis and Neesima, a Japanese. On the other hand, the hope that the advances made in the 1880s would continue exponentially, and that Japan would be evangelized within a generation, did not survive the upsurge of nationalism which began in 1889. The fact that there were unequal treaties still in place proved that conversion to Christianity was not enough to loosen the hold of Western imperialism. The traditional religions, Buddhism and Shintoism, returned to favor, encouraged by the government, which was promoting the mystique of the emperor. Protestantism itself, divided by sectarian squabbles, became less attractive, particularly since in the West itself powerful movements were challenging religion, which was losing its prestige to science. From 1890 on, the American missionary movement in Japan was at a standstill, doing little more than keeping its head above water, before being submerged by the ultranationalist wave of the 1930s. As a consequence, its influence was progressively weakened, compared with that of the more energetic China lobby.

As was the case in the economic domain, all hopes focused on China, with its masses of humanity and its high level of civilization.[23] If China adopted Christianity, the triumph of the true faith over its adversaries would be sure to follow shortly, but if, on the other hand, it opted for irreligion, this would pose a serious threat to the West and its Judeo-Christian values. So it is hardly surprising that China became the primary field of activity for young students, as well as for missionaries of the old school. Inland from the ports opened by the treaties lay the vast spaces of rural China, to which legal access was not granted until the treaty of 1903 (though missions had already been trying to make its spiritual conquest for several decades, despite local resistance and the risks involved). The hostility of the educated class required a different strategy from the one followed in Japan. On the one hand, the ordinary people were not forgotten, and attempts were made to reach them by preaching in the streets or in churches erected in the towns. Religious tracts and Bibles translated into the various Chinese dialects were distributed, often with little impact on illiterate masses unfamiliar with Christian symbols (like Pearl Buck's hero, Wang Lung, completely mystifed by the scrap of paper given him by a European, on which was depicted "a man, white-skinned, who hung upon a cross-piece of wood"[24]). Such methods of evangelization were still essential to those Protestants for whom a literal reading of the Bible and the quest for personal salvation were paramount. The Chinese Inland Mission (CIM)—mostly British but with Americans providing a third of its personnel—actually considered this the only way to achieve its objective of converting the heathen. A Southern Baptist like the fanatical and bigoted Reverend Tarleton P. Crawford, was quite prepared to sacrifice the school which his wife had managed to keep going in order to pursue, with no success whatsoever, his chimerical dream of saving Chinese souls from eternal damnation.[25]

In fact, the Chinese who attended missionary schools often did so with a purely practical purpose in mind. They valued Western medicine highly, as they did modern scientific knowledge and the English language, which opened doors to employment in commercial dealings with the industrialized countries. The Protestants who were preaching a social gospel shared this interest in things secular, hoping that education and medicine would open Chinese eyes to the benefits of the true faith. Schools, and boarding schools in particular, removed young people from their traditional environment, making it possible to make converts of them, who would in their turn bring the gospel to their fellow countrymen, in their own language. At the 1877 Shanghai conference, Calvin Mateer, a Presbyterian clergyman, defended the educational effort against a critical majority of his colleagues. He was the founder of a secondary school and then, in 1882, of the first post-secondary college, at Dengzhou in Shandong province.[26] Establishments of this nature became increasingly numerous until the Boxer insurrection of 1900, when the missionaries paid dearly for their association in people's minds with Western imperialism.[27] They only acquired real importance after the failure of this nationalist uprising, which had been encouraged by the xenophobic and threatened imperial authorities. The Manchu government then became convinced, rather late in the day, that its salvation lay in modernizing the country and that the missions, instead of being treated like foreign intruders, could assist in bringing about the needed changes, particularly since the Americans were opposed to any carving up of the empire. So it was that during the next twenty years a relatively favorable climate prevailed for the creation of a variety of schools, hospitals, and clinics. The number of students quadrupled between 1907 and 1920, rising to almost 200,000 at the primary and secondary levels—i.e., between 4 percent and 8 percent of the total enrolled in the public education system. In higher education the Protestant colleges enrolled an increasing number of students (2,000 in 1920, and 3,500 five years later—very nearly a tenth of the total enrollment), in such prestigious institutions as Yenching College in Peking, Jinan College in Shandong, and Nanjing.[28] When compared with the Chinese population of many hundreds of millions, these figures seem discouragingly low and even more so if the ultimate objective of conversion is taken into account, for it never occurred to ever-increasing numbers of students to adopt Christianity, since the nationalist cause seemed to them to contain much more promise for the future.

A Religious Depression

Enthusiasm for overseas missions fell into decline in the years following the Great War. The decade of prosperity was characterized by a profound disillusionment which was the ultimate consequence of the crusade for democracy. Any ambitious project was discredited in a society which had fallen prey to the seductions of mass consumption and become thirsty for quick profits. The fruits of evangelization were too slow to ripen. After 40 years of effort in China the institutions responsible for the missions had very little to show. Endless repetition of their slogan had dulled its impact on a public which was now diverting its money to different purposes, and on students who were becoming more and more attracted by the well-paid professions in preference to the austere—though hardly monastic—life in a distant mission on the other side of the Pacific. The churches often lacked both funds and men, though not single women who, although they were excluded from ordination, continued to volunteer as teachers, caregivers, and preachers of the gospel message.[29] The great economic depression that set in during the fall of 1929 made the situation even more critical by reinforcing the effects of the religious depression that had gone before. The foreign missions were no longer a priority, even for those who had always considered them important, with the result that they now had financial limitations to contend with.

Even more serious was a challenge to the very principle of the missionary effort which came from within American Protestantism itself. In the 1920s, a conflict which had been brewing between liberals and conservatives was brought into the open. For the conservatives, Christianity was unique, unlike any other religion. It offered the only true path to salvation, and the missions had a duty to bring the light of the gospel to the benighted heathen. For liberals, on the other hand, such a cut-and-dried distinction was unacceptable. They believed Christianity was indeed the best of the "superior" religions, but it was just one on a scale of several, on which it indeed held a pride of place. This did not mean it should attempt to displace the others, but rather, to make allies of them in the struggle against the real enemies of religion in the modern world, namely materialism and secularism. Consequently, the liberals urged that the missionaries seek dialogue rather than confrontation with the Buddhists or the Confucians, in order to discover religious truth through a process of mutual exploration. Such was the position expounded in 1932 in a ground-breaking report by Harvard philosopher William E. Hocking, who had been asked to "re-think" the missions after a century of activity.[30] Between the two extremes, a moderate majority accepted the idea that all the great religions offer a genuine knowledge of God, so that the proper approach was to study them sympathetically instead of criticizing them aggressively. However, they still refused to consider Christianity as just one religion among many, for they still held it to be the only true revealed religion, able to hold out a hope of salvation in this life and in the next. To other faiths it offered "the steel hand of truth encased in the velvet glove of love" as the Presbyterian James S. Dennis put it. The missionary's duty was not so much to spread a doctrinal message based on the Bible as to demonstrate the goodness of Christ by himself leading an exemplary life which might inspire the heathen to follow his example. All these differences, which had long been kept below the surface by the common enthusiasm at the time the missionary movement was having such success, reemerged in the 1920s, when there was a definitive rupture between the fundamentalists and the more moderate liberal establishment.[31] The quarrel was extensive enough to do serious damage to the unity of the Protestant effort and to the credibility of its message. The fundamentalists brought back nineteenth-century theology in full force, while the liberals ultimately made Christianity into a flabby, spineless, doctrinal muddle, quite incapable of inspiring the enthusiastic spirit of a few years earlier. The missionary spirit was undermined from within by the deep differences between the different persuasions. In the circumstances, the aim of evangelizing China within a single generation became no more than an empty slogan.

But even if American Protestantism had remained united it seems doubtful that it could have attained its objective after the First World War, when it came face-to-face with a formidable opponent in the form of nationalism. Whatever efforts it had made to adapt to the culture, it had been unable to avoid being identified with the Western imperialism for which it paved the way culturally. To stop Christian evangelization from being identified with Westernization, and indeed with the "occupation" boasted of by the Continuity Committee of the World Missionary Conference in 1922, indigenous Chinese churches would have had to take over. However, if there was some movement toward the creation of a national organization, it would not have been possible without the financial and human support provided by the United States. This suggested that Americans still exerted at least some direction and control—something which was becoming more and more intolerable to the nationalists, who were becoming increasingly impatient of any foreign yoke, whether they were on the political right, and attached to tradition, or leftist apologists of Communist revolution.[32] During the unrest in China, in 1926 and 1927, the work of the missions was endangered by a wave of violence. John E. Williams, vice president of the University of Nanjing, was assassinated by the soldiery, and the churches decided to withdraw their people from the interior of the country, moving them to the ports or bringing them back to the United States until order was reestablished under Chiang Kai-shek, whose conversion to Protestantism in 1930 offered a glimmer of hope in an overall rather discouraging picture.[33] To keep in touch

with events in China during the 1930s, the missionaries turned their attention to the countryside, lending their support to programs of rural
development. They considered this more productive in the long-term
than the traditional evangelism which the fundamentalists of the CIM
were still practicing. This change of policy was given little time to
achieve any significant results, for the Guomindang was opposed to any
extensive agrarian reform, and very soon the war started by Japan made
the nationalist struggle for the country's liberation a priority. Missionaries who until then had been hostile to communism came to appreciate the contribution made by Mao Zedong's partisans to the struggle
against a merciless invader, while others believed the only solution for
Christianity in China lay in unwavering support for Chiang Kai-shek,
despite the obvious inadequacies of his Nationalist regime.[34]

The religious depression had indeed slowed down the missionary
effort considerably, but the movement had been able to achieve sufficient impetus around 1920 to be able to make a fairly strong showing
during the two following decades, particularly in China, which had
become its favorite child. In 1938, fifty-one missionary societies were
active in China, and thirty-six in Japan. That year, China received
$3,700,000 in funding, and Japan $1,350,000. There were three times as
many American missionaries in the Republic of China as in Japan—
2,500 compared to fewer than 700.[35] The evangelization of the Pacific
countries would no more be achieved in two generations than in one.
The cultural failure was evident, but the campaign did have important
strategic implications, for a powerful lobby group resulted in American
foreign policy playing the China card against Japan, in spite of the much
closer links which had been established with the Japanese economy.[36]

DANGER FROM THE WEST?

For Europeans, Asia is in the east. For Americans, particularly those
living along the Pacific Coast, it lies to the west, beyond the vast expanse of the Pacific. This relative geographic position did not appear to
be an entirely positive thing when, at the turn of the century, certain
learned minds put forward the theory of the historical march of civilization from east to west. Of course, this theory did hold out the
promise that the United States would take over the torch from western
Europe, but it also implied that it in turn would have to surrender it to

a Far East populated by masses of humanity quite capable of dominating the entire world. So, the Pacific came to be viewed by large sectors of American opinion as a frontier fraught with danger—however unrealistic such a view may have been.

Anthropological Jousting

From the eighteenth century on, Oceania has often been described as a kind of heaven on earth by certain Western intellectual milieus which quite early on discovered it had certain charms which seemed to be lacking in Judeo-Christian civilization or in a capitalist economy. The multitude of microsocieties inhabiting the islands provided countless examples of different social organizations and bore witness to the diversity of human structures. In the United States, early in the twentieth century, a doctrine of biological determinism (taken to an extreme by the eugenics movement), seemed to have taken control of the social sciences. It is not surprising, therefore, that at the same time racism and racial segregation, as its institutional manifestation, were at their height. But this whole discourse, originating in a supposedly scientific anthropological discipline, was actually based more on fancy rather than on any conclusions which had undergone the rigorous experimental proof which is fundamental to the scientific method. A movement opposing this trend grew up around Franz Boas, a professor at Columbia University, and led to a broad debate which was to last for at least two decades, from 1910 to 1930, regarding the respective importance for the development of the individual personality of innate, as opposed to acquired, characteristics. It seemed that no compromise was possible between these two opposing factors of "nature" and "nurture." Boas, with his disciples Kroeber and Lowie, together with the behaviorists, argued for cultural instead of biological determinism, claiming that the individual personality is shaped, not by its genes and its heredity, but by the environment in which it exists from birth. But they still needed proof to support their theory. It was noticed that in all societies adolescence was accompanied by a crisis, by a questioning of the values imposed by adults. Was this not an indication that a biological upheaval was taking place at this age, resulting in some degree of rebellious behavior? Or should it not rather be blamed on the dominant culture itself and the tensions arising from it? To settle the problem, in 1925 Boas sent Margaret Mead, one of his students, aged twenty-three, to study the lives of adolescents on Samoa, which, as an American colony, was preferred to the Tuamotu Islands, a French possession. After nine months of research, spent mostly on Tau, the most easterly of the Manua group, the young ethnologist had a solution to offer: "Much of what we ascribe to human nature is no more than a reaction to the restraints put upon us by our civilization."[37] Among the two dozen young Samoans she interviewed, Mead found no real evidence of any crisis of revolt. Unlike young Americans, they enjoyed almost total sexual liberty prior to marriage, and lived as members of large family groups, without any particularly close relationship with their biological parents. It was true that they did not have many choices, for Samoan culture, compared to that of the United States, appeared to be very simple, with one day followed another in an almost unchanging pattern. Nevertheless, the general impression given by this island society was one of happiness. Here was the "negative example" which would invalidate all the conclusions of the biological determinists. Samoa was the battleground where the great nature versus nurture debate would be settled. It might even provide some principles on which to base an American liberal education to make transition through adolescence less painful. The success of Margaret Mead's book—an anthropological best seller published in 1928—ensured the triumph of cultural determinism. For at least until the recent research by Derek Freeman, which was poorly received by anthropologists, no one questioned the quality of the evidence collected or the lack of any comparisons with the many previous studies which, by contrast, had insisted on the prudish morals of the Samoans and described a considerable amount of tension within Samoan society. In short, Mead's Oceania had little to do with reality. Underlying the seductive picture she painted was the rivalry between two parties of zealots around a dispute which it was impossible to settle in the 1920s and on which very little further light has been cast since then.[38] Nevertheless, the new cultural anthropology did have the merit of combating one of the shortcomings of American society at the time, namely, its racism, from which the Oriental peoples suffered, among others, particularly where immigration was concerned.

Unwelcome Immigration

In spite of the emotions it aroused, immigration from the Pacific countries between the Civil War and the quota laws of 1924 was only a drop compared to the flood entering the United States, which was transforming it into "a nation of immigrants"—it represented only 2 percent of the total immigration numbers, i.e., around 600,000 individuals between 1853 and 1914, compared with 26 million European immigrants. The Pacific Latin American countries contributed almost nothing to this percentage, for the Chileans and Peruvians had been put out by their poor reception in California at the time of the Gold Rush. There were also very few Australians, for there was no lack of space remaining to be conquered on their own frontier, nor of jobs in their port cities. The Filipinos, who after the annexation were granted the status of "nationals," halfway between foreigners and citizens, filled the void left by the Asians once these had been excluded.[39] This means that the problem of immigration from the Pacific was actually limited to the Chinese, who made up 60 percent of the total numbers, and to a lesser extent the Japanese, who amounted to a third. Even though they arrived in limited numbers, the Asian immigrants were a cause of uneasiness because their arrival took place in the space of only a few years, and was concentrated on California: almost a third of the Chinese entered the United States within a ten-year period, and the same fraction of Japanese arrived between 1900 and 1908.[40]

The same factors which had induced the Europeans to cross the Atlantic also played a part, though to a lesser extent, in attracting immigrants from the two Asian empires. Both the pull of America and the repulsive force of Asia led people to move from one shore of the vast ocean to the other.[41] The great majority of the Chinese came from the Canton region, where population pressure was leading to poverty, and provoking tension between the original inhabitants and the Hakkas, recent arrivals from the north. Numbers of men were obliged to leave temporarily—or at least so they hoped—to earn enough money to enable the family they left behind to survive. They usually emigrated in groups, organized by Canton merchants, who attempted to keep control over the immigrants after they arrived on the other side of the Pacific. A number of societies, some of them official, and others secret, maintained family unity and helped immigrants adjust to a deeply foreign, and even hostile, environment. The celebrated Six Companies of San Francisco, founded in 1862, were typical of this concern to organize immigration along purely ethnic lines, in a way which was basically similar to what was happening in certain city neighborhoods where European immigrants tended to concentrate.

Contrary to the latter, however, there was only a very small proportion of women among the Chinese—only 8 percent, and in addition more than half of these were engaged in prostitution. Since interracial marriage was both little practiced and was frowned upon, the Chinese immigrant community in the United States was unable to give birth to a new generation, or to grow in numbers, falling from 107,000 in 1890 to 62,000 in 1920.[42] In Japan also, poverty, and the lack of any brighter economic promise, caused peasants and artisans to set out, to Hawaii initially, where, from 1900 on, they constituted the largest ethnic group, and later to California.[43] At the end of the nineteenth century, they could expect to earn from five to ten times more in the United States than in their country of origin. In addition, they were immune from conscription, and enjoyed a freedom of expression unknown in Japan (the Japanese Socialist party was founded in Seattle). It is true that during the Meiji era Americans enjoyed an extremely positive reputation, and that the Japanese government, uneasy at its industry's inability to provide employment for the rapidly increasing population, encouraged emigration, while maintaining control over it. In the case of the Japanese also, the predominance of male emigrants in the period before 1908 had an effect on the gender distribution of the group. This lasted as late as 1940, three decades after Japanese immigration had been halted, at which point only 43 percent was female.

Much earlier than immigration from elsewhere, immigration from Asia encountered violent hostility from nativists of all stripes. Antipathy was initially focused on the Chinese, who had been the first to cross the Pacific, but subsequently was extended to the Japanese, who had formed the second wave. The Californians, living on the farthest fringe of the United States, and made uneasy by what seemed to them the first stages of an invasion, were at the forefront of the resistance, but, as Stuart C. Miller has shown, diplomats, merchants, and missionaries were also responsible for spreading an image of China and its people which was negative enough to mean that the working class in San Francisco and the West should not be considered the only ones responsible for the exclusionary measures.[44]

These were introduced in spite of the fact that in 1868 the American statesman Anson Burlingame, on behalf of the Chinese government, had persuaded the United States to sign a treaty granting the emperor's subjects the right to immigrate. But unrest soon grew in California, which had a great deal of influence on the Washington administration because of its propensity to swing easily from the Democrats to the Republicans, or vice versa. The economic crisis between 1873 and 1878, and the resultant unemployment, obliged Congress to act by abolishing the reciprocal rights pertaining to immigration—a unilateral action vetoed by President Hayes, but which led to a renegotiation of the treaty with the Peking authorities, who were forced to agree to allow the federal government, if it considered public order in danger, to "regulate, limit, or suspend" the "coming or residence," but "not absolutely prohibit" the immigration of unskilled Chinese laborers alone. Other categories of immigrants, and those already settled in the United States, were to continue to enjoy the right to freedom of movement and be protected against mistreatment. This was too good an opportunity to be missed, and the law of May 6, 1882, passed with a large majority, suspended the further immigration of Chinese manual laborers, whether skilled or not, with all others being required to obtain a certificate from the imperial government. Renewed for a further ten years in 1892, this suspension was made permanent in 1902, despite the treaties which had been signed with China. Furthermore, the Supreme Court was liable to restrict the rights of people not affected by this measure, but whose legitimate entitlement could be questioned when they returned to the United States from a trip home. There was no guarantee of fair procedures being followed in such cases. A sworn statement from at least one credible white witness was required, for it was thought that the Chinese were incapable of understanding the significance of such a solemn act. In general, the court gave as free a hand as possible to the executive branch, and it was up to any foreigner to prove the legitimacy of his case to immigration officials.[45] In 1927, the same authority forced Martha Lum, an American citizen of Chinese origin, to attend a black school, for it considered that the fundamental distinction was between whites and all the other races and agreed that a Southern state was entitled to segregate schools.[46]

Initially escaping these discriminatory measures, Japanese immigrants were affected in turn in the twentieth century, despite the support given them by their government, which was militarily much stronger than the government of China, and despite the fact that, wishing to protect its image, Japan selected prospective emigrants. Once again, the first restrictive moves were made in San Francisco. In October 1906, the school board required children of Asian origin to attend an "oriental" school, separate from the schools for young whites. In fact, this was an attempt to provoke an incident which would justify excluding the Japanese. The strategy was successful, for President Theodore Roosevelt,

conscious of strategic realities and of the wave of indignation sweeping
through Japan, was obliged to intervene. On the one hand he persuaded
the city of San Francisco to abrogate its segregation decision, while on
the other he reached a gentleman's agreement with the Tokyo government in February 1907, the precise terms of which have never been revealed, and which did not require approval by Congress. Japan promised
to discourage emigration by manual laborers, unless they were former
residents, family members of residents of the United States, or farmers
already established there.[47] But once the flow of immigrants was halted,
there still remained the problem of land. In 1913, the Californians, jealous of the efficiency and industriousness of peasants from across the Pacific, passed a law prohibiting strangers not entitled to citizenship from
owning agricultural land. They were only allowed to sign leases for less
than three years—a concession that was abolished in 1920, for it prevented the achievement of the desired objective.[48] The final piece of the
puzzle was put in place when the Japanese were refused any possibility of
acquiring American citizenship. In 1922, the Supreme Court revoked
the naturalization granted to Kakao Ozawa, who had entered the
United States as a student in 1894, graduated from the University of
California, and settled in Hawaii, where he took out American citizenship. The court had only one argument against him, and this was a racist
one: he was not white, whereas the naturalization laws of 1790 only
made mention of "free white persons." The door was slammed shut in
the faces of the Japanese, as it had been for the Chinese in 1882. The law
of May 26, 1924, established quotas for Asian immigration similar to
those which already existed for Europeans, but with the important difference that each Asian country was allowed only the minimum number
of 100 immigrants per annum, compared with 51,227 for Germany,
34,007 for the United Kingdom, and 28,567 for the Republic of Ireland,
or even the 3,954 awarded France. The Japanese saw this unjust treatment as a profound insult, which explicitly designated them as racial in-
feriors. Excluded from naturalization, Asians could now hardly even
hope to obtain the status of permanent residents.

The "Yellow Peril"

The legislation to exclude Asian immigrants was a sign of the fear
felt by American society at the end of the nineteenth and at the beginning of the twentieth centuries. The phrase "yellow peril" was made

popular by Kaiser Wilhelm II in 1895, but it only achieved general acceptance because the fear already existed, in a more or less latent form. For several decades the Far East was viewed in people's minds as a threat to Western civilization. This was all the more surprising because at the time Europe was at the peak of its ability to subject the other continents to its imperialist domination. Such a fear might perhaps have been comprehensible on the part of Europe, divided as it was between systems of alliances which represented a threat to peace, but in the United States, with its isolationist tendencies and the remarkable prosperity it enjoyed, there was no reason for the notion of the "yellow peril" to gain such wide acceptance. The Pacific Ocean was coming to be seen as the last line of defense between civilization and barbarism. In spite of a mass of often contradictory arguments used to support this truly preposterous idea—which was in addition completely contradictory, since on the one hand it expressed a racist view of Asians as an inferior breed of human being and on the other hand a fear of the competition they could offer in many areas, whether demographic, economic, cultural, or military— there was no rational explanation for the emergence of this basic fear which had always lurked beneath the surface.[49]

The propaganda aimed at the "yellow peril" conjured up a vision of swarming Oriental hordes, preparing to cross the Pacific and pour into the relatively empty spaces of the American West, the real frontier of Western civilization. It capitalized on totally irrelevant historical memories of the waves of invasion from the steppes of Asia which had constantly menaced the medieval world. The influx of Oriental immigrants was represented as a catastrophe, since any mixing of the races would surely lead to the degeneration of the supposedly superior white race. Dressed up in false science, the racism of Madison Grant or Lothrop Stoddard gave an unwarranted importance to the notion of racial purity, and predicted a gigantic "conflict of color," in the twentieth century.[50]

Economically, the threat was perceived as a direct challenge to white interests. The first, most immediate, and most urgent danger arose from the fact that Asian immigrants were prepared to work longer hours for lower wages, since they needed less to survive than did "Caucasian" workers. In California, the accusation was constantly leveled against the Japanese that they offered unfair competition. In his novel *The Pride of Palomar*, Peter B. Kyne turned his hero, Don Miguel Farrel, the scion of a Californian Hispanic family, into a virulent spokesman against the Japanese invasion, as he argues with a rich New York

banker, John Parker, who is intending to make a lot of money by selling land to a certain Okada, a Japanese "potato baron":

"So you do not believe it possible for a white man to compete
economically with these people, Farrel?"

"Would you, if you were a white farmer, care to compete
with the Japanese farmers of this valley? Would you care to live in
a rough board shack, subsist largely on rice, labor from daylight to
dark and force your wife and daughter to labor with you in the
fields? Would you care to live in a kennel and never read a book or
take an interest in public affairs or thrill at a sunset or consider that
you really ought to contribute a dollar towards starving childhood
in Europe? Would you?"

"You paint a sorry picture, Farrel." Parker was evasive.[51]

The second economic threat came from an industrialized Asia
where low manufacturing costs, due to the low wages, made it impossible to compete. At the end of the nineteenth century the American
proponents of a bimetallic standard saw this as an argument against the
gold standard, since a devaluation of silver (the only standard in Japan
and in China) against gold would give an advantage to exports from
those countries. This was no longer to see Asians as inferior, but as cunning, well-suited to factory work, and capable of competing with
Westerners in those sectors of production in which they excelled. In
the United States, which was very protectionist up to 1913 and from
1922 on, this argument struck home. The Pacific countries would take
jobs from American workers and deprive them of all their hard-earned
benefits. There is no need to stress the blinkered nature of such a view,
totally permeated by the notion that human races are endowed with
specific immutable, unvarying characteristics.

The notion of a cultural threat was rooted in the same prejudices.
Stereotypically, various types of Asians were all endowed with identical
characteristics. Up to 1905, the Chinese seemed to represent the major
threat. They were depicted as scheming, deceitful, cruel, and vicious,
devoid of all morality, with innate criminal tendencies—in short, creatures quite incompatible with the democratic civilization of America.
The terrible Dr. Fu Manchu, in Sachs Rohmer's novels, was the incar-
nation of the perverse Oriental mentality. In comic books bloody
tyrants were often given almond eyes, the most typical physical feature

of Far Eastern people. In addition, the Chinese, because they were thought to be unclean, were seen as a danger to public health—for an inferior race can even conquer a superior one by spreading the germs of an epidemic. The image of the Japanese, who after 1905 replaced the Chinese in American nightmares, often took a similar form, but they were considered an even greater threat because of their greater intelligence and discipline and their attachment to the emperor, traits which made them, consequently, even more difficult to assimilate.

In the minds of those who warned of the "yellow peril," the combination of so many demographic, economic, and cultural threats could only lead to armed conflict, all the more dangerous to America since the Asians would probably form an alliance based on racial solidarity, or perhaps prepare the way for czarist Russia's ambition to extend its empire. As early as 1879, the Californian Pierton W. Donner, in a work intended as part analysis and part prophecy, described the inevitable development of a situation which, early in the twentieth century, would lead to the "last days of the Republic."[52] Chinese immigration was paving the way for the armed hordes which would take advantage of America's lack of foresight and of its internal divisions to conquer the entire country and raise the imperial dragon flag over the dome of the Capitol. This invasion from the Pacific would transform the "Temple of Liberty" into the "Western Empire of His August Majesty the Emperor of China, and Ruler of all Lands." Thirty years later, Homer Lea, in *The Valor of Ignorance*, would outline the different stages which he believed would lead to an almost equally disastrous outcome taking the form of a future war between the United States and Japan. He argued that American democracy was too ill-prepared to withstand an attack by Japan. Japanese troops would land in California, capture the entire Pacific Coast, and establish an impregnable line of defense along the Rocky Mountains. Such would be the fate a country which prized individualism and "commercialism" much too highly would suffer at the hands of a nation where everything was subordinated to the interests of the state.[53] For many geopoliticians, a major Japanese-American war for control of the Pacific lay in wait over the horizon of the twentieth century. Beginning around the time of World War I, a number of books analyzed the tensions and attempted to predict the results of a hypothetical war which most believed would end in a United States victory, after some initial setbacks. In 1925, the British writer Hector Bywater, in a prophetic work, actually prophesied a Japanese raid on Pearl Harbor. Reality would imitate fiction,[54] for before long the imaginary "yellow peril" would be replaced by the real threat of Japanese imperialism.

Nevertheless, it would be unfair to mention only the expressions of racism in the United States. In the cultural domain, fruitful exchanges took place on both sides of the ocean. To take just one example, the so typically American architecture of the Prairie Style owed a great deal to Japanese influence. Phoenix Hall, the Japanese pavilion at the Chicago Exposition of 1893, was in strong contrast with the neoclassical style of the official buildings. It was to inspire a new generation of architects, such as Frank Lloyd Wright who, with his "organic" style of architecture, as he himself pointed out, had more in common with the East than with the West. This type of architecture developed from the inside out, rather than the reverse. His study of Japanese houses led Wright to pierce walls to let in air and light, and to join rooms together, using wooden screens to create a more subtle interior space than was provided by the traditional walls. Emphasizing horizontal lines, cantilevered flat roofs, and compatibility with the natural setting, Wright's houses in Oak Park and in Chicago during the first decade of the twentieth century, at Taliesin in Wisconsin, or the Fallingwater House, of 1936, display a community of spirit with Far-Eastern architecture.[55] On the institutional level, numerous movements in the United States advocated cooperation and goodwill between the peoples of the Pacific. In these circles it was hoped that a spirit of friendship and understanding would grow out of the public discussion of both common interests and controversial topics. Such was the objective of the Institute of Pacific Relations (IPR), under the direction of a Council of the Pacific with representation from the various countries, both independent and colonial powers. The IPR had its headquarters in Honolulu, the site of its first International Conference in 1925, which was to be followed by others at two-year intervals.[56] The institute's cultural activities helped the participants to develop a better knowledge of one another, to appreciate diversity as well as things in common, and encouraged further studies. Another organization, the Pan-Pacific Union, founded in 1917, avoided any topics likely to become controversial. It promoted conferences on scientific or educational matters, and in Honolulu in 1922 it held an important conference dealing with commercial subjects, in an attempt to find some common ground between divergent interests. The first Pan-Pacific

Women's Conference was held in 1928, followed by others in 1930 and 1934.[57] By the time the Great Depression began, it was beginning to seem as if the gulf between East and West might be bridged. The desire for international understanding seemed strong enough to successfully stand up to chauvinism and racism—but it was too recent a development to bring about any profound change in attitudes which still labored under the dead weight of the past.

SEVEN

Imperialism and the Open Door

The increase in cultural interchanges and the intensification of cultural relations between 1870 and 1940 could not be a matter of indifference to Washington and the different forces in society with influence over its decisions. At a time when the European powers were reaching their tentacles into every corner of the planet, the Americans could hardly be content to watch them divide up the spoils from afar, unless they were to lose any advantages they had gained at the cost of so much effort. And yet, for a former colony which had won its freedom from British rule after a bitter struggle, it was not easy to overcome the psychological barriers to ruling over other countries in its turn, so that there was indeed some reluctance to set out on the path of imperialism, and a tendency to seek other solutions more acceptable to the conscience of its people.

THE CONQUEST OF EMPIRE

The word "imperialism" has so often been used polemically that it has lost any precise meaning. We will use it here with its original meaning, defined as the desire to acquire an empire by extending a country's rule beyond its own frontiers. The United States was indeed an imperialist power at the end of the nineteenth century, no less so than the major European nations at the time. Was this a temporary aberration in a tradition considered fundamentally anticolonialist, or was it the inevitable outcome of the growth of its productive capacity? Historians, depending on the school to which they belong, have debated the question for decades, but, if the motivations are in doubt, the facts are clear: in 1865 the United States possessed no territory in the Pacific outside its own continental boundaries, but by 1899 it had taken over Hawaii, Samoa, and the Philippines, together with other islands of various sizes which gave it strategic control over the Pacific shipping lanes. This "new empire" only seems a new development relative to the preceding phase of expansion within the continental landmass, but in fact it was no different from the European colonial ventures of the same period.[1]

The Empire Debate

If, at the end of the nineteenth century, those who advocated the annexation of overseas territories won out over their opponents, it was because their arguments were more in tune with the circumstances and mood of the time. Favoring their opponents was the long-standing tradition that there could be no question of a country such as America behaving like a vulgar colonial power. The United States, they held, had room enough for its population, both then and in the future. It was relatively self-sufficient, and whatever could not be produced domestically could easily be acquired through international trade. For development to take place, there was no need for trade to depend on political domination. In any case, colonies were a heavy financial burden on the mother country, for, apart from the administration costs, their defense would require a huge Navy, constantly needing to be updated. Liberal democracy could very well be dragged down by such a burden, and by the militaristic mentality it would encourage. Racial considerations also made it more desirable to refuse to annex any new territories, since the United States already had enough problems with its black population and its Asian immigrants without increasing the proportion of colored people in its midst—for, indeed, how could colonial citizens be prevented from coming to settle in the mother country, modifying its ethnic composition as a result? Morality, the efficient use of economic resources, and racism all provided reasons to shun any colonial ventures.

But these arguments were carrying less and less weight in public opinion, particularly during the recession years, between 1873 and 1879, and above all between 1893 and 1897. Economic and strategic considerations, far from being mutually exclusive, were closely linked, enabling powerful lobbies to gain greater influence over the decision makers. If the local resources of the countries which might be candidates for annexation seemed limited, and if their population seemed too small to consume a significant proportion of American agricultural and industrial surpluses, they could still serve as stepping-stones, or even as springboards, to markets with enormous future potential, such as China or Australia. What had to be done, therefore, was to obtain a firm hold on the principal shipping routes before the Europeans took all the best locations. With steam replacing sail, it was necessary to have coaling stations where steamships could refuel. The development of commercial navigation required the creation of a modern Navy, capable of protecting it from enemy depredations in wartime. The interests of merchants and of the navy went hand in hand. But it was not enough just to possess a simple naval base, for it was necessary to control the surrounding territory in order to prevent a surprise attack. As a result, the search for naval bases led progressively to the occupation of the entire country in which they were situated, if only in order to deny access to future enemies. The whole logic of this strategy was expounded by Alfred Mahan, whose fundamental work, *The Influence of Maritime Power on History, 1660–1783*, published in 1890, became a bestseller on naval strategy, with equal fascination for the graduates of the U.S. Naval War College at Newport and the Washington politicians who dreamed of the United States as a future great power, supported by a modern fleet of battleships.[2] After being hobbled for many years by the inertia of the antiannexationist tradition, the case for conquest finally gained the upper hand in 1898–1899, advancing American power across the Pacific to the very gates of Asia.[3]

The Samoan Tangle

It may seem surprising that the first colonial-style adventure by the United States in the Pacific took place in the Samoan islands, situated in the southern hemisphere at a latitude of 14 South, 4,650 miles from San Francisco.[4] Neither the local resources nor the purchasing power of the inhabitants was such as to justify such an intrusion. However, the island of Tutuila offers one of the finest natural harbors, Pago Pago, in an ocean lacking in easily accessible ones, and is situated on the shipping lane between North America and Australia and New

Zealand.[5] It was an ideal port of call for steamships needing to refuel with bunker coal. Since 1870 there had been regular liner services between San Francisco and Sydney or Auckland (the Hall Line, succeeded by the Webb Line, between 1871 and 1873; the Pacific Mail Steamship Company, from 1875 to 1885; and the Spreckels brothers' Oceanic Steamship Company, from 1885 to 1907, all supported by the postal service).[6] This was reason enough for the United States to take an interest in Samoa—particularly since the Germans and the British also had their eyes on the islands.

In 1872, on his own initiative, Richard W. Meade, commanding a frigate, negotiated a treaty with the chiefs of Tutuila, granting the United States the exclusive right to establish a coaling station, and offering the inhabitants the protection of the powerful American republic—a treaty which the Senate, hostile to the Grant administration's expansionist ambitions, refused to consider. But this did not deter the administration, which gave the go-ahead to the mission of Colonel Albert B. Steinberger, who managed to gain the confidence of the islanders to such an extent that he became a sort of "Prime Minister" for the islands from 1875 to 1876, when he was expelled by the German, British, and American consuls, who were envious of his influence. It looked as if the German Reich would emerge as the winner, for it had the strongest economic interests, but the native chiefs were afraid of becoming a German colony, and proposed to Britain and the United States either to annex Samoa or establish a protectorate over it. In Washington, in January 1878, Secretary of State Frederick W. Seward, the son of Lincoln's minister and just as expansionist in his views as his father, leaped at the chance to negotiate a treaty which would give the United States the nonexclusive right to establish a naval base and coaling station at Pago Pago, without its having to make any commitment in return apart from providing assistance should conflict arise with a third party—an agreement which the Senate immediately ratified by a unanimous vote. From that point on, the Americans were gradually drawn into the complicated affairs of Samoa. In 1879, their consul, also acting on his own initiative, negotiated an agreement with his German and British colleagues giving the three nations the right to jointly govern the town of Apia and, de facto, the entire archipelago. This kind of arrangement—a "tridominium"—was absolutely unheard-of to the Americans who, until then, had always rejected any kind of alliance with the Europeans which would tie their hands.

Tensions between the Germans and the natives soon revealed the weaknesses of the "tridominium" arrangement. When Bismarck's envoy attacked the local monarch in 1885, the Cleveland administration expressed its outrage and came to the defense of the unjustly treated islanders. When the American consul proclaimed a temporary protectorate, this increased the risk of confrontation between the two powers, whose objectives were too opposed for any compromise to be reached in the short term. In August 1887, an intervention by the German Navy raised the tension created by the civil war in the islands by another notch. It was an electoral year, so Cleveland, the Democratic president, could not be seen to back down, and his defeat in the election of November 1888 made him even less inclined to accept a kind of imperialist behavior he abhorred. In January 1889, an armed confrontation became imminent when three American warships came face to face with the same number of German vessels, and public feeling reached fever pitch. But Bismarck, whose position in Europe was weak, was reluctant to initiate hostilities with the United States, and called a conference in Berlin to settle the problem. The situation was suddenly resolved, and tempers were cooled when, on March 15, a typhoon wreaked havoc on the six ships off Apia, with the loss of 150 sailors. With a military solution no longer possible, it was up to the diplomats again. The Act of Berlin (June 14, 1889) recognized Samoan independence, but subject to the supervision of the three powers who would have the final say, as arbitrators, in any internal disputes in the islands.

In practice, this solution turned out to be unworkable because of the rivalries between tribal groups and the divergent interests of the protectors. In 1898, on the death of the monarch, Germany supported one pretender to the throne, Britain and the United States another. Tension rose once again in March 1899, when the German consulate was damaged by joint military action by Britain and the United States— a completely unprecedented phenomenon, which heralded the closer relations between the two Anglo-Saxon powers. But a peaceful solution was found quite rapidly, at the expense of the islanders. The monarchy was abolished, and partition agreed on. The treaty of December 2, 1899, divided the Samoan archipelago into two groups with the island of Tutuila, and Pago Pago, going to the United States, and the westerly islands to Germany, while the British renounced their claims, accepting compensation elsewhere. So, after an unsatisfactory twenty-year trial of a joint protectorate, the Americans had finally come to accept the colonial solution. They were now the sole possessors of a group of islands in the South Pacific—although in the short term this was of little practical use to them, and still did not make them a power in the region, until the naval base would be built at Pago Pago.

The Inevitable Annexation of Hawaii?

If the Samoan islands seemed marginal to the fundamental interests of the United States in the Pacific, the same could not be said for the Hawaiian archipelago, lying on the Tropic of Cancer, and providing a natural line of defense for the California coast. The Americans could never have allowed a foreign power to take control of Hawaii, for this would have been a direct threat to their coast, and allowed traffic through the projected canal across the Central American isthmus to be easily cut off. But there was no need to annex the Polynesian kingdom, for it would have been enough to develop closer ties with it and to snuff out any thought of its annexation by European powers, particularly the British and French, who had been very active in the region before the Civil War. However, this purely defensive view was not to the taste of the annexationists, who considered Hawaii a necessary springboard to markets in the Far East. So the importance of the archipelago consisted less in the market it would itself provide (which had in any case been in a relative decline since the collapse of whaling) than in its strategic value, both economic and naval. This was why it would inevitably be brought under American control, until its ultimate annexation—although this met with resistance for a long time.[7]

As a first step, Seward had negotiated a reciprocal treaty with the Hawaiian monarchy in 1867, under which sugar and rice from the island would be imported duty-free to the United States. The national mood, hostile to any expansionism, persuaded the Senate to delay ratification of this treaty, and to throw it out three years later, since it had aroused hostility in the southern states, where the same commodities were produced, as well as in the North, where refiners felt they would be put at a disadvantage relative to their Californian competitors. The kingdom of Hawaii wanted reciprocity to lay a foundation for economic expansion, even at the risk of becoming increasingly dependent on a single major client. In the United States, on the other hand, reciprocity was opposed, not only by those who felt it would be damaging to their interests, but also by those who doubted its constitutionality, saw it as a dangerous exception to the official policy of protectionism, and considered it a fool's bargain. To overcome this reluctance, in 1873 King Lunaliho tried to tempt the Grant administration by proposing to give Pearl Harbor as a naval base—an offer which was soon withdrawn in the face of hostility from the Hawaiian nationalists. Finally, his successor, Kalakahua, during a visit to Washington, succeeded in having a reciprocity treaty signed (January 30, 1875), with a commitment not to offer Pearl Harbor, or similar commercial privileges, to any other nation. These conditions made Senate approval easier, but it was more difficult in the House of Representatives. From this point on, Hawaii would be drawn increasingly into the American orbit, mainly because of its links with the west coast. Large sugar plantations, set up as public companies, took over the best land. The substantial investment they required was within the means only of whites—mostly Americans, but also British and Germans. Between 1875 and 1890 cane production boomed, almost all for export to San Francisco. When the treaty expired after seven years, the Hawaiian government requested its renewal. In the United States the same debate was thrashed out once again. On one side were the refiners and sugar beet producers, and on the other Claus Spreckels, the sugar king of the West (who had been opposed in 1875 but had opportunistically changed sides, having himself become a major sugar planter in the islands), the advocates of trade expansion into Asia and Australia, and anybody who could appreciate Hawaii's strategic value.[8] This debate explains why the new treaty, signed in December 1884, was not ratified by the Senate until January 1887, after the addition of an amendment granting the United States the exclusive right to establish a naval base at Pearl Harbor.

The sugar monoculture brought about profound changes in Hawaiian society. The plantations needed a cheap workforce, which was provided by bringing in indentured laborers from abroad—Chinese, Japanese, and Portuguese. The indigenous Polynesians, whose numbers were in constant decline, found themselves in the minority. The Asian immigrants, whose numbers continued to grow, had no political rights, while the whites, eager to protect their property rights, demanded a greater share in the kingdom's administration. In 1887, a peaceful revolution, conservative in nature, obliged Kalakahua to surrender some of his powers to a House of Nobles, elected, on a tax-based suffrage, not only by Hawaiian subjects but also foreign residents. This gave the

Americans an even greater say, and created discontent among a large proportion of the islanders. In 1890, the situation began to evolve even more rapidly. In that year the McKinley Tariff Act allowed sugar into the United States duty-free, and granted a subsidy to domestic producers. This deprived the reciprocity treaty of all its value for the Hawaiian planters, who now found themselves obliged to sell their produce at a world price 40 percent lower than what they had enjoyed. The resultant financial difficulties made it difficult for them to afford the taxes and the expenditures of the monarchy. Annexation seemed a solution, even though it was far from certain that the subsidy would be permanent, and since in any case it would probably have to be shared equally with the cartel established by the New York refiners. Historians have argued a lot about the role of sugar in the republican revolution of 1893 in Hawaii. We know that the major planters, Spreckels among them, disapproved of it and, with the exception of the American A. S. Wilcox and the Englishman Alexander Young, were excluded from the new political establishment. This opposition arose from the fact that they wanted the influx of Asian indentured workers to continue, something ruled out by American law. However, numerous businessmen and members of the liberal professions, whose wealth depended on the sugar exports and on the efficient management of public affairs, adopted an increasingly hostile stance toward the monarchy, particularly when Queen Liliuokalani, a Polynesian nationalist, succeeded to the throne in 1891. The question of the Asians seemed not to play any role in the revolutionary movement, which concerned only two races, the "Caucasians" and the native islanders.[9]

When, on January 14, 1893, Liliuokalani illegally abolished the 1887 constitution, her enemies from the Annexation Club jumped at the opportunity to organize a Committee of Safety with the direct support of American Ambassador John L. Stevens, a convinced annexationist who considered the fact that his letters to Secretaries of State Blaine and Foster did not provoke any reprimand to be an indication that they found this solution acceptable. On January 16, a landing by marines from the *Boston*, in President Cleveland's words, "upon false pretexts respecting the danger to life and property" of American citizens and the need to help maintain public order, forestalled any resistance from the monarchists. The next day the queen gave in, hoping that Washington would restore her to the throne. The provisional government—which included not a single native Hawaiian and which was dominated by Hawaiian-born Americans, like the lawyers Lorrin A. Thurston and Stanford B. Dole—hastily negotiated an annexation treaty with the Republican administration, which was about to transfer power to Cleveland's Democratic administration, who had defeated Harrison in the preceding fall. The new president, a strict moralist, refused to submit this treaty for ratification, since the revolution had succeeded only with the help of American military force, and enjoyed no popular support. But in that case should the United States intervene to restore the queen to her throne? Cleveland might have considered this if she had not further damaged her cause by threatening to behead the rebels. Washington's official attitude was therefore to let events take their course. In the end, this restraint worked in the revolutionaries' favor, for in 1895 they snuffed out a last monarchist rising, and took a firm grip on power. The reciprocity treaty was still in force, and when in 1894 the tariff on sugar was revised this gave new impetus to cane production, and further encouraged the immigration of Japanese laborers, who became the largest ethnic group in the population. The whites in power now became aware of the growing racial problem, which was all the more dangerous because the Empire of the Rising Sun, having acquired a new self-confidence from its victory over China, was determined to protect its emigrant subjects, and insisted they should enjoy equal political rights. If Hawaii was to remain part of the Western world, and not be absorbed by "oriental civilization," it would soon have to be annexed by the United States, the only power capable of protecting the archipelago against Japan. As soon as the Republicans returned to power in 1897, an annexation treaty was negotiated, on June 16, despite protests from Tokyo and the opposition of American sugar interests and trade unions. These opponents had enough influence to delay approval by the Senate, making it necessary for its supporters, to eliminate any risk of failure, to follow the precedent set in the case of Texas by proposing a joint resolution of both chambers of Congress, for which a two-thirds majority was not required. But the decisive event was the Spanish-American War and Dewey's victory at Manila. The Republic of Hawaii, which opened its doors to troops in transit to the Philippines, showed itself to be an indispensable staging post in the new geostrategic configuration of the Pacific. In spite of ferocious opposition from the antiannexationist minority, who were mostly Democrats, the imperialists—Republicans in the great majority, in a reversal of the party split of the 1850s—finally carried off an easy victory on July 6, 1898. Hawaii was officially proclaimed a territory of the United States on August 12. It was not a particularly glorious event, following such lengthy trials and tribulations, but it now gave the United States mastery over the eastern Pacific.[10]

Fallout from the Spanish–American War

The year 1898 is generally considered a major turning point in American history. The war against Spain heralded the arrival of the United States on the world scene as a new imperialist power. It was indeed an impressive development in its role in the Pacific, for, with the conquest of the Philippines and Guam, the annexation of Hawaii and Wake Island, and the partition of Samoa, its front line, which until then had not extended beyond the West Coast, was abruptly brought forward to the very threshold of Asia.

Yet, designs on the Pacific originally played no part in such a radical change, for the Spanish–American War broke out over Cuba. Once war was officially declared on April 25 it was only to be expected that the United States would try to strike at its adversary wherever it was weak. This was on the ocean, far from its European bases. Spain had strengthened its presence in the Pacific during the 1880s, adding the Caroline and Mariana Islands to its long-standing colony in the Philippines in order to forestall any designs German imperialism might have had on them. But, as in Cuba, Spanish colonial rule in the Philippines would not go unchallenged. In 1896 the Spaniards executed the Filipino nationalist José Rizal, and a revolutionary party, the Katipunan, was founded. In the following year a rising headed by Aguinaldo failed, despite the weakness of the Spanish forces. The American plan of action initially assumed that most of the operations would take place around Cuba. However, two months earlier, on February 25, Assistant Secretary of the Navy Theodore Roosevelt, a convinced disciple of Mahan, had sent the following telegram to Commodore George Dewey, commander of the Asiatic Squadron:

> Order the squadron except *Monocacy* to Hong Kong. Keep full of coal. In the event of declaration of war with Spain, your duty will be to see that the Spanish squadron does not leave the Asiatic coast, and then offensive operations in the Philippine Islands.[11]

It was indeed essential to prevent Spanish ships from destroying American trade in the Pacific. This explains how Dewey, who had been obliged to leave Hong Kong, a neutral port, arrived on May 1 in Manila Bay, where Admiral Montoyo's entire fleet lay at anchor. In the space of two hours, Dewey's six ships utterly destroyed the Spanish fleet, and imposed a blockade on the Philippine capital. Such a stunning victory created a totally new situation. It made it quite impossible for Dewey to leave, for he might well have had to do battle with another Spanish fleet which had set out from Cadiz (though this would turn back before reaching the Indian Ocean). Most important, he had to wait for military reinforcements to capture Manila and prevent the imperialist powers—principally the Germans—from taking advantage of the Spanish defeat. The arrival of American troops from California, who on their way had taken Guam without firing a single shot, brought the United States one step closer to the point of no return.

When the armistice of August 12 brought the "splendid little war" to an end, President McKinley had still not decided what do with the Philippines. Would it suffice to establish a naval base there? Manila Bay, which was as good a harbor as Pearl Harbor or Pago Pago, would suit very nicely: its proximity to China would facilitate trade with that country, and it would give the United States the edge in countering German, Russian, and French ambitions. But no base can be properly defended unless there is also control over the surrounding territory. So how far was it acceptable to go? Would the island of Luzon be enough? In that case, wasn't it possible that the other powers would take over the remainder of the archipelago, and so threaten the American foothold in Manila? In the final analysis, was it not the best solution to take over all the Philippines? But in that case should it be annexed, or made a protectorate? The latter solution had the disadvantage of making it necessary to deal with a local government which would probably only cause endless problems, as had happened in Samoa. In October, McKinley, after taking military advice, opted for the annexation of the entire archipelago, and for a base in the Marianas. During the negotiations leading up to the Treaty of Paris, signed on December 10, he had no difficulty persuading a humiliated and isolated Spain to accept his demands. He acquired Guam and the Philippines for a payment of $20 million.[12] He could have obtained all the Carolines and the Marianas, but the Americans had no interest in having them all. Germany took the opportunity to buy them, thus becoming a potential rival, until this role was taken over by Japan, who took them during World War I. If the United States had established their rule over all of Micronesia in 1898, the North Pacific could well have been transformed into a kind of American lake many years earlier. By limiting their pretensions they weakened their strategic position in the Philippines, allowing potential future rivals to occupy vital positions along the great shipping lanes.[13]

The ratification of the treaty was the occasion for a major debate between annexationists and anti-imperialists in the Senate and in the entire country, for this was the first time that the United States was taking possession of a populous country far from its own soil. The old arguments still prevailed. For the imperialists, the Philippines were a jumping-off point to the Chinese market, which could be expected to absorb American surpluses, if not immediately, at least in the future. To miss such an opportunity would leave the field open to the European powers and to Japan, and create a risk of world war. Another consideration was the United States' moral responsibility—for how could it refuse to assume the "white man's burden" when the Filipinos were incapable of governing themselves? The anti-imperialists, a heterogeneous coalition made up of pacifist businessmen like Carnegie, the sugar interests, ethnic groups like the German-Americans, socialists, unionists, and reformers, were still loyal to the antiannexationist tradition. Conquest, for them, would inevitably lead to militarism. Colored peoples were ill-fitted to become citizens of a democracy like the United States. The country did not need colonies to sell its goods abroad, and its security would only be weakened by extending its lines of communication across the ocean. In general, the Republicans, with a few exceptions, followed their leader, McKinley, while the Democrats flocked to the opposing camp, until their leader, Bryan, advised the senators from his party to ratify the treaty for the sake of world peace, with the intention of granting the Philippines independence shortly after. On February 6, 1899, the Senate voted its approval by 57 votes to 27—one over the required two-thirds majority.

In this entire matter, the American government had given little thought to those most affected, the Filipinos themselves. It so happened that the overwhelming majority of them, led by Aguinaldo, believed they had won their independence from the Spanish colonial yoke. A congress meeting at Malolos put forward a constitution for the new country. Relations with the American military commanders, which had been excellent under Admiral Dewey, deteriorated rapidly when he was replaced by Army generals, Merritt and, later, Otis, both of whom considered the Filipinos to be savages, not unlike the American Indians. Early in 1899 tension mounted because of the American officers' attitude. An incident could have triggered hostilities at any time, and this is what happened on the night of February 4, two days before the Senate vote. Otis was convinced that the conflict would be a brief one, particularly since certain *ilustrados*, made uneasy by the revolutionary aspects of the nationalist movement, agreed to temporary rule by the United States. But Otis was proved wrong, for the Filipinos fought hard, waging a merciless guerrilla war over the difficult mountainous terrain. Otis was too timid to achieve a rapid victory, and made little headway. Volunteer reinforcements came from the United States The government did not want to Filipinize the conflict, and enlisted only 5,000 Macabebe scouts. In May 1900, Otis was replaced by Arthur MacArthur, who was more aggressive. When Aguinaldo fell into a trap and was captured in March of the following year, it marked a decisive step toward peace, for the nationalist leader asked his troops to surrender and he himself took an oath of allegiance to the United States. Nevertheless, a few pockets of resistance at Batangas and on the island of Samar continued a desperate struggle, while the Moslem Moros on Mindanao rose up in revolt. Not before 1902 was General Adna Chaffee able to proclaim the conclusion of operations. Like the Boer War, the Philippines war had lasted more than three years. To win it, the United States had put 200,000 soldiers in the field and had lost 4,200 men in the fighting, while the cost to its opponents was 16,000 dead, plus about a hundred thousand civilians, victims of famine and disease.[14] Innumerable atrocities on both sides make this guerrilla war one of the less glorious pages of American military history. Curiously, however, it left very little resentment against the colonial regime among the Filipino people.[15]

In Search of Bases and Special Preserves

In the nineteenth century, the American Navy had needed ports of call which could be used by its two squadrons, the Asiatic and the Pacific. But after the Civil War, the coal needed by steamships could be purchased in foreign ports and, if need be, repairs could even be carried out. But what would happen in wartime if neutral countries closed their ports to them? The solution was to acquire naval bases where fuel could be stockpiled and docks constructed, and which could be fortified against attack.

In the Far East, the American government obtained a perpetual lease from Japan in 1864, allowing it to establish a coaling station at Yokohama, which it shortly afterwards subleased to the Pacific Mail Steamship Company until 1900.[16] The Navy therefore became dependent on the facilities available at Hong Kong and Shanghai, an arrangement which would prove unreliable when war broke out with Spain. So, after taking Manila, the United States, following Mahan's theories, lost no time in looking for bases of its own. At least one was necessary between the Yangtze and Manchuria to protect American interests in China. Since the English had already taken over Weihaiwei, an acceptable substitute might have been Zhifu, in Shandong Province, close to the region where cotton goods were marketed, or else the Zhushan Islands, ninety miles from Shanghai. However, these alternative solutions were unacceptable to the rival nations, jealously protective of their spheres of influence, and the search proved fruitless. It must be remembered that early in the twentieth century the Pacific was still considered to be of only secondary interest compared to the Atlantic, where the battleships of the new fleet were stationed. The thinking was that the creation of bases difficult to defend in the Pacific would weaken the Navy's offensive capacity. For the time being, therefore, some thought was given to taking advantage of the possibilities made available by the recent conquests, at Subic Bay in the Philippines, Guam, or Pearl Harbor. However, almost nothing was put in place before World War I broke out.

On the Pacific Coast of the American continent, the United States was wary of Chile, the major power in the region and the "Prussia of South America," which had an excellent army and a relatively up-to-date fleet with which to back up its expansionist ambitions. This was why the United States generally supported Peru, as it did during the war of the Pacific, though clumsily and ineffectively in that particular instance. During the Chilean revolution of 1891, which was a struggle between President Balmaceda and the parliamentarians supported by England, Washington made no secret of its sympathy with the former, though it was unable to help him avoid defeat. As a result, strong anti-American feelings developed in Chile, ready to flare up at any moment into armed conflict. The situation was not helped by the fact that the

American ambassador in Santiago, Patrick Egan, committed one blunder after another. On October 16, the captain of the *Baltimore* was rash enough to allow his sailors ashore in search of entertainment in the port of Valparaiso. A fight broke out, in the course of which two Americans were killed. President Harrison, with an eye to reelection, treated this as an attack on the national honor, while the Chilean minister of foreign affairs accused him of lying. In spite of Blaine's attempts to stall, the crisis took a dramatic turn between January 20 and 25, 1892, when Chile demanded Egan's recall. Hostilities would probably have broken out had the South American republic, aware of its military inferiority and the lack of any European support, not rapidly capitulated and offered its excuses.[17] However, conflicts over disputed frontiers were always simmering among the countries on the Pacific Coast of Latin America. Peru was conscious of its weakness, and several times attempted to ensure its security by offering the United States a lease enabling it to establish a forward base at Chimbote, but this led to nothing because the tensions always blew over. Besides, the Peruvian people were opposed to the surrender of any part of its national territory. Similarly, the United States gave up the idea of a base on the Galapagos Islands, a possession of Ecuador, which it had been eyeing for some time, as it would have provided protection for the isthmus of Panama. When the Panama Canal was opened to traffic in 1914, the Pacific Fleet was no longer so isolated, and could be quickly reinforced by battleships from the Atlantic. Panama became the principal base for operations along the South American coast, so that it was no longer necessary to look for other bases which would prevent the United States acting as arbitrator in conflicts such as the Tacna-Arica dispute between Chile and Peru, or in those between Peru and Ecuador.[18] For the same reason, the idea of a base on the Gulf of Fonseca in Central America (which had been given strong consideration after the signing of the Bryan-Chamorro Treaty in 1913) was abandoned when Costa Rica and El Salvador raised objections to the concessions offered by Nicaragua.

American imperialism was also evident where the ocean itself was concerned. In the nineteenth century, international law set the limit of territorial waters at three nautical miles, beyond which the sea belonged to everyone, even in partly enclosed waters. For example, the Bering Sea is only connected to the North Pacific through channels between the Aleutian Islands, but the Americans rejected the czarist proclamation of 1821 declaring it an "enclosed sea." However, after the

Alaska Purchase, around 1890, the Republican administration did not let this prevent it from making the same claim as the Russians, for the most worthy ecological reasons. In the summer the fur-bearing seals migrate from the south to breed on Saint Paul and Saint George, the Pribilov Islands, in the Bering Sea. In 1870, Congress had granted a monopoly over the seal hunt to a company which professed its concern to maintain the numbers of the species. The increase in fur prices had attracted Canadian hunters who, being forbidden access to the islands, massacred the animals indiscriminately in the open sea during their migration, putting this rich resource in serious danger. The American government, also wishing to protect the company's monopoly, therefore decided, in 1886, 1887, and 1889, to arrest ships engaged in this seal hunt in defiance of international law. The result was a crisis involving Great Britain, which was concerned to protect the interests of its dominion. A naval confrontation was avoided in 1890 when, after two short-lived compromises, the matter was put before an arbitration commission which met in Paris in 1893. The commission dismissed the American claim that the Bering Sea should be considered an enclosed sea and reaffirmed the three-mile limit, but counterbalanced this ruling by outlawing the seal hunt in a zone extending sixty nautical miles out from the islands, and imposing a total ban on seal hunting in the open sea between May and July. This decision, which reconciled respect for international law and the need to protect the seals, was confirmed by the conference held in 1911.[19]

So, the end of the nineteenth and the beginning of the twentieth centuries saw the United States, as a power, reach out far beyond its continental borders and make an overt claim to be one of the masters of the Pacific. It appeared to be joining the ranks of common nations, and renouncing what until then it had considered its uniqueness. Was this the end of American "innocence"? Not quite, for Americans still were less at ease than the Europeans with their new role as a colonial power.

THE RELUCTANT COLONIALIST

Taking possession of the Philippines raised very few problems of conscience for the American government, but very soon, unlike France for example, it came to think of its rule as a temporary one, lasting just long enough to teach the Filipinos how to live under a democratic, liberal system.[20] In any case, was this not the best way to win the support of the local elite, the *ilustrados*, which it was essential to do as long as Washington had no intention of opening up its purse to the new colony? By tempting them with the prospect of independence within the foreseeable future—since, in any case, for racial reasons there could be no question of their ever being granted American citizenship—they could be won away from the forces of insurrection. In the meantime they would be offered a role in governing the archipelago, and allowed to remain active in its political life.

In May 1899, three months after hostilities began, the colonial power organized the first municipal administrations, elected on a tax-based suffrage, and as the country was pacified, the same measures were offered to the provincial governments. The initial plan was to construct a locally based democracy, but this idea soon had to be abandoned in favor of oligarchic social structures. The instructions handed down in 1900 to the members of the Taft commission reminded them that:

> the government which they are establishing is designed, not for our satisfaction or for the expression of our theoretical views, but for the happiness, peace, and prosperity of the people of the Philippine Islands, and the measures adopted should be made to conform to their *customs, their habits, and even their prejudices*, to the fullest extent consistent with the accomplishment of the indispensable requisite of just and effective government.[21]

The "prejudices" referred to were the respect for the traditional order, based on the power of the local chiefs, despite all its faults. In the provinces, control was entrusted to a triumvirate, composed of the governor, elected by indirect suffrage by municipal councilors, and two members nominated by the central authorities. In the initial years these were usually Americans who, it was hoped, would provide models of integrity and efficiency—a hope which was rarely realized, for their untrustworthiness, corruption, and poor use of tax revenues made these individuals very little different from their Filipino colleagues.[22] To limit abuses, the commission decided, starting in 1903, to partially reduce municipal autonomy and to strengthen central authority, while at the same time encouraging Filipinization at the local level, as a way to consolidate the power of the *ilustrados,* with their advantages in culture and in wealth. Within a decade, Filipinos occupied 71 percent of positions in the civil service, 99 percent in the municipal government, and over 90 percent in the autonomous provinces.[23]

The year 1907 saw further progress in Filipino involvement. Until then, executive and legislative power had been in the hands of a seven-member commission, comprising four Americans, including the governor, and three *ilustrados*. The reform shared the legislative functions between the commission and a legislative assembly elected by tax-paying voters literate in either English or Spanish. The political parties were allowed to pursue their activities in freedom. The Nationalist party, under Sergio Osmeña, which emerged victorious from the first elections, opposed the commission, finding it too representative of American interests for his liking, while the commission was exasperated by its adversaries' cronyism and lack of concern for public interest. As W. Cameron Forbes, the former governor, noted:

> The Commission, with their American sense of justice and dislike of delay, display, sham, and subterfuge, were turned loose upon a world of medieval mismanagement and abuse like a group of knights-errant looking for wrongs to right and abuses to end. They found plenty of these and literally worked themselves sick in their efforts to bring into the Islands the blessing of the kind of administration to which Americans have become so accustomed that they take it as a matter of course.[24]

Under the circumstances, the Republican administration in Washington considered that the Filipinos were not ready for independence, since this would not be to the advantage of the common people. In 1909, therefore, Congress approved a free-trade agreement which granted commodities and raw materials from the Philippines considerable advantages over their other competitors in the tropics, and allowed American manufactured goods exported to the Philippines the same privileges. Philippine independence seemed to have been postponed indefinitely.

Wilson's election in 1912 brought new hopes. The new president was favorably disposed toward the emancipation of the Philippines within a short time. The governor-general, Francis B. Harrison, made approaches to the nationalists. On August 29, 1916, Congress approved a comprehensive bill put forward by William H. Jones, the Democratic representative from Virginia. In its preamble, it affirmed that the United States had never had any intention of conducting a war of territorial conquest against Spain, and that the time had come for it to renounce its sovereignty over the Philippines "as soon as a stable government can be established therein."[25] Until the appropriate opportunity arrived, the Filipino people should be readied for independence by allowing them greater control over their own affairs. Filipino citizenship was introduced—a measure which excluded any possibility of granting American nationality to Filipinos. As under the American Constitution, legislative power was given to a bicameral legislature elected by tax-qualified voters literate, not only in English and Spanish but also in an indigenous language. The governor-general, appointed by the president of the United States, held the reins of executive power and had a veto. In short, the form was American in inspiration, but the autonomy granted was still subject to control by Washington. All these concessions did not fully satisfy the nationalists, who still adhered to their demand for immediate independence. It is true that their rhetoric was mainly meant to impress the public, as they vied for leadership of the party. Their demagoguery assured the leaders of their hold over the masses, while not preventing their being much more ready to compromise behind closed doors. Thus Manuel Quezon would say on the one hand that he would prefer to have damned bad government as long as it was run by Filipinos and on the other: "Hang the Americans! Why don't they tyrannize us even more than they do?" In any case, the disagreements among the leadership were short-lived, for all of them wanted to preserve the social structures inherited from the past, inequalities and all, so that Quezon and Osmeña were soon reconciled in 1922, after an interval of squabbling, as they would be once again in 1935.

The harder line taken by the Republicans returned to power during the years of prosperity—among them the governor, Leonard Wood, fiercely hostile to granting independence prematurely—obliged the nationalists to join forces. Events took a new turn during the 1929 crisis. The Philippines were considered an encumbrance. They would be difficult to defend should war break out with Japan, and the local oligarchy seemed too firmly ensconced for there to be any hope of transforming the archipelago into an American-style democracy. The commodities it produced—sugar and hemp—were in competition with certain sectors of United States agriculture which were already suffering from overproduction. The cause of Philippine independence made considerable progress in American public opinion, less out of idealism than out of an explicit desire to defend the United States' material interests. In March 1930, the Osmeña-Roxas mission to Washington resulted in a bill setting a date for independence. This was adopted in January 1933, over Hoover's veto.[26] Contrary to everyone's expectations, Quezon was against it, supposedly because it was excessively generous in accommodating American military bases, and because it eliminated the free-trade advantages too quickly. In actual fact he was merely trying to prevent his two henchmen and rivals from getting all the credit. Under the New Deal he renegotiated the agreement in almost identical terms, happy to be the one to lead his people into independence. A law passed on March 24, 1934, brought forward by Senator Tydings from Maryland and Congressman McDuffie from Alabama, promised complete independence in ten years, counting from the date the Constitution was put into effect, with the United States retaining only its naval facilities. In the meantime, the United States, through its high commissioner, would keep a say in the affairs of the Commonwealth of the Philippines, particularly where the public debt, the currency, overseas trade, and foreign affairs were concerned. Free trade would be done away with in stages. Above certain allotted import quotas (50,000 tons of refined sugar and 800,000 tons of raw sugar), Filipino sugar would be subject to the same duties as foreign sugar. From the sixth year on, the Philippines government would levy an increasing export tax in order to progressively reduce the advantages its goods enjoyed on the American market, the proceeds of this tax going into a fund to pay down the debt. These provisions were all motivated by the desire to eliminate competition, just as the clause abolishing immigration rights was inspired by the wish to halt the influx of people of color, considered undesirable.

In 1934, colonization seemed to the vast majority of Congress to offer nothing but disadvantages. Unlike France or Britain, the United States had no intention of using the crisis as an excuse to fall back on its colonies. Since it was in November 1935 that the Commonwealth of the Philippines had come into existence, with Quezon as president and Osmeña as vice president, the Americans were supposed to lay down their colonial burden in November 1945, as had always been their intention. In fact, however, this event would be postponed by a few months, until July 1946, because of World War II.

THE OPEN DOOR

John Hay's Two Notes

In the strict meaning of the term, the principles of the "open door" policy are defined in two circular letters from Secretary of State John Hay, responding to the situation in China. In the first of these, dated September 6, 1899, Hay requested the German, English, and Russian, and later French, Italian, and Japanese governments, to give "formal assurances," regarding the spheres of influence they had recently developed or would subsequently develop in China, that each country:

> First. Will in no way interfere with any treaty port or any vested interest within any so-called "sphere of interest" or leased territory it may have in China.
>
> Second. That the Chinese treaty tariff of the time being shall apply to all merchandise landed or shipped to all such ports as are within said "sphere of interest" (unless they be "free ports"), no matter to what nationality it may belong, and that duties so leviable shall be collected by the Chinese Government.
>
> Third. That it will levy no higher harbor dues on vessels of another nationality frequenting any port in such "sphere" than shall be levied on vessels of its own nationality, and no higher railroad charges over lines built, controlled, or operated within its "sphere" on merchandise belonging to citizens or subjects of other nationalities transported through such "sphere" than shall be levied on similar merchandise belonging to its own nationals transported over equal distances.[27]

On July 3, 1900, at the height of the Boxer insurrection, and during the siege of the Peking legations, Hay broadened his proposal in a letter addressed, in addition to the former recipients, to Austria-Hungary, Belgium, Holland, Spain, and Portugal:

> We adhere to the policy initiated by us in 1857, of peace with the Chinese nation, of furtherance of lawful commerce, and of protection of lives and property of our citizens by all means guaranteed under extraterritorial treaty rights and by the law of nations. . . . [The] policy of the government of the United States is to seek a solution which may bring about permanent safety and peace to

China, preserve Chinese territorial and administrative entity, protect all rights guaranteed to friendly powers by treaty and international law, and safeguard for the world the principle of equal and impartial trade with all parts of the Chinese Empire.[28]

In neither of these notes does the expression "open door" appear, but this was the name immediately given to American policy in China at the turn of the century, particularly in the light of its favorable connotations—for, as the historian Marilyn Blatt Young comments, "an open door is, prima facie, more fair, more desirable than a closed one—at least to those who would like to get inside."[29] Originally, it was a British notion, for the English had been made uneasy by imperialist encroachments into China, starting in 1897, when the empire granted Germany (whose example was soon followed by Russia and France) the exclusive lease to territory, the port of Jiaozhou, and also certain rights in mining and railroad construction in Shandong, its "sphere of influence." The integrity of the Chinese economic space seemed to be under threat, with the risk that instead of becoming a huge potential market it would be carved up into regions enclosed by walls of discriminatory customs barriers. Just as much as such a solution was reassuring to protectionist nations who felt uncertain of their ability to compete in world markets, it was a threat to the interests of those which remained attached to free trade, like England, or which were confident of succeeding thanks to their dynamism, like the United States. Keeping an open door in China meant, initially, that goods and capital could circulate freely, without foreign powers having any special privileges in the zones where their major interests were concentrated—i.e., Shandong in case of Germany, Manchuria in that of Russia, Yunnan in that of France, Fujian in that of Japan, and the Yangtze valley in that of Britain. Very soon, however, the London government was obliged to reduce its demands. In the spring of 1899 it accepted implicitly that investments should be exempt from the principle of the open door, and that the spheres of influence constituted regions within which investments made by the dominant power should enjoy a quasi-monopoly. In return, it attempted to obtain equal treatment in trade, with the same customs duties, the port dues and railroad freight rates applying to all, so that in trade competition all merchants and industrialists, of whatever nationality, would enjoy a level playing field. This was the position taken by Hay in his first note. In it, there was no mention of investments, but only of the conditions applying to trade and the movement of goods. Furthermore, even if he put the term in quotes, he never questioned the existence of the spheres of influence, but simply wanted to prevent their being an obstacle to expanding the market for American goods in China. As for defending the integrity of the empire against attempts to carve it up, he makes no reference to this, except in the letter addressed to London, in which he urges joint action by the foreign powers "in favor of the administrative reforms so urgently needed for strengthening the imperial government and maintaining the integrity of China in which the whole Western world is alike concerned."[30]

In this espousal of the British attitude, even though he expressed it in his own terms, some saw an indication that the Department of State, headed by the Anglophile John Hay, was being influenced or even manipulated by the English, who were looking out for their own country's interests rather than for those of the United States.[31] Indeed, Lord Charles Beresford, in his book *The Break-up of China*, published in 1899, stressed the extent to which British and American interests were the same, and put forward a plan to reform the Chinese Empire, under foreign control, in order to prevent its disintegration. In July of the same year, William W. Rockhill, the expert on Chinese affairs in the Department of State, had received a letter from Alfred E. Hippisley, an Englishman serving with the Imperial Maritime Customs, drawing his attention to the urgency of demonstrating to the imperialist powers that the United States was determined not to be deprived of its trading rights, which would very soon come under threat if the spheres of influence were to apply to merchandise as well as to investments. On August 21, Hippisley sent Rockhill a memorandum containing all the substance of the September 6 note—"yes" to equal treatment, but nothing on the subject of a more general equality of opportunity, and nothing on the territorial integrity of China. But this sequence of events should not be taken to mean that, without the suggestions of well-meaning Englishmen, the Department of State would not have expressed its views in exactly the same terms. It was a matter of common interest rather than of cause and effect. The United States still had little interest in investments overseas, since its situation as net debtors left it without substantial enough financial resources. On the other hand, the rapid growth in its exports of manufactured goods seemed very promising, and this was all the more welcome because the American economy was producing more goods than the home market could regularly consume.

On the other side of the Pacific, China, with its some four hundred million consumers, represented an enormous potential market for American agricultural and industrial surpluses. A whole literature expressed the excitement inspired by this fabulous promise, exaggerated by the frequently extravagant propaganda put out by pressure groups such as the Asiatic American Association (AAA), controlled by New York businessmen engaged in trade with the Far East and by southern cotton manufacturers, delighted at the success of their wares in Northern China and in Manchuria particularly. It is easy to show how many negative factors were overlooked in this enthusiasm—the weak purchasing power of the Chinese, further diminished by the devaluation of silver against gold, the magnitude of domestic consumption, and the lack of any will on the part of American producers to devise a serious export policy which would take the needs of their clients into account. Nevertheless, if not the present, then at least the future seemed to belong to American industry, and people allowed themselves to be intoxicated by the carefully cultivated fantasies which were continuously put before them. It was believed that China would become the major market in the twentieth century, and that by dominating it the United States would be assured commercial supremacy. The United States owed its advantage to a number of factors: its direct access to the Pacific Ocean, its productive capacity, and, after 1899, to the position it occupied in the Philippines, at the very gates of this future Eldorado. To achieve its objective, it only needed to prevent the other nations from getting a stranglehold on China, and excluding goods "made in the USA" from their spheres of influence.[32]

But if this was to be achieved, the European and Japanese imperialists would have to be receptive to the American demands. The relevant capitals received the September 6 note without any enthusiasm. Saint Petersburg procrastinated, feeling it was being targeted because of its policy in Manchuria, and demonstrated reluctance to make any written commitment. But since Russia did not question the principle of equal treatment, it did agree in the vaguest terms to the three points made by Hay. It only remained for the secretary of state to act as if he had brought off a coup. In the directive he sent his ambassadors on March 20, 1900, he declared that, since the conditions of acceptance had been fulfilled by all, the Washington government considered their assent to be "final and definite"—a euphemism which would give rise to much debate.

Three months later, events in Northern China evolved so rapidly, with the Dowager Empress Cixi's complicity, that Hay was forced to give renewed attention to considerations which he had previously dismissed regarding the territorial integrity of China. The United States was afraid that the Europeans and Japanese would take advantage of the military operations launched to relieve the besieged legations to carve up the empire between them, as they had done to Africa and southern Asia. Hay was therefore obliged to maintain the fiction that the armed intervention was aimed at the Chinese soldiers and rebels, and not at the Peking government, particularly since the central and southern provinces, controlled by energetic and prudent viceroys, remained peaceful all along. There was no justification for dismembering China once the revolt had been put down. Hay's second note, dated July 3, 1900, contrary to his first, did not ask for a response from the various capitals, but took the form of a declaration of the principles of American policy. Since no one protested at its contents, Washington considered it had been accepted, de facto, and that unanimity prevailed on the subject of preserving Chinese territorial integrity and independence. One consequence was that the United States came to consider itself the only real friend the Chinese had, and the only nation making no territorial claim on her whatsoever, a fact which established a "special relationship" between the two countries.[33]

The Open Door in Action

From the end of the nineteenth century, a pressure group made up of diplomats, certain economic interests, and missionary groups, encouraged the idea widespread among the American public that China enjoyed a special relationship with the United States, differentiating it from all the other Pacific countries, and from Japan. This arrangement was supposed to imply mutual responsibilities, not only with the more powerful nation helping a protégé nation to develop but also, in return, respect by the latter for the—"unequal"—contracts and treaties it had signed.[34] For its part, China considered the United States to be different from the other countries which had achieved a similar level of development. The Peking authorities and the intellectuals realized that the Americans had no designs on their territory. Sometimes the government, as was its wont, attempted to use Washington to play off the "barbarians" against one another—though this was usually unsuccessful, since the United States still did not possess adequate military and naval resources, or the will to make use of them. Thus, on four occasions, Li Hongzhan called in vain on American assistance to extract himself from difficult negotiations (over the Ryukyu Islands in 1879, when ex-President Grant's mediation with Japan proved fruitless; over Korea in 1880–1882, in which he failed to obtain recognition of Chinese sovereignty; during the war with France over Tonkin in 1884–1885; and, finally, during the Sino-Japanese war in 1894–1895).[35] At the beginning of the twentieth century, both of Li's successors, Zhang Zhidong and Yuan Shikai, made the mistake on several occasions of thinking that in the name of the open door policy the United States would protect them from European and Japanese pressure. To their great disappointment they found that this was not to be.

This was because, in spite of sentimental professions of goodwill from certain well-meaning souls, and from missionaries in particular, the powers in Washington and on Wall Street were primarily concerned with the interests of their country and of their shareholders. The open door, as far as they were concerned, was not an expression of idealism—which is to say of Sino-American friendship—in international relations, but rather a statement of American economic and strategic interests in the western Pacific. Depending how they were viewed in Washington, the United States was ready, to different degrees, to make an active commitment to defend the principles expressed in the notes of 1899–1900.[36] Hay, made uneasy by the Russian intrusion into Manchuria and the list of exclusive rights being claimed, reiterated his attachment to the open door on four occasions, between 1901 and 1905, even giving the appearance of broadening it to include investments. However, he did not take any retaliatory measures.[37] Theodore Roosevelt, on the other hand, was much more circumspect, being a realist who perfectly realized that the Chinese market was a mirage, that international relations are determined by power rather than by feelings, and that in any case the American Navy, still in the process of being built up, was in no position to assert mastery over the Pacific for the time being. In the circumstances, it was impossible for the United States to challenge the spheres of influence. In any case, the breakup of China had been forgotten now that the foreign powers had discovered that nominal integrity of the empire was more to their advantage, guaranteeing as it did a balance between them which prevented any one of them from attempting to achieve hegemony. On the other hand, if the Chinese failed to respect any treaties, Roosevelt took out the big stick. During the boycott of American goods in 1905–1906 to protest the closing of the United States to Chinese immigration and the unfair harassment of nonmanual laborers attempting to immigrate, he brought pressure to bear on the Peking authorities to punish the ringleaders. The atmosphere changed completely when he was succeeded by Taft. "Dollar diplomacy" required the Washington administration to actively support foreign ventures by American capitalists. In 1909, equal treatment of imports was no longer enough, as it had been ten years earlier, for discrimination against investment by other nations in the spheres of influence was an impediment to trade. The way sales of American cottons collapsed in Manchuria once the Japanese had expelled the Russians from its southern part and taken their place was pointed to as proof, but this was mistaken, as we have already seen, since the real blame lay with the southern manufacturers' lack of interest in selling abroad. Taft and Secretary of State Philander Knox demanded equal opportunities for all nations. American investment in the construction of railroads and factories and in mining operations should be allowed within any sphere of influence—indeed, the plan they put forward for the neutralization of the Manchurian railroad system actually went as far as to deny the validity of these spheres. But all these efforts came to nothing, for the return on capital in China, once the risk factor was taken into account, was less than could be expected in the United States at the same period, and, in addition, the fact that Taft had no intention of using force to enforce compliance with the open door principle was perfectly obvious to the other powers. In 1913, the Democrat Woodrow Wilson repudiated "dollar diplomacy" in the name of the idealism which was supposed to animate American diplomacy. The Chinese republicans placed great hope in this vision, so favorable to the preservation of China's territorial integrity, but they, too, like their imperial predecessors, were to be disappointed. Wilson's declarations remained purely theoretical in the face of Japanese intrusions during World War I. In spite of their proclamations of friendship for China, the United States was not prepared to come to its assistance when it was in danger. In any case, was there not a contradiction between the decision to close America to Chinese immigration and the desire to open China not only to American goods but also to the activities of missionaries and of reformers and modernizers from across the Pacific? In 1914, according to Marilyn Blatt Young, "the Open Door would mean only the right of all to exploit China with equal ferocity."[38]

The limits of the open door policy were just as evident in Korea's case. Immediately after the Civil War the Washington government would have liked to repeat Perry's achievement, this time in the "hermit kingdom." The massacre of the *General Sherman*'s crew in 1866 had persuaded Washington to dispatch an expedition of five gunboats five years later, when the American Navy destroyed five forts and killed 350 Koreans at Inchon, but failed to have any trade treaty signed. In 1880, Korea was still closed to economic relations with the West. Concerned for the fate of shipwreck victims, the Department of State then sent the ambitious Commodore Robert W. Shufeldt to negotiate the opening of Korea to trade. He had to wait for some time to get Li Hongzhan's approval, for China wished to maintain nominal sovereignty over the peninsula, though it was prepared to allow the Americans in, in the hope of countering Japanese expansionist designs. Finally, in May 1882, Shufeldt's mission ended in success when he signed a treaty with Korea which was more favorable to that country than the treaties the United States had previously signed with China or Japan. Tariffs were set at 10 percent instead of 5 percent, and the Americans promised their good offices should Korea be wronged by a third party. Under Kojong's reign, the "hermit kingdom," weakened by internal unrest and the intrigues of its powerful neighbors, Japan, Russia, and China, sought American support to maintain its territorial integrity, expecting the principle of the open door to provide them with a guarantee against these threats to its independence. But Korea was only marginal to Washington's concerns, for trade with it was still insignificant and investments remained very small, while the Protestant missionaries still ministered to a flock of fewer than 40,000 in 1905, and in any case the United States was completely lacking in the naval resources it would have needed to thwart its neighbors' designs. So, despite American protests of friendship, the United States abstained from all interference, and maintained strict neutrality whenever a crisis arose. Theodore Roosevelt had no intention of picking a quarrel with Japan with so little at stake. The Taft-Katsura memorandum of July 29, 1904, explicitly recognized the sovereignty of Japan over the Korean peninsula, heralding its establishment of colonial rule five years later.[39]

Apart from its origins in the Chinese situation at the turn of the century, the open door policy, which is sometimes seen as the second pillar of United States foreign policy, after the Monroe Doctrine, is considered by the historians of the Wisconsin school—William Appleman

Williams and his disciples—to be "the strategy of American foreign policy" during the first half of the twentieth century. Conceived by a coalition of businessmen, intellectuals, and politicians opposed to traditional colonialism and favoring an "informal empire" which the United States would establish over the underdeveloped world, thanks to its economic power, it was neither as naive nor futile a policy as the "realists" like to think. Rather than being categorized by legalism, moralism, or idealism—reproaches heaped on it by its detractors—it aimed at commercial expansion, but did so with much more subtlety than imperialism proper, as we defined it earlier. According to Williams, "if it ultimately failed, it was not because it was foolish or weak, but because it was so successful," for its very success stimulated the antagonisms which were eventually to prove fatal to imperialist ambitions.[40] It is possible that its beginnings go back to well before the end of the nineteenth century, and indeed we can find traces of it in the foreign policy of the Jacksonian era. But it is in the nineteenth century that the ascendancy of the great bureaucratic organizations gave birth to the "corporatist" state, defined by a close coordination between decision making by the major business sectors and by influential politicians.[41] The modern state has no need of colonies to establish the world economic order on which the proper functioning of the system depends. The free flow of merchandise and of capital—the fundamental idea of the open door policy—can be brought about without any need for direct political control. To impose its will, the United States did not need an empire in the formal sense. In the Pacific, the achievement of empire and the open door policy were in the final analysis the two faces of the same dynamic thrust, of the same drive for economic supremacy, with the second being a sign of greater self-confidence. However, it did involve a risk of conflict with other powers whose attitude was more dog-in-the-manger. It was the open door policy, and not imperialism, that led to war with Japan.

EIGHT

The Pacific Problem

— "What is the Pacific problem?" asked André Duboscq in the article on this topic he contributed to the *Dictionnaire diplomatique*, around 1930. His answer was as follows:

The problem of the Pacific is simply the problem of China—this country with its prodigious riches, both above and below ground, this enormous market of more than 400 million consumers, so attractive to the covetous attentions of the great powers. In principle, and apart from a local evolution which has already begun, but which will certainly advance very slowly, whoever has mastery over the Pacific should enjoy an exceptional position in coastal Asia. But in the race for the Chinese Eldorado—a race which China itself is attempting to slow down—there is a rivalry between two competitors which is particularly worrisome for world peace. On the one hand there is the United States, prepared to use every means at its disposal to win the good will of the Chinese and the various advantages which will ensue, and on the other Japan, which, thanks to the affinities between the two yellow races, and with its advantage of proximity, has year by year strengthened its position in China, in spite of Chinese resistance.[1]

This French observer clearly underlines the fact that during the first four decades of the twentieth century the relations existing between the

United States and Japan would decide whether there would be war or peace in the Pacific, with China as the principal bone of contention between them. Was conflict inevitable? Were the seeds of Pearl Harbor already present in these earlier tensions?

THE UNITED STATES, JAPAN, AND THE OTHERS

Until 1931, a war between America and Japan was only one of several possibilities. A number of commentators on both sides explain the momentum leading to an inevitable conflict by invoking the implacable rules of geopolitics. But the path of history, with its sometimes unpredictable twists and turns, can always lead to a number of different outcomes. In fact, the United States and Japan enjoyed excellent relations before 1905, with only a small cloud emerging briefly on the horizon in 1897, because of Hawaii, when McKinley hastily put forward his annexation treaty in order to preempt Japanese pressure on the republican government. However, Tokyo made no protest when the United States occupied the Philippines in the following year. During the Russo-Japanese War in 1904–1905, American sympathies were with the little yellow soldiers in their victory over the Russian bear, and loans from New York helped Japan finance its land and sea campaigns. However, the brilliant Japanese victory marked a turning point, for the goodwill soon turned into fear and deep hostility. Japanese public opinion was upset at the part played by Theodore Roosevelt in the Portsmouth negotiations, blaming him for having allowed Russia avoid paying an indemnity, and giving up all of Sakhalin Island.[2] The first anti-American demonstrations were an indication of this change of mood, and of a new distrust toward a country which had previously been considered friendly. For their part, the Americans, obsessed by their fear of the "yellow peril," suddenly became aware that this danger was assuming the very specific guise of an empire with an excellent army and a powerful navy, posing a threat to their lines of communication, but which was nevertheless their best customer and foremost source of imports in the whole Pacific zone. Henceforth, the protagonists were clearly identified, and, depending on whether a policy of confrontation or of cooperation prevailed, the Pacific either grew dark with the storm clouds of their disputes, or enjoyed a spell of brighter weather.[3]

Speaking of the quarter-century between 1906 and 1931, the Pacific problem cannot be reduced to the ongoing drama between Japan and the United States, nor can the Pacific be viewed in isolation if there is to be any hope of arriving at a valid explanation of events. What occurred in the Pacific region had repercussions in the Atlantic and in Europe, and vice versa. The European powers were by no means absent from the debate. Among the major powers, France certainly played the most minor role, for its interests were more in the South Pacific or in Indochina, on the fringe of the North Pacific where the tensions arose. However, it was involved because of its alliances. Prior to World War I, Germany was perceived by American strategists as a very real threat. They were alarmed by its designs on the west coast of Latin America, obliging it to concentrate the fleet of battleships in the Pacific. The purchase of the Carolines and Marianas in 1898 strengthened the German presence in Micronesia, next door to Guam and the Philippines. If Germany made an alliance with Japan, the United States would have to face them on two fronts. However, this uncomfortable situation lasted only 15 years, for the Reich, having lost its colonies under the Treaty of Versailles in 1919, was temporarily excluded from the equation. Czarist Russia, before its defeat by Japan in 1905, was considered the most dangerous potential adversary. American exporters complained of the difficulty of gaining entry to the sphere of influence Russia had established in Manchuria. Commentators were afraid that Russia would gain control over China and establish a powerful continental stronghold—the "heartland" spoken of by the geopolitician Mackinder—which the surrounding maritime nations would be unable to penetrate.[4] But Russia, after 1905, weakened by internal dissent, seems to have lost any influence except as an ally of one of the other protagonists. The Bolshevik revolution, once its victory was secure, turned the new Soviet Union into a source of problems, for its ideology was seductive to peoples suffering under foreign domination, particularly in China, where the Communists gained considerable influence in the ranks of the Guomindang, until they were wiped out by Chiang Kai-shek in 1927.[5]

If Germany and Russia were occasional players in the Pacific, this was not the case for the British, with their three dominions, Canada, Australia, and New Zealand, and their numerous colonies, ranging from Malaya to tiny Polynesian islands. The British Empire was also the possessor, at least until 1922, of the most powerful fleet in the world. Its position in the North Pacific was indeed relatively weak, but it had to defend its lines of communication between the western coast of Canada and the rest of the empire. For the United States, even though by the twentieth century the days of confrontation seemed to be over, relations with Britain were far from always being rosy, particularly since the United Kingdom had signed a mutual defense treaty with Japan in 1902. The real nightmare would have been for the United States to find itself at war with an Anglo-Japanese alliance. But this seemed rather unlikely, if only because of the racial prejudice in the white dominions, which were just as exclusionist as the United States. This treaty, later renewed until 1921, was in fact a sign of England's weakness in the Pacific, for it was being obliged to concentrate its fleet in European waters, where it was involved in a ruinous naval arms race with Wilhelm II's Germany. Despite the importance of its economic and cultural interests it was no longer a key player in the North Pacific, and this was even more so after World War I, which finally put paid to its pretensions to world leadership.

By default, there remained only two powers capable of deciding the region's fate, either individually or by way of the alliances they had formed. At first sight, they were of unequal strength. The economic power of the United States, its enormous natural resources, and its more numerous population, seemed to ensure its supremacy over a Japan whose remarkable growth was not enough to conceal the fact that it was still lagging behind in development, for it was only recently that it had been opened to world trade. A sense of realism should have led the emperor's subjects to seek, as far as possible, a way to coexist with the giant across the ocean, rather than look for confrontation. Certain Japanese leaders, like Shidehara, were in favor of such a policy of moderation, while refusing to simply knuckle under. They felt that their country did indeed need to turn outward. If confined to their own islands, the fast-growing Japanese population would lack the land and employment opportunities it needed to feed it and provide it with a decent standard of living. Overseas settlement was the solution for overpopulation. The sparsely populated white nations—the United States or the British dominions—would have been ideal outlets, but they did not allow immigration from Asia, so that other solutions had to be found, whether on the Pacific islands or on the Asian continent, without launching out on a vast program of conquest. Korea, or even Manchuria, would fill the bill. Likewise, they felt, Japan, poor in natural resources and in sources of energy, could satisfy its needs, and find markets for its goods, without establishing a formal empire in the western Pacific, as long as it was assured that it would not be discriminated against by favorable treatment granted its European or American competitors. Such a conception of Japanese expansion was not incompatible with a peaceful international order. This was far from the case for those at the opposite end of the political spectrum in Japan, who fomented a warlike nationalism. A raw nerve was touched whenever they saw their country refused equal status by other powers. The United States, invoking the Monroe Doctrine, had given itself the right to exclude any external political influence from the American continent, so why should Japan not have the same right to assert her special interest on the neighboring mainland of Asia? Prejudice against the yellow races aroused deep-seated feelings of humiliation. Everything which could be interpreted as a slight fueled an anger liable to result in irrational behavior. An exalted conception of national honor encouraged defiant posturing and a hardening of attitudes which were a threat to peace in the Pacific, for they made compromise impossible. This kind of attitude was very common among certain Army and Navy officers, who considered themselves the heirs of the ancient samurai, but it was also propagated among the general population by an education system which emphasized the imperial myth as a way of bringing the nation together.[6] If Japan was to resist the temptation of military adventurism, the political authorities would have had to exercise effective control over the armed forces, so as to prevent any gaps in the chain of command. But, unfortunately for the Japanese, their institutions functioned poorly, leaving the Army and Navy too much freedom to make decisions for themselves. Whenever tension heightened, the government was made aware of its weakness in dealing with the nationalist expansionists, for whom the solution was the conquest of a vast region which would be brought under Japanese domination and subordinated entirely to the service of the empire. This was in spite of the latter's outward advocacy of cooperation between the yellow races against the white, whom it considered *persona non grata* in the region. But the political authorities held one trump card in the fact that the two armed forces had rather different objectives, the Army favoring expansion across the Asian mainland, either into China or Siberia (thus ensuring the major role for itself), while the Navy, to the contrary, preferred a push into the archipelagoes to the south. At times, therefore, these two pressure groups neutralized one another, with as a consequence some degree of initiative being left to civilians less intoxicated with military glory and more conscious of the balance of power in the Pacific.

The Americans never doubted their superiority over the Japanese, and yet they feared them. The notion of the "yellow peril" awakened irrational fears.[7] It created an enduring feeling of distrust in many sections of the population, and in certain influential quarters, such as the Navy. Sometimes this came close to paranoia, when it was suspected that the Japanese were planning to invade the American continent. But even if the picture it had of the other nation had been less alarming, the United States could not have avoided occasional tension in its relations with the Japanese Empire. After all, it, too, was an expanding country; its Pacific policy was governed by the principle of the open door, and it considered, more or less correctly, that the markets in the region provided an indispensable outlet for its surpluses, as well as an indispensable source of raw materials. There could be no question of leaving the western Pacific to Japan, for this would be damaging to the American economy, even more in the long than in the short term, and would give even more strength to a potential enemy, whose cunning and even treachery it feared, and which had shown itself ready to appeal to feelings of solidarity among the yellow races in its effort to stir up antagonism toward the envied and detested white man. Yet there was no shortage of "realists" in the United States, who were ready to recognize that within a certain distance of its own territory Japan did indeed have special interests in the Far East, and that it was legitimate for it to exercise a preponderant influence within this zone, on condition it did not practice any discrimination contrary to the principle of the open door. At the other extreme, the "idealists" were trying to bring about an international order without any taint of colonialism, under which every country would benefit from a level playing field in competition with each other. Japan seemed all the more threatening to them in that it was heavily militarized and lacking in democracy. Depending on whether the realists or the idealists held the reins of power in Washington, the United States was more or less favorable to the compromises required if peaceful relations were to be maintained with Japan. It was actually a matter of degree rather than of kind, for the Pacific was only one of the spheres—and at times one of secondary importance—where American foreign policy was at work, and America had no desire to go to war as long as there was no threat to its vital interests.

The other nations of the Pacific were not really considered major players. The Latin American nations were of little significance in the geostrategy of the Pacific, for their underdevelopment, their small populations, the mediocrity of their naval power, and their position on the fringes of the main sphere of activity, all deprived them of the opportunity to play an important role. On the opposite shore, China, despite its masses of humanity, reacted to events rather than directed them. The foreign powers retained privileges which limited its sovereignty, and its territorial integrity was endangered by foreign ambitions, particularly by those of its Japanese neighbor. The civil wars which raged after the republic was established in 1912, and which continued even after Chiang Kai-shek's apparent victory, turned China into the "sick man" of the Pacific. What prevented China from being completely at the mercy of others was the emergence of a public desire to be free of foreign domination, the importance of which became evident with the May Fourth Movement of 1919. In the rivalry, latent or open, between the United States and Japan, it was to the former that China turned for support, for the Americans did not make any claims for a sphere of influence or for territory. China placed its hopes in the powerful pressure group of Protestant missionaries, who were emotionally committed to its modernization and opposed to the designs Japan had on it. A great majority of the Chinese saw Japan as the major threat to the integrity of their country, and were far from convinced by the propaganda calling for the yellow races to unite against the white, since they were perfectly aware that exploitation by the Japanese would very likely be worse than any European domination. They resisted Japanese intrigues by appealing to the conscience of the Americans who, if they were sometimes reluctant to become involved in China's extremely complicated affairs, were unable to completely forget the objective of their mission, which had been to awaken the slumbering giant—not least since such an endeavor was in tune with their economic and strategic interests. This explains why China was the crux of the Pacific problem, even if the Chinese were not themselves one of the major protagonists. It was mainly because of China that the two expanding powers, the United States and Japan, found it impossible to establish a durable basis for cooperation and mutual trust.[8]

STORMS AND A BRIGHTER SPELL (1906–1931)

Between war and peace there lies a wide, gray area in which recriminations, intemperate actions, fear, and distrust are to be expected, but where restraint and prudence are also to be found, as well as a desire to keep events under control in case things should escalate too far—in short, everything that represents a victory of clearheadedness over the emotions. The compromises which emerge—often considered a sacrifice of principles—prevent disagreements from turning into open war, while the brighter spells always seem unlikely to last.

The First Quarrels with Japan (1906–1921)

Coming a year apart, Japan's crushing victory over Russia and the racist school policy in California injected an element of suspicion into Japanese-American relations. Despite its alliance with England, Japan was not strong enough to attack the United States directly, but in Korea and Manchuria, Japanese interests were able to push out American competition. For his part, Theodore Roosevelt was aware of his country's strategic inferiority at a time when the plans to modernize the Navy had still not been put fully into effect. With the annexation of the Philippines, the lines of communication across the Pacific had been lengthened, resulting in a weakening of the empire. In the circumstances, it seemed preferable to play for time, reaching agreements to safeguard what was essential, even if this meant sacrificing principles. In the long term, the United States' position could only improve. This way of thinking led to the signing of the Taft-Katsura memorandum, on July 27, 1905, and the Root-Takahira agreement, on November 30, 1908. The first of these was often presented—and Japanese propaganda never missed an opportunity to do so—as a win-win proposition. The Japanese minister of foreign affairs declared that his country had no aggressive designs on the Philippines and made the case for an Anglo-American-Japanese Triple Alliance to defend the principle of the open door—something to which the United States could not agree. Taft, as secretary of war, speaking only on his own authority, was quite prepared to accept that Korea had no right to make treaties with foreign powers without Japanese consent—which was not the same thing as giving a green light to Japanese expansionism. But apparently this had already been done just a month before, when Roosevelt, wishing to bring off a diplomatic coup in the Portsmouth negotiations, had sacrificed Korea in order to induce the Tokyo government to give up its insistence on an indemnity and its claim to the northern part of Sakhalin, which it wanted to take from the Russians.[9] Three years later, following discussions in Washington, Japanese Ambassador Takahira, and Secretary of State Elihu Root, noted that their two countries, which both possessed islands in the

Pacific, were "animated by a common aim, policy, and intention" in the region, namely, "to encourage the free and peaceful development of their commerce in the Pacific Ocean," the maintenance of the status quo, and the defense of the open door principle in an independent China. But this was a commitment without any substance, for in case of a crisis, there was merely provision for consultations.[10] Theodore Roosevelt displayed similar caution when, on December 22, 1910, he wrote to his successor, Taft, that, since Japan had "vital interests" in Manchuria and in Korea, "[it] is therefore peculiarly our interest not to take any steps as regards Manchuria which will give the Japanese cause to feel, with or without reason, that we are hostile to them, or a menace—in however slight a degree—to their interests."[11] This did not prevent him, during his second term as president, from indulging in a show of force to discourage Japan from considering any warlike moves. This was one of the intended effects of the circumnavigation of the globe (December 1907–1909) by the "Great White Fleet" of sixteen battleships from the Atlantic Fleet, even if its major purpose was really to test the maneuverability of modern vessels and study the logistic problems of naval war in the Pacific.[12]

In relations with Japan, Taft and Secretary of State Philander Knox adopted a legalistic approach quite different from Roosevelt's, which had been based on the balance of power. In the name of "dollar diplomacy" they could not allow Japan to carve out its own special preserve in Manchuria as it had done in Korea, which became a Japanese colony in 1910. It was better, Knox thought, to "try to bring Japan's policy in China up to the level of ours, where we may differ, than to lower our policy to the level of theirs."[13] However, the weakness of the United States became evident when its plan to neutralize the railroads failed. It still lacked the power to take effective action in the western Pacific.[14] Japan was so aware of this that early in World War I it did not hesitate to join in with the aim of getting the German possessions in China and Micronesia into its clutches, and to take advantage of the fact that the European powers were busily expending their resources in a fratricidal conflict to exert its dominance over the Chinese Republic and gain an unassailable foothold there. Such were the aims of the "Twenty-One Demands" made by Japan in January 1915. These demands fell into five groups. One of these concerned the transfer to Japan of the German rights in Shandong province, a second the preponderant position of Japanese interests in Manchuria and in eastern

Mongolia (notwithstanding the open door principle), a third with China's obligation to cooperate with Japan in mining in the Yangtze valley, a fourth the prohibition on China to cede or to lease ports to other powers, and, lastly, a fifth formulated demands (which would later be presented as merely desiderata) having no less an objective than the establishment of a kind of protectorate in China. Japanese advisers would be installed at the highest levels of state, there would be joint policing in large cities, China would be required to buy at least half its armaments from Japan, and concessions would be granted for railroad construction in the Yangtze valley and Fujian province, and the right be given to spread Japanese religious doctrine.[15]

For the Wilson administration, accepting such a catalog of demands would have meant abandoning the principle of the open door. Even if he rejected "dollar diplomacy," Woodrow Wilson was too firm a believer in free trade as the basis for a peaceful international order to accept such a fait accompli. His close ties to the missionaries made him more sympathetic to China than to Japan. Instinctively, he felt closer to the Chinese lobby, represented by experts such as Paul Reinsch, the ambassador in Peking, and those responsible for the Far Eastern Division of the Department of State, than to the pro-Japanese groups. Nevertheless, he was also aware that the American people would never go to war to save the Chinese Republic. So he contented himself with taking a firm moral position, contrary to the advice of Secretary of State Bryan, who was more ready to give way on the basic points, and refused to recognize the concessions the Tokyo government had been able to extract from Peking in May 1915, when the Chinese had negotiated from a position of weakness.[16] Among ever larger sections of American public opinion, Japan was being perceived as a dangerous potential enemy. It was not surprising that Germany had tried to make it an ally, despite their differences in the Pacific. In January 1917, the Zimmermann telegram proposed, in addition to an alliance between the Reich and Mexico against the United States, an understanding between Germany and Japan—a proposal which came to nothing, since, indeed, quite to the contrary, in April Washington and Tokyo found themselves fighting on the same side, without any formal alliance having been made. Henceforth, American attention would be directed primarily toward the Atlantic, so that tensions in the Pacific had to be temporarily reduced. To counter Japan's attempts to take advantage of the situation to gain more territory, Secretary of State Robert Lansing, reached an agreement with Viscount Ishii on November 2, 1917. The Japanese envoy wanted Washington to recognize the "paramount interests" of his country in China, but was only able to obtain recognition of the fact that "territorial propinquity creates special relations" in the Chinese territories adjoining its Manchurian possessions—no small matter, though it was not held to be in any way prejudicial to the sovereignty, independence, and integrity of China, or to the open door.[17] The Chinese lobby, with its considerable power in Washington, saw this as a betrayal of its protégé and as a recognition of the principle of spheres of influence. As for Wilson, he hoped to win back as much as he could, later on, at the negotiating table. Likewise, during the Bolshevik revolution in far-eastern Siberia, he made every effort to keep the Japanese in check. To protect its interests, Tokyo was prepared to send an expeditionary force, but Washington, anxious not to allow its opposite number to act independently, found a good excuse to intervene in July 1918 to prevent the Germans from getting their hands on the arms dumps, and so that the United States could assist the evacuation through Vladivostok of the Czech Legion, composed of former Czechoslovak prisoners-of-war in Russia, which was retreating across Siberia in order to continue fighting on the Allied side in Europe. The American intervention, with only 7,000 men, was given limited objectives. Once these were achieved, the troops withdrew in April 1920, while Japan kept its forces (which were ten times this number) in the Maritime Province for a further two years, until it acknowledged the Communist victory. In Siberia, Japanese appetites were in the end contained by Bolshevik resistance rather than by Washington's action.[18]

In negotiating the peace treaty in Paris in 1919, Wilson attempted to block Japan's eager attempts to secure the German rights in China and in Micronesia (which were, however, granted it by the other allies) and to obtain recognition of the principle of racial equality. On the latter point, however good his initial intentions may have been, the American president ultimately adopted Australia's exclusionist policy, deciding to drop an amendment which would have opened the door to Japanese immigration—something to which politicians in the South and in the West were opposed—in order not to delay ratification of the treaty in the Senate. This made it much more difficult for him to stand in the way of Japan's territorial claims. To the great distress of his admirals, he had to accept Micronesia being made a "C mandate" of the League of Nations, with a ban on the construction of fortifications.

Worse still, he had to allow the inclusion of Shandong in the Japanese sphere of influence—a concession which the Treaty of Versailles' adversaries were quick to criticize hypocritically, assuming a sanctimoniously altruistic air in rejecting it.[19] When Wilson left the White House in March 1921, relations between the United States and Japan were at a low ebb, with considerable acrimony on from both sides. The enormous naval buildup launched by Washington in 1916 put the United States in a position to envisage war in both the Atlantic and Pacific oceans, while Japan, with its limited power, made every effort to avoid an imbalance of power which would be fatal to its long-term ambitions. Was it almost time for a Pacific War?

A Brighter Spell (1922–1931)

Contrary to the expectations of geopoliticians such as Bywater and Golovin, no Pacific War was to break out in the 1920s.[20] In the United States, the election of the Republicans and the growing influence of the pacifists led by Senator Borah demonstrated an awareness of the folly of the naval armaments race. At the same time, in Japan, liberals like Hara and Shidehara managed to make their voices heard against the imperialist and militarist tendencies. The way was suddenly open to establish good relations between the two major powers in the North Pacific. An era of imperialist diplomacy gave way to one of cooperation, aiming to promote a new international order.[21]

The Washington Conference, held between November 12, 1921, and February 6, 1922, established a framework for the postimperialist "system." Organized with consummate skill by Secretary of State Charles Evans Hughes, it dealt simultaneously with complex, involved problems, and resulted in the signing of three treaties. The first, the Five-Power Treaty, dealt with partial limitation of naval armaments. Hughes, wishing to halt an arms race, potentially disastrous for all, proposed to Britain, Japan, France, and Italy a massive overall reduction in the tonnage of large warships (battleships and battle cruisers), even if it meant scrapping vessels under construction. A maximum tonnage in the ratio of 5-5-3 would be imposed on the major fleets—American, British, and Japanese—together with a maximum caliber for naval artillery. Likewise, there would be restrictions on auxiliary warships (light cruisers and destroyers), submarines, and aircraft carriers, and strict rules would govern the replacement of outdated vessels. It was a sign of returning confidence that this ambitious disarmament plan was largely accepted, except where auxiliary warships and submarines were concerned. The Japanese delegation, led by Admiral Kato, wanted to substitute the more favorable ratio of 10-10-7 for the 5-5-3 proposed, so as to set the overall tonnage of Japan's capital ships at 70 percent of what was allowed each of the two Anglo-Saxon powers, instead of the 60 percent proposed by Hughes. This proposal was rejected, since Japan could provide no real justification for it, its interests being confined to the western Pacific, whereas the United States had two coasts to defend, and the British had to protect the lines of communication of their widely scattered empire. In return, to avoid the possibility of a surprise attack launched from a forward position, Kato won an extension of the status quo where fortifications and naval bases in the Pacific were concerned. Despite advice from his naval experts, Hughes agreed to halt construction in Guam and in the Philippines, but not in Hawaii, which was considered a defensive base, offering no threat to Japan, while Japan renounced the fortification of any islands which could be used as staging areas for an attack on the American colonies, i.e., the Kurile Islands, Ryukyu, Formosa, the Pescadores, and even the archipelagoes (the Bonin Islands and Amani-Oshima) which were part of Japan proper.[22]

The second treaty, which had four signatories (since Italy was not concerned), dealt with insular possessions and dominions in the Pacific. The Anglo-Japanese alliance, which had caused the United States so much uneasiness, was not renewed. The four powers undertook to respect their mutual rights and to reconvene the conference should any controversy arise over a Pacific "question"—a word Hughes preferred to "problem."[23] But no mechanism was laid down to force compliance, in conformity with the diplomatic policy followed by the United States. Since the United States had not ratified the Treaty of Versailles, Hughes made it clear that his country was not implicitly recognizing the League of Nations mandates over the former German archipelagoes, which had been granted to the British Empire and Japan, while he requested freedom for missionary activity and equal treatment where tariffs were concerned. The main discussion point was the notion of "insular possessions." Did it include the four large Japanese islands? Shidehara could not accept that it did, for in that case the treaty would have applied to Japanese soil but not to that of the Anglo-Saxon powers, thus underlining the unequal status allowed Japan. Conversely,

Balfour, speaking for England, had no intention of allowing it to seem that Australia and New Zealand were inferior to Japan. For Hughes, it was not a matter of extending the treaty to continental landmasses in the Pacific region—i.e., to China, Siberia, Chile, etc.—for this would have meant including too vast an area. In the end, a supplementary accord stipulated that in Japan's case the expression "insular possessions" applied to the southern part of Sakhalin Island, Formosa, the Pescadores, and the Micronesian mandates. One last sticking point was settled by the American-Japanese treaty of February 11, 1922, under which Japan recognized the American right to install transpacific telegraph cables on Yap, in the Caroline Islands, without discrimination.

In addition to the original five nations, the third, so-called Nine Power Treaty involved Belgium, Holland, Portugal, and also China, since it addressed the problem of this country.[24] It reaffirmed the principles of the open door and rejected spheres of influence and discrimination where railroads were concerned. Cooperation would replace international rivalry. For the Chinese, the benefits were minimal. They failed to have all the special rights and privileges granted foreigners abolished, despite the efforts of certain of their diplomats, who were, in Hughes' words, subjected to pressure from "agitators devoid of any common sense."[25] The most they could obtain was a number of undertakings: that customs duties would be revised to an effective level of 5 percent ad valorem, that a conference would be called to study the abolition of the *likin* (an internal tax) and that customs surtaxes would be imposed. The possibility of abolishing extraterritoriality was also raised, "as soon as Chinese law permits," once a commission had reported on the justice system, as was that of the withdrawal of foreign troops whose presence was not authorized by treaty as soon as China showed itself to be capable of protecting the lives and property of foreign residents. The country was encouraged to reduce its own armed forces so as to devote its energy to economic development and political stability. Obedient to the spirit of this treaty, Japan withdrew from Shandong, although it retained solid economic interests in the province.

The Washington Conference showed that Japan had drawn back from its World War I objectives, and the Japanese liberals, who did not espouse these, did not feel they had emerged as losers. It is true that in naval matters the empire was not granted a status comparable to that of the United States, but its strategic position did emerge somewhat improved. Even in China it retained its economic strongholds in the north. The spirit of optimism and faith in international cooperation, which permeated the conference, seemed to herald a new era. It would, however, soon run into difficulties, for the Chinese Nationalists could not tolerate the great powers making compromises at their expense. They still had the support of the Soviet Union, which finally enabled Chiang Kai-shek to launch his northern campaign in 1926. Faced with such a tide of nationalism, the Americans and the Japanese, together with other Western nations, could have cooperated to protect their interests. In fact, their entente was far from perfect, so that initially there was a failure to act, and then a rush to preserve the individual interests of each in discussions over customs tariffs and extraterritoriality—leading to the failure of the meetings called for in the Washington treaty. To best care for its own interests, each party put more trust in its bilateral relationship with the Chinese than in a multilateral approach—the Americans in particular making fine-sounding declarations of friendship toward China. When the Communists were wiped out by Chiang Kai-shek in 1927, Japan took over from the Soviet Union as a driving force in the Far East. Tanaka took a harder line toward China than Shidehara had done, sending two military expeditions into Shandong. More serious was the fact that the Kwantung Army pursued an increasingly independent policy in Manchuria, disregarding any decisions taken in Tokyo, which showed the slightest trace of moderation or prudence.[26] In 1928, it assassinated Chinese Proconsul Marshal Zhang Zuolin whom it considered excessively nationalist. But just at this very moment China was being reunited under Chiang Kai-shek. With an apparent order replacing the earlier chaos, it demanded the abolition of unequal treaties. In July 1928 the United States was the first to recognize its freedom to set tariffs.[27] Japan, with Shidehara returning to Foreign Affairs in 1929, followed this example, and, in turn, recognized the Nationalist government.

Good relations between America and Japan reached their highest point in 1930, at the London Naval Conference (January 21–April 22). As it had done in 1922, the American government sought to reduce military expenditures and achieve some progress toward genuine disarmament by limiting tonnage—by category of ship and not overall, as France would have preferred—for a period of six years.[28] It obtained an agreement on this principle with Britain and Japan, under which the three naval powers would not construct the projected replacement tonnage for battleships, the number of which was limited to fifteen for the

United States and Britain and nine for Japan. The Japanese government argued strenuously for a more favorable ratio for auxiliary vessels than for battleships. The Anglo-Saxon powers agreed to a ratio of 10-10-7 for cruisers and destroyers. This was to the considerable disappointment of the Japanese admirals who, after the Washington conference, had lobbied for the construction of cruisers, but now found they had reached the ceiling imposed, while the Americans were allowed to make up their shortfall. Where submarines were concerned, the Japanese were granted parity, but at a relatively low level. The London Conference thus established the conditions for a genuine balance of power in the Pacific, with each nation feeling secure because of the inadequacy of its potential adversaries' aggressive capabilities. These good relations might have endured if the world economic crisis which set in at the end of 1929 had not upset the international order. It made each country more disposed to turn back into itself rather than cooperate with its partners and competitors. In Japan, where exports were heavily affected, the Depression destroyed the prestige of the civilian elite, while the military became uncontrollable, grasping this opportunity to finally put into effect the aggressive plans they had been hatching all through the decade with the aim of creating a self-sufficient empire devoted to preparations for war. All the Kwantung Army needed to do was organize a raid on the southern Manchuria railroad for this "incident," on September 18, 1931, in order to provide the excuse for intervention in the Chinese province of Manchuria. This finally brought the brighter spell in relations between the United States and Japan to an end.

Planning for a Pacific War

For the men in charge of the United States Navy this brighter spell had never been real. Since 1906 they had considered Japan their principal enemy in the Pacific—a sentiment entirely reciprocated by the Japanese Navy. For it to have come to this, each of the two powers had had to equip itself with a fleet of modern warships.

In the nineteenth century, the United States had not possessed a unified fleet in the Pacific. Its two squadrons, the Pacific and the Asiatic, did little more than show the flag in their own areas of the ocean in order to impose respect for American commercial interests.[29] After the Civil War, the decline of a fleet which only a short time before had been comparable to England's was the reflection of a policy based on the two basic principles of continentalism and commercialism, implying the rejection of any ventures outside the North American continent and a concern to protect the country's foreign trade.[30] To achieve this, there was no need for battleships, but only, at the most, of cruisers capable of protecting commercial shipping from piracy. By 1883, however, the Navy seems to have lagged so far behind technically that Navy Secretary William E. Chandler took the decision to create a "new" Navy, with a nucleus of modern, steel-hulled vessels. This program, pursued by his successors, Benjamin Tracy (1889–1893), Hilary Herbert (1893–1897), John D. Long (1897–1902), and expanded by that impenitent "navalist" President Theodore Roosevelt, provided the United States with a battle fleet second only to the British.[31] Mahan laid down how it should be used. It would be concentrated in the Atlantic, where the major potential threat was located, with the Pacific remaining a secondary theater, with the result that the squadrons stationed there did not include a single battleship, the primary warship of the time. This doctrine remained in place when in 1898 the United States came into possession of an empire stretching from California to Hawaii and the Philippines. The conquest of the former Spanish archipelago weakened the Navy's lines of communication, for to reach Manila they had to pass close to the German possessions in Micronesia. Nevertheless, no effort was made to construct strong, fortified bases at Pearl Harbor, Guam, or Subic Bay. The Navy was prepared to accept the temporary loss of the Philippines should war arise, for the most important thing was to regain mastery of the seas with the help of its fleet of battleships, after a great, decisive naval battle in which the adversary would be annihilated. The opening of the Panama Canal in 1914 made it even less necessary to consider a two-ocean Navy, capable of conducting simultaneous campaigns in both the Atlantic and the Pacific, for the capital ships stationed in the Atlantic could now intervene more rapidly in the Pacific. The 1916 naval program, which finally was to bring the United States up to naval parity with the British, would have allowed the fleet to be divided into two if it had been possible to bring it to completion, but this was prevented by the restrictions agreed to at the Washington conference. The Navy therefore continued to adhere to Mahan's principle of the concentration of forces. But the Pacific was becoming less and less secondary. In 1919, Secretary of the Navy Josephus Daniels decided to move a number of large battleships to bases at

Bremerton, Mare Island, and San Diego.[32] Indeed, since Germany's defeat, and despite tension with the British, it was Japan that represented the major threat.[33]

From 1906 on the Naval Command and the U.S. Naval War College at Newport had, as an exercise, been working on plans for a hypothetical conflict with Orange, the code name for Japan. Until the end of 1940, a succession of "Orange Plans" were to reflect American thinking about ways to defeat a Japanese aggression. Indeed, the basic hypothesis was that the United States would never be the first to attack, but would merely respond to operations launched by the enemy. Furthermore, it would have to confront Japan on its own, without the help of any allies. A surprise attack against the Philippines was to be expected, in the Japanese tradition, for Hawaii seemed out of reach of the Japanese fleet. The Philippines would be lost, for their defenses were poor, the construction of fortifications having been forbidden in 1922. The problem for American strategists was therefore to know how the archipelago could be retaken, and the Japanese fleet so soundly defeated that it would leave Japan no alternative but surrender. The principal difficulty was the ability of the Navy to cross such a vast expanse of ocean without sacrificing too much of its strength, for the farther ships travel from their home base the more their effectiveness is diminished. The halfway mark between Hawaii and Japan was between the Marshall and Caroline Islands. The closer American vessels got to the Philippines, the more they would be at a disadvantage against the Japanese navy. They would be vulnerable to a war of attrition in the form of nighttime attacks by destroyers, cruisers, and submarines, or aerial bombardment by planes based in the Micronesian Islands that Japan had inherited after World War I.[34]

Nevertheless, the Americans never doubted their ultimate victory. Out of range of the Japanese, their immense economic resources would provide them with superior numerical strength, after a longer or shorter initial period. The debate focused on the best strategy to adopt. On the one hand the "thrusters" advocated moving the whole fleet west, from Hawaii to Mindanao, the large southerly island in the Philippines, usually by way of the Micronesian archipelagoes, but sometimes taking a more southerly route to avoid Japanese counterattacks. It was the equivalent, in naval terms, of a cavalry charge with drawn sabers. The "cautionaries," on the other hand, were very conscious of the risks involved in such a daredevil attack. They proposed a slower, gradual advance. They would take the "Royal Road" to Japan, capturing the intervening groups of islands one by one, without necessarily occupying all of them. Bases would be established in the most suitable lagoons—Wotje, Eniwetok, or Truk—from which attacks could be launched on the Philippines, or at the last Japanese line of defense. Both the thrusters and the cautionaries thought the war would end with a massive naval battle involving the battleships of both fleets. American strategists felt that Japan, battered and isolated from the rest of the world, would have to admit defeat without it ever being necessary to send in land forces.

Before 1934 it was usually the views of the "thrusters" that prevailed.[35] Their strategy did not seem altogether unreasonable as long as the Germans occupied Micronesia and the Americans could hope to create a powerful base on Guam, the "Gibraltar of the Pacific." This was no longer the case after 1915, by which time Japan had become the dominant power in the western Pacific. The treaty of Washington now prevented Guam, and also Manila, from being fortified. The closest staging-point for an attack was Pearl Harbor, but construction did not really get under way there until after 1919. It seemed suicidal to dispatch the fleet with a "through ticket" for the Philippines. Finally, in 1934, the "cautionaries" were able to impose their point of view, studying the landing techniques needed to take the atolls, and perfecting the logistics—all of which would prove extremely useful when hostilities eventually broke out. Six years later, the very idea of the Orange Plan no longer seemed relevant, for the United States would not wage war alone, but act in conjunction with its British, Dutch, and Chinese allies. The Orange Plan was replaced by a "Rainbow Plan," five successive variants of which were developed, "Rainbow 5" providing the framework for the impending Pacific War.

These changes were made necessary by disruptions in the naval balance of power. The London Conference of 1930 had laid down the relative strengths which would enable each nation to feel secure. The Japanese Navy disapproved of these provisions, claiming parity with the Anglo-Saxon naval powers, demanding that ratios it considered discriminatory be abandoned, and that vessels it considered "offensive"—i.e., battleships and aircraft carriers—be eliminated. All of these guarantees failed to conceal its desire to enjoy supremacy in the western Pacific, with the objective of establishing a "Greater East Asia Co-Prosperity Sphere." The Americans, who had neglected their fleet during the years of the Great Depression, under Hoover, found in Franklin D. Roosevelt, with his New Deal, a president sensitive to naval issues but one who nevertheless could not allow himself to act too much in defiance of an isolationist public opinion. In spite of this, in 1934 Representative Carl Vinson was able to obtain an increase in the Navy budget, allowing it to grow to the limits permitted under the London agreement. However, this American rearmament was not what caused the Tokyo government, in December 1934, to denounce the restrictions imposed in 1922 and 1930. In the circumstances, the second London Naval Conference (December 1935–January 1936) was destined to end in failure, despite the touching but fruitless efforts of the British, who came up with compromises equally unacceptable to the Japanese and the Americans. The latter would agree neither to parity—for they had two coastlines to defend—nor to the elimination of so-called offensive armaments, which would have meant leaving the field open to Japan in the Far East, and giving up the open door.[36]

In 1936, Japan relaunched the naval arms race. It began rebuilding much sooner than its future enemies, having nothing to fear from an internal pacifist opposition which had been successfully muzzled by terror and totalitarian indoctrination. It rejected all of the previous qualitative limitations, and began to build superbattleships of 70,000 tons, with a firepower vastly superior to anything sailing the oceans at the time. Roosevelt, on the contrary, was hampered by a population which remained mostly indifferent to foreign affairs. The second Vinson Bill, in 1938, did little to make up the ground being lost to Japan. As for Britain, it lacked the economic and financial ability to conduct two separate campaigns, one against Nazi Germany in the Atlantic and another against Japanese imperialism in the Pacific. It considered leaving the defense of the Pacific to the Americans. This was considered a questionable honor by the latter, who felt ill-prepared, after their late start. When, in 1940, after the fall of France, the United States finally put a gigantic rearmament program into effect, it could hardly count on this having a noticeable effect before early 1943. In the meantime, a window of vulnerability was left open. By the autumn of 1941 Japan enjoyed almost complete numerical equality with the combined navies of the United States, Britain, Holland, and Free France in the Pacific, while as far as the quality of its armaments was concerned, it was much superior.[37] But this situation would not last forever, and the Japanese admirals knew it. Their only chance of winning the

Pacific War was to strike soon, and hard. The plans they had begun to make ten years before, in 1931, were therefore put into effect.

Was Conflict Inevitable?

In 1931, when Japan, in search of a solution to its internal problems, chose the path of militaristic imperialism in China, the last thing the United States was prepared to do was rush to the assistance of the victim. Secretary of State Henry Stimson would have liked his country to take a strong line, but neither President Hoover, nor American public opinion, which was preoccupied with the deepening economic crisis, nor its potential European allies, wished to take such a step. Stimson therefore had to be content with a doctrine of nonrecognition of military conquest. The United States did not accept the Japanese seizure of Manchuria nor, later, its creation of the puppet state of Manchukuo— though these declarations of principle were not accompanied by any effective measures or sanctions.[38] China did enjoy the sympathy of the American public, which lent an ear to the Protestant missionary movement, and Japan was rapidly using up the little moral credit it had left, but these sentiments found little response in the Department of State, which was quite prepared to admit that Japan had a special interest in Manchuria by virtue of the fact that the latter was next door to its Korean colony.[39]

In July 1937, an incident on the Marco Polo Bridge, on the outskirts of Peking, quickly degenerated into a new Sino-Japanese war. This time, the Japanese imperialists and military men intended to bring all of China into their sphere of influence. They thought it essential to control it if they were to obtain its raw materials and gain access to the markets from which they considered they had been excluded by the more highly developed Western nations. To survive in a Darwinian world, there seemed to be only one solution for the Japanese Empire, namely to create an East Asian regional bloc under its aegis, rich in resources and in manpower, and well able to thrive on its own. This project called for the expulsion of the Americans and Europeans from the western shores of the Pacific. As it was most unlikely that they would willingly give up their interests in the region, Japan would have to prepare for total war, mobilizing all its material and human resources in order to achieve an unassailable strategic position, even if this meant allying itself with the group of totalitarian nations led by Nazi Germany and Mussolini's Italy—which is precisely what it did when it signed the Tripartite Axis Pact in 1940.[40] It is clear, whatever may be said by historians who argue about the relative responsibility of the United States and Japan for the origins of the Pacific War, that it was Japan which constantly took the initiative, while the American government merely reacted.[41]

Was imperialist conquest the only path open to Japan? This can be questioned in the light of the extraordinary economic development achieved, after 1945, by a country deprived of its colonies and relying only on the ingenuity and discipline of its people. It is true that the United States encouraged this new direction by substantially opening its markets to Japanese goods and encouraging the emergence of a liberal international order. It is also true, on the other hand, that prior to 1941 growing protectionism in the United States restricted Japan's possibilities for growth through exports. However, the possible directions which might have been taken in purely internal development were still far from all having been explored. As far as the raw materials lacking within the Japanese Empire itself were concerned, it would have been possible to obtain them by reaching bilateral agreements with producer countries. Why would owners of mines, oilfields, or any other natural resource refuse during a crisis to sell the commodities which were their principal source of income? Control over raw materials is important only from a military point of view. But in the 1930s neither the United States nor Britain represented a real threat to Japan. By opting for aggression, the Tokyo government was revealing its true mentality, for the rhetoric it used in claiming to be a victim of the Anglo-Saxon powers was no more than a smokescreen to conceal its desire to dominate the nearby countries. Believing themselves superior to the other Asians, the Japanese set out to establish their own rule over the region, much as Nazi Germany planned to do in Europe. To achieve this, they were able to call on a military strength without equal in Asia, and in 1937 they put it to use in an attempt to bend Chiang Kai-shek's China to their will, before it had a chance to rebuild after the havoc wreaked by the "warlords."

An objection to this interpretation is that it takes no account of the shades of opinion among those in power in Japan. In fact, the American experts at the time were constantly trying to gauge the relative influence of the moderates and extremists. The former were considered favorable to the international order which had resulted from the Washington conference in 1922, and the latter hostile to anything which did not contribute to the greater glory of the empire. The failure of American policy can in large part be attributed specifically to an unjustified confidence in the influence of the Japanese moderates. Terrorized by the very real risk of assassination, and rendered powerless by ultranationalist propaganda, the moderates were barely able to confront the fanatical militarists squarely, and were reduced to fighting a rearguard action in which their successes were very limited and short-lived, for they were incapable of putting forward a fundamental critique of their opponents' case. A sense of despair sometimes overwhelmed them, as they witnessed a drift toward belligerence which was making conflict with the United States inevitable. They were too aware of the latent power of America not to foresee that the war would end in a disaster for Japan.[42] But the moderates had no control over Japanese policy, and even less did they have any over the armed forces—the Army and the Navy—who were laying claim to considerable independence from the civilian authorities. The military—younger officers for the most part—were imbued with imperialist ideology and confident of their worth in battle. They had no hesitation in confronting an enemy which appeared much stronger economically. What did give an appearance of indecision to Japanese government policy was the fact that the Army and the Navy had serious differences about the precise expansionist policy to be pursued. The Army preferred a thrust into the continent of Asia, in which it would play the leading role. It believed it could easily control eastern China, the most populous region, and install a puppet regime which would allow it to exercise the real power, after which it would turn its attention to the Soviet Far East, using an attack on Bolshevism as an excuse, with the objective of capturing its rich natural resources. This continental adventure would require all the resources of the empire, and implied giving up the idea of conquests in Southeast Asia, where they would have to confront the Americans, the British, the French, and the Dutch. The Navy, on the other hand, saw that China, with its swarming humanity, would not be so easy to control, and that there was a risk that the empire would exhaust itself in its effort to do so, making it less able to confront its powerful maritime rivals. To conduct such a war, Japan would in any case have to depend on raw materials—oil among others—which were not available either on its own soil or in the regions under its control, and which the Anglo-Saxons could cut off,

obliging it to evacuate China. Expansion into Southeast Asia was a necessary preliminary to the continental thrust, and in this the Navy would
play the leading role, since it alone had the ability to protect the
empire's lines of communication.[43] But in either case, whether continental or maritime imperialism prevailed, Japan would be the aggressor,
the first to attack.

In spite of its stunning initial success, the invasion of China soon
ran into trouble. Chinese resistance did not collapse, and was even intensified by civilian massacres such as the odious "sack of Nanjing."
Some of the Army's best divisions were tied down on the mainland
carrying out policing duties. In 1939 at Nomonhan, the Japanese troops
suffered a severe defeat at the hands of the Soviets, making it clear that
the taking of eastern Siberia would certainly be no military cakewalk.[44]
In April 1941, therefore, Matsuoka, the Japanese foreign minister, passing through Moscow, signed a neutrality pact with the USSR, which
provided that should war break out between one of the two signatories
and a third power, the other would remain neutral. In addition, the
USSR recognized the independence of Manchukuo, and Japan that of
Outer Mongolia, two puppet states carved out of the former Chinese
Empire.[45] The Soviet setbacks at the hands of the Nazis in the summer
of 1941 encouraged the leadership in Tokyo to take advantage of the
opportunity to revive their designs on Siberia, but the Soviets had too
many troops in the Far East for this to be an option. Only when it was
certain there was no longer any risk of a Japanese attack did the Soviet
chiefs of staff remove large numbers of troops from the Siberian front
to take part in the battle to save Moscow, with the onset of winter. But
in fact, since July 1941, the thrust into Southeast Asia had become
Japan's main priority, as a result of the effective embargo on American
petroleum exports, itself imposed in reaction to the Japanese occupation of French Indochina.

As for the United States, it could have abandoned the entire western Pacific to Japanese influence, retaining only control over the eastern half for itself. America for the Americans, Asia for the Asians—the
ocean was broad enough to allow two relatively self-contained spheres
of influence to coexist in peace. The Philippines did present a problem,
but the archipelago was supposed to achieve independence by 1946 at
the latest, and America had no very important interests there. However, such a solution was never seriously envisaged in Washington. The
United States rejected any division of the world into self-contained

blocs. Cordell Hull, secretary of state from 1933 to 1944, advocated the creation of an international order based upon free trade, the lowering of customs barriers, and multilateralism. There could be no question of recognizing any kind of a Japanese sphere of influence in Asia, for this would mean a return to the age of mercantilism, whereas only a liberal economic system offered a way to recover from the financial crisis. Putting these principles into effect called for a firm response to Japanese imperialism. But, in practice, Hull tried to avoid any confrontation with Japan, and resisted pressure from the anti-Japanese lobby within the Department of State. He had no intention of going to war over Japanese aggression in China, and at most he occasionally agreed to provide economic and financial aid to Chiang Kai-shek's government. The United States, despite its expressions of goodwill toward China, certainly did not fly to its assistance, and indeed Hull made no effort to bring the two antagonists to the negotiating table. Japan, he felt, had painted itself into a corner it would have to get out of by its own devices. In the meantime, while a million soldiers were tied down on the Asian continent, the risk of a Japanese-American confrontation was reduced, and America's situation could only improve in the long term.[46] Hull did not really favor a policy of appeasement, but American interests in China did not seem to him significant enough to merit going to war. But, on the other hand, much more was at stake in Southeast Asia, with its invaluable resources of rubber and tin. A Japanese conquest of Dutch Indonesia, British Malaya, and French Indochina would destabilize the balance of power in the region and provide Japanese ambitions with solid material support.[47]

If he kept a firm hand on American diplomacy until the summer of 1941, Hull constantly had to deal with alternative views expressed within the Roosevelt administration. The president himself displayed a certain sentimental attachment to China. Three months after the Japanese aggression, on October 5, 1937, he condemned the evermore frequent breaches of international treaties, suggesting that the 90 percent of peace-loving nations should "quarantine" the infected 10 percent, so as to prevent the disease from spreading. However, in the actual conduct of affairs he usually supported his secretary of state's policy of nonconfrontation, and as events evolved rapidly in Europe he accepted the opinions of his advisers, who told him that the Atlantic would be the main theater of action. The fall of France in June 1940 left no further doubt that Nazi Germany was the principal adversary, against whom the essential effort—primarily aid to the British—should be directed. The Pacific now became a secondary arena, though inseparable from the Atlantic in the overall strategy. By moving their forces— the Navy in particular—east of Panama, the United States risked creating a vacuum which Japan might be tempted to take advantage of in China, Siberia, or Southeast Asia. A coordinated defense with the British, centered on the recently fortified naval base at Singapore, would have some chance of giving the Japanese imperialists second thoughts, but considering the enormous distances to be covered this could not be put into effect right away.[48] Also, in the short term, any such signs of cooperation might well provoke the contrary reaction, encouraging the Japanese to take the initiative and attack in the south to preempt any threat of encirclement. From the moment it decided that the German Reich was the main enemy, the United States needed to gain as much time as possible in the Pacific: the longer it was possible to postpone hostilities with Japan, the better the American strategic position would become. This was the thinking of Cordell Hull and of General Marshall, the chief of staff. But should nothing be done? The Chinese lobby, in which Stanley Hornbeck played a leading part, pointed out that if, in refusing a confrontation with Japan, the United States allowed China to collapse, Japanese divisions would be freed up for other military ventures. And who could tell just how much it would take to satisfy the appetites of totalitarian regimes?

In the short term, the only effective weapon the United States possessed to bring the Japanese war machine to a standstill was an economic one. Japan needed scrap metal to construct ships, tanks, and bombs, and its planes could not fly without high octane fuel. It so happened that America was the only source which could supply these commodities in sufficient quantities. American observers of the military operations in China, were scandalized to see that the Japanese aggressors were using so much materiel originating in the United States. They demanded an embargo. But neither Roosevelt nor Hull would consent until rearmament entered an active phase. In July 1939, the Washington government abrogated the Japanese-American commercial treaty of 1911, without this being enough to halt the soaring Japanese strategic purchases. When America began serious preparations for war, some kind of embargo seemed more justified, to ensure that the Department of National Defense had the first claim on scarce resources. After the export of airplanes, scrap metal was affected in its turn in

1940, when Japan occupied the northern part of French Indochina, to be followed by certain machine tools. As for oil, the embargo enforced in August 1940 applied only to an octane rating of over 86, which did not cause much of a problem for the Japanese air force, whose engines were less powerful than those of the Americans. The ineffectiveness of this partial embargo exasperated the anti-Japanese leaders and bureaucrats. So, when the occupation of French Indochina was completed in July 1941, and Operation MAGIC, which decrypted the Japanese secret codes, left no further doubt about Japan's intentions in Southeast Asia, it was hoped that a total embargo on petroleum would be enough to bring the Japanese forces to a standstill.[49] Instead of proclaiming the embargo officially—the consequence of which would have been to accelerate the Japanese offensive in the Dutch East Indies—an indirect approach was taken. Japanese assets in the United States were frozen, and allowed to be used only for purchases which required individual approval by Washington. Assistant Secretary of State Dean Acheson, who was responsible for putting this policy into effect, transformed it into a wholesale embargo, whether on his own initiative or with the tacit approval of his superiors—a point debated by historians.[50]

By this time war had become inevitable, for the Japanese Navy could not allow its fuel reserves to become exhausted, which would very quickly make it cease to be operational, particularly since Tokyo's official decision came down on the side of an offensive strategy. In the fall of 1941, renewed negotiations between Hull and Ambassador Nomura failed to hold out any hope for peaceful relations, not because the secretary of state demanded a comprehensive solution, including the evacuation of China, while the Japanese envoy only proposed a limited compromise, but because the American leadership, kept fully informed by MAGIC, rightly placed no confidence in the offers made by the Konoye and, later, Tojo governments. These were in any case morally unacceptable, since their only purpose was to satisfy the remaining scruples of the apprehensive moderates.[51] So, at the end of November, a squadron set sail, under orders to destroy the American fleet at Pearl Harbor.[52]

Was conflict between America and Japan inevitable? Even if Roosevelt had met Prime Minister Konoye in the fall of 1941, as had been contemplated, it seems unlikely that anything would have been different, since the president could not have taken a different position from that of Hull, who could hardly be suspected of seeking a confrontation with Japan. If the petroleum embargo had not been applied in August 1941, Tokyo, it is sometimes said, would not have been obliged to take the Dutch East Indies, or to try to destroy the fleet in Pearl Harbor in order to cover the flanks of its lines of communication to the south. Instead, well-supplied with gasoline, it would have had the resources it needed to launch a major offensive in China (as it did in 1944), and if Chinese resistance had collapsed—which is far from certain, for the Communists would have carried on an effective guerrilla war against the invader—it would have been able to lay the foundations of a powerful, self-sufficient, and virtually impregnable empire.[53] But even if the Americans were indifferent to the fate of the Chinese, they could never have allowed the creation of such an inherently hostile bloc on the far shore of the Pacific, headed by a nation noted for its warlike mentality and willingness for sacrifice. In the summer of 1941, Japan was reaching a certain boundary which the United States could not allow it to cross without putting its long-term security at risk. Not, of course, that it had to fear an invasion of California—but of Hawaii, perhaps. Who could tell?[54] It would also have been risking a contraction of the international economic zone which it felt was essential to its future prosperity. This invisible boundary line passed through Singapore, Sumatra, Manila, and, further south, Australia.[55] Any Japanese incursion to the south, beyond French Indochina, was charged with latent conflict. What would have happened if Japan had again changed strategy in the fall of 1941, and attacked Siberia, gambling on the fact that its Axis partner would be victorious in the USSR? One can imagine that the Soviet troops, attacked on two fronts, would have cracked, and the USSR would have found itself occupied by the two aggressor powers. The United States had no immediate vital interest in Siberia, so that Roosevelt would have found it difficult to persuade the country to accept a war against Japan. However, the American people would have been, willy-nilly, obliged to play a larger part in the Battle of the Atlantic, if only to safeguard Britain, its first line of defense. War against Nazi Germany was unavoidable. If it had wished, Japan could have disregarded the provision of the Tripartite Treaty which required it to intervene on Hitler's behalf if it was the United States who began the hostilities. In this way, it would have had an opportunity to consolidate its rule in continental Asia and, by making some judicious concessions, been able to exercise an unparalleled amount of influence once the war in Europe was over. Japan, like Germany, committed a major miscalculation by attacking the liberal democracies before turning its attention to Soviet Russia.[56] Like the Nazis, the Japanese ultranationalists had nothing but contempt for liberals, whom they considered to be cowards, devoid of all manly pride. They felt certain they would easily defeat them or, at least, involve them in a lengthy struggle in the course of which they would become discouraged and reduced to a state where they would accept any compromise offered them.[57] Totalitarian mentalities, in their arrogance, are led to underestimate their opponents, and to misjudge the moral resilience and determination of free peoples who, even though they do not spontaneously resort to force as a solution, are nonetheless prepared to stand up for their rights whenever a certain limit seems to have been reached. By the end of 1941, for many Americans, this limit had been very nearly crossed, in both the Atlantic and the Pacific oceans.

PART III

THE "AMERICAN LAKE" (1941–1994)

— We have seen how, between the Civil War and World War II, the original outlines of the Pacific frontier progressively lost their clarity, without being completely obliterated. The most obvious indication of this was the increasing importance of political and strategic interests. The Pacific had become less and less marginal in Washington's calculations. Nevertheless, the Atlantic was still the primary preoccupation. It was there that major economic activities were conducted and where the major military powers were situated, as also were the cultures with which the Americans felt the most affinity. In commercial and financial relations with the Pacific countries, on the other hand, an imbalance remained, and the cultural gulf was far from being bridged. Despite the Japanese threat, which had obliged the American government to move its battleships to the West Coast in 1919, Nazi Germany seemed to pose a much greater threat to the vital interests of the United States than did Japan in 1940. The Pacific was no longer the "great frontier" of earlier days, but it was still far from attaining equal status with the Atlantic. The third and last stage of the frontier's evolution, in the course of which it would disappear, was already beginning, but this process was certainly hastened by World War II. The military victory over Japan turned the Pacific into an "American lake." After 1945, the United States' line of defense was established on the archipelagoes close to the Asian continent and sometimes on the mainland itself. This meant that the vast expanse of the Pacific had become an integral part of America's defense strategy in confronting the communist world. Japan, now included in this defensive perimeter, was, just as much as Western Europe, a key element in the geostrategic vision underlying the stationing of American forces throughout the globe. Likewise, the economic boom in the Far Eastern and Southeast Asian countries, resulting in part from their exports to the United States and facilitated by the revolution in transportation and in communications, wiped away all earlier inequality with the Atlantic. Under the aegis of the American eagle even cultural contacts became less superficial, in spite of some resistance. Less than four decades were required for the proper conditions to be created in the American lake for the frontier to disappear almost entirely.

It was only during this final stage that the westward advance of the American population was to play a part in relations with the Pacific. As late as 1940, the five states and territories on the ocean were home to 10 million inhabitants, or 8 percent of the total population. Even if one thinks that reciprocal interests between nations are a function of their geographical interface, it is difficult to attribute a decisive voice to California in decisions concerning the Pacific until just before Pearl Harbor. In the past half-century, however, things have changed, for 40 million Americans, or 16 percent of the total population, now live along the Pacific seaboard, and California has been the most populous state in the Union since 1963. However, this new importance did not just emerge from relations with other Pacific countries, but rather from the overall development of the United States and the intrinsic conditions of its growth. The major decisions are still not made in Los Angeles or San Francisco, but on the East Coast, in Washington and in New York. In ceasing to be a frontier, the Pacific certainly encouraged the growth of the five western states—but not to the extent of paving the way for them to dominate the future.

NINE

War in the Pacific

— Early in December 1941 the United States controlled only the eastern half of the Pacific. It seemed unassailable within the triangle constituted on one side by the American continent and, on the ocean itself, by the lines joining the Aleutian Islands to Hawaii and Hawaii to the Panama Canal. This area was made secure by the enormous distances, considered too great to be crossed by the Japanese fleet. Farther to the south, between Panama and Tierra del Fuego, the fact that the coastline angles back by almost 50° of longitude relative to California placed the shorelines of Peru and Chile well out of reach of any serious threat from the Far East, but at the same time made these countries useless as bases in any campaign of reconquest. It seemed that the Pacific coast of the South American continent would be marginal in the coming conflict. Apart from these safe zones for naval communications, the United States did not have many aces in their hand—a few atolls close to the international date line (Midway, Johnston, Howland, and Baker) and Eastern Samoa. Forward of these, their possessions were a liability, for they were surrounded by territory under Japanese control. Wake Island was within reach of the Marshalls, and Guam seemed completely isolated on the edge of the Marianas. As for the Philippines, their westerly position left them encircled by Japanese forces on the Carolines, in Taiwan, and in French Indochina.

Against the Americans, Japan controlled the western half of the North Pacific. Its continental conquests in Korea and China protected it from any threat from Asia, particularly since the USSR was involved in a desperate struggle against the Nazi invader. Japan's strategic position, whether offensive or defensive, was an excellent one, if its intention was to expand its empire toward the southwest or the southeast, and be in a position to defend it subsequently. To the southwest, British Malaya and the Dutch Indies appeared to be as weak as the Philippines, while to the southeast each of the islands taken from Germany during the First World War was a potential launching pad for incursions into the area ruled by Britain and its Australian and New Zealand dominions (eastern New Guinea, the Bismarck Archipelago, the Solomon and Gilbert Islands). Japan's problem was to know where to stop, for the farther it ventured across the ocean, the longer its supply lines became, increasing their vulnerability. It needed to find the most useful perimeter, and one which would be relatively easy to defend. This is why Japanese strategists had little intention of carrying hostilities beyond 100° S, which meant leaving not only Australia and New Zealand out of the war, but also the archipelagoes extending from New Caledonia and the New Hebrides to French Polynesia. Nevertheless, Japan's ambitions transformed the Pacific into a vast battlefield, for the first time in its existence, with the two dominant powers, one on its American and the other on its Asian shorelines, contending across it to decide the ocean's future. Was it to be divided into two by a maritime frontier or remain an open sea?

THE JAPANESE OBJECTIVES

The Japanese military commanders, who had provided the principal inspiration behind Japanese policy before taking it over completely in October 1941, under the Tojo government, pursued a double objective, namely to break Chinese resistance and to gain control of Southeast Asia's raw materials. The first of these objectives depended on the achievement of the second. The Navy was determined to rid itself of American pressure on its oil supplies. In addition, the conquest of Southeast Asia would isolate Chiang Kai-shek's China by cutting off the Burma Road. These aims implied war with the British, the Dutch, and, probably, the Americans.

The Japanese strategists considered several plans, one of which was very risky, for it included a direct attack on the Dutch East Indies. Another, supported by the Army, took the form of a counterclockwise advance from Malaya to the Philippines, through Sumatra and Borneo. The Navy, on the other hand, preferred to move in the opposite direction, in order to eliminate any threat to its offensive from American forces. They finally opted, in August 1941, for an audacious and complex compromise solution, consisting in a surprise attack on all points simultaneously. The plan had three stages. The first, a centrifugal offensive, required the imperial ground and sea forces to fan out between Thailand and the Gilbert Islands, destroying the enemy forces. This would be followed by a phase of consolidation, involving only an attack on Burma, with elsewhere the emphasis being placed on reinforcing the support positions and the occupied zones. Finally, the empire would take up a purely defensive position to await an enemy counterattack, which it would be able to annihilate close to the defense perimeter by taking advantage of the most favorable conditions for battle.[1] By the end of 1941, only the first phase had been prepared in detail. One point which had provoked much discussion was the question of a preemptive strike against the American naval base at Pearl Harbor. This idea had been put forward in early 1941 by the commander-in-chief of the combined fleet, Admiral Yamamoto, who had previously been hostile to any rapprochement with the Axis, and who was extremely aware of the United States' potential power. He considered that Japan's only chance in the coming war was to take a gamble: the destruction of the Navy ships which, at Roosevelt's insistence, had been stationed at Pearl Harbor since early May 1940. Roosevelt had viewed this as a way of exerting pressure on the warmongers in Tokyo, disregarding the advice of Admiral Richardson, who was much more conscious of the disadvantages of exposure to the enemy in such a forward position. If they could take the United States by surprise, the Navy planes of the Japanese fleet would have a good chance of destroying the threat to the left flank of the centrifugal offensive in one fell swoop and would gain precious time, during which the enemy might become discouraged, or find itself confronted with an impregnable Japanese defense. The Japanese Navy was highly conscious of the risks entailed by Yamamoto's plan, for the surprise would have to be total and refueling ships in the rough waters of the North Pacific in December was a difficult operation. Would it not be better to rely on Japan's traditional strategy of waiting for the enemy to enter Japanese waters, where it would be far from its bases and therefore more vulnerable? It was by threatening to resign that Yamamoto tipped the scales in favor of his plan.

Between December 1941 and April 1942, the Pacific war evolved according to plan. The limits of the defensive perimeter were reached with such apparent ease—the taking of Singapore being a case in point—that the Navy considered pursuing its offensive without a pause. Initially, Rabaul, in New Britain, had been captured to provide protection for the Fourth Fleet's base, at Truk. Now, to protect Rabaul, it would be necessary to take the northeast coast of New Guinea and capture the Solomon Islands, from where the Japanese could push on into New Caledonia, Fiji, and Samoa, severing the lines of communication between the United States and Australia, the staging area for the Anglo-Saxon counteroffensive. Yamamoto's staff had grandiose dreams of capturing Hawaii, Ceylon, or even Australia itself. But these suggestions were rejected as too ambitious, the preference being for the capture of Midway, northwest of Hawaii. The objective was to draw the American fleet into a decisive battle, even though the location was not particularly advantageous to the Japanese. As he had done over his plan to attack Pearl Harbor, to overcome objections Yamamoto put his reputation, then at its height, on the line. Japan seemed set to dominate the Pacific.

On June 4, 1942, in the space of only five minutes, planes from the aircraft carriers *Enterprise* and *Yorktown* shattered the Japanese fleet's offensive. The Battle of Midway marked an early turning point in World War II, well before Stalingrad and El Alamein. From that point on, Japan would fight an essentially defensive war, delaying the American and, at the same time, the British advance as much as possible and achieving its only major advance in China in 1944.[2]

Japan attempted to portray its aggression in a different light. Japan's slogan of "Asia for the Asians," which portrayed it as freeing the yellow peoples from the white colonial yoke, had been part of its ideological armory for several years. After its merciless exploitation of the so-called "liberated" peoples, Japan put forward the idea of a fundamental community of interest between them, within what was termed a "Greater East Asia Co-prosperity Sphere." In November 1943 the Greater East Asia Conference ended with a declaration every bit as universalist in tone as the Atlantic Charter. The countries of the region would cooperate so as to ensure its stability and create an order based on the principles of coexistence and coprosperity. They would respect their mutual autonomy and establish friendly relations among themselves, and with other nations, in a spirit of progress, while refusing racial discrimination and guaranteeing free access to natural resources.[3] In fact, there was a considerable distance between the words and the reality, and there was every reason to doubt the sincerity of such declarations coming from a government which had invaded country after country and never hesitated to vaunt the unique, and therefore superior, character of the Japanese people.[4] Nonetheless, this propaganda did erode the foundations of European colonialism and paved the way to eventual decolonization.

This gave Japan the impression that it had not completely lost the war. In 1944–1945 it prolonged hostilities because of its unwillingness to accept unconditional surrender. The Tokyo government wanted to preserve the imperial regime as the personification of the nation's historical continuity—something a growing number of American leaders were prepared to concede. It clung desperately to the futile notion that the USSR might serve as mediator and that an agreement with China might allow Japan to concentrate all its forces on the struggle against the Anglo-Saxons, whom it considered its principal enemy. But these maneuvers were a useless waste of precious time, the direct consequence of which was the dropping of the atom bombs on Hiroshima and Nagasaki.[5]

THE PACIFIC IN AMERICAN STRATEGY

In 1940, Washington considered the Pacific secondary to the Atlantic as a theater of war. The Orange Plans were replaced by the Rainbow Plans, the ultimate version of which, "Rainbow V," set out two primary objectives: the defense of the "Western Hemisphere"—by which was meant the American continent above 10° N—and the dispatch of American forces to Europe in order to inflict a decisive defeat on Germany and Italy. If war were to break out on both ocean fronts, the United States would have no alternative but to adopt a defensive posture toward Japan from within the Aleutians-Hawaii-Panama triangle. The Atlantic was given priority where the production of war materials and the deployment of troops were concerned. These general principles, endorsed by the British, were never put in question, for it seemed evident which enemy represented the greater danger. Nazi Germany possessed a military, industrial, and scientific potential far greater than Japan's and had to be defeated first. All through 1941, Roosevelt and his high command moved naval units from the Pacific to the Atlantic to combat the U-boats of the German Navy. They desperately sought to avoid any confrontation with Japan so as to avoid fighting a much-feared war in both oceans. In fact, the United States was ill-prepared. However, propelled into the war against its will by the attack against Pearl Harbor, it did not allow its desire for revenge to lead it astray. At the London Conference, in April 1942, the Allies decided to limit their efforts against Japan, making it their priority to strengthen Britain and provide aid to the USSR. This strategy would bring about the German surrender on May 8, 1945.

Nevertheless, the Pacific could not be ignored. This was particularly true for the Americans, who were responsible for operations throughout the entire ocean, since the British were much too pre-occupied with Europe, Africa, and the Middle East to have any interest in such a role and were quite content to control operations in Burma, Malaya, and Sumatra. To respect the sensitivities of their two major armed forces, the Washington authorities divided the Pacific into two zones. The larger of the two, allotted to the Navy under Admiral Nimitz, included the north, center, south, and southeast of the ocean, on either side of the equator. The other, allotted to the Army under MacArthur, included Australia, New Guinea and the adjoining archipelagoes, the Philippines, and the Dutch East Indies, minus Sumatra. China was considered a separate zone, but Generalissimo Chiang Kai-shek made the American General Stilwell his chief of staff. Clearly, the Navy had the lion's share, and its highest ranking officer in Washington, Admiral King, was determined to see its role would not be reduced to a merely secondary one. For his part, MacArthur would never agree to play second fiddle, for he hoped to reap as much glory in the Pacific as his colleague Eisenhower had done in North Africa and Europe. This made strategic coordination between the Army and Navy somewhat difficult, but in the end the problem was satisfactorily resolved by the high command in Washington, when General Marshall and Admiral King worked out the necessary compromises with the top political leadership.[6]

From early 1942, King, a believer in offensive warfare, thought it would be a mistake to remain within the boundaries of the Aleutians-Hawaii-Panama triangle, for he felt this would leave the field open to the Japanese and deprive the Navy of the forward bases it needed to launch a counteroffensive. He considered the defense of Australia, as well as its supply lines, to be essential. Also, since it would take time to organize landings in North Africa or France, some of the available troops could, in the meantime, be transferred to the Pacific. On February 18, 1942, King proposed attacking the Solomon Islands, but this was somewhat premature, since the Japanese centrifugal attack was still meeting with success in all quarters. He did succeed in having numbers of American troops stationed in Australia, New Caledonia, and the New Hebrides, as well as on Bora Bora and the atolls used as way-stations, so as to oppose any possible Japanese offensive. The result was that, a month after the Battle of Midway, on July 2, 1942, the high command in Washington found itself in a position to develop a plan to retake the southwest Pacific. After much discussion, it managed to combine King's suggestion of a step-by-step advance launched from the Solomon Islands with MacArthur's impetuous plan for a lightning advance through New Guinea, striking at the Japanese base at Rabaul.

The first successes, at Guadalcanal (August 1942–February 1943) and in the Owen Stanley Mountains (September 1942–January 1943), finally provided the opportunity, in April 1943, to finalize a strategy for the defeat of Japan which would be approved the following month at the Trident Conference in Washington. This relied not only on a naval blockade to force Japan into unconditional surrender, but also on a very extensive aerial offensive. For this purpose, bases close to Japan would be required, since the range of the bombers was still limited. With the USSR still neutral, Siberian airfields could not be used, and Formosa appeared to be too well-defended. This left only China and brought about renewed interest in Chiang Kai-shek's regime. Roosevelt wished to include him among the great powers after the war, despite his government's obvious incompetence. Hong Kong was considered the ideal port for assembling the necessary materiel—a curious idea, since the South China Sea is a kind of cul-de-sac, quite easy to defend. To take Hong Kong, the high command envisaged a coordinated effort by all the Allied forces, with the Chinese attacking from the provinces under their control, while the British, after recapturing Burma, would bypass the Straits of Malacca and liberate French Indochina. The Americans would mount a two-pronged attack on the Philippines, with MacArthur leading one prong along the northern coast of New Guinea toward the Moluccas and on to the Celebes and Sulu Seas, and Nimitz the other, across the central Pacific from Hawaii to the Marshalls and then the Carolines, joining up with the first somewhere around the New Guinea "bird's head." The naval operations would follow the Plan Orange strategy for the liberation of the Philippines after a sweep through Japanese Micronesia.

This strategy of May 1943 was not completely realized in this form. In March and April 1943 Japanese forces again went on the offensive along the Indian frontier, and General Slim's troops of the British 15th Army were unable to get the better of them until March 1945. In China, the Ichigo offensive launched by the Japanese Army in the center and south of the country from April 1944 to January 1945 met with remarkable success. It thus basically fell to the Americans to carry through the attack. The strategy of an offensive through the central Pacific came back into favor. This was where the American Fleet enjoyed an ever greater superiority over its Japanese rival, and it was also the shortest route to the Philippines, with only the capture of a few atolls being required to advance along it. Furthermore, these did not present the same health risks as the shores of the southwest Pacific, meaning that fewer troops would be needed. A decisive victory in this sector would make a direct assault on Japan possible, without there being any need to take a roundabout route through the Philippines and China. Of course, MacArthur was opposed to any such plan, which would have reduced his role. The high command in Washington therefore decided to attack from both directions, particularly since, at the end of 1943, it now had the necessary human and material resources to do so. The two-prong offensive had the advantage of obliging the enemy to spread its resources thinly, not knowing where the next blow might come, so that each of the prongs made the other stronger.

Nevertheless, it was in the central Pacific that the outcome was decided. The capture of Tarawa, in the Gilbert Islands (November 1943), and then of the Kwajalein, Majuro, and Eniwetok atolls in the Marshall Islands (January–February 1944), pushed back the Japanese defensive perimeter. Their base at Truk, in the Carolines, was put out of action by bombing. As King wished, and as was confirmed by the conferences of Cairo and Teheran (December 1943), the offensive was switched farther to the north, toward the Marianas, which the new B-29 Flying Fortresses could much more easily use as a base to bomb Japanese cities than airfields on the Chinese mainland. On March 12, 1944, the Washington command revised its plan of the preceding year, replacing Hong Kong as an objective by the Marianas to the north on the one hand, and, on the other, by the island of Mindanao to the south, from where a choice could be made between Formosa and Luzon. The first objective was achieved between June and August with the capture of Saipan, Tinian, and Guam. When Peleliu in the Palaus fell in September, MacArthur could finally glimpse the fulfillment of his dream of a return to the Philippines, which he did in October 1944, first to Leyte, and then to Luzon. A short time before, the high command, against King's advice, had given up the idea of taking Formosa, for it seemed that invading it would be very costly.

At the beginning of 1945, with Nazi Germany collapsing, the strategy to defeat Japan became clear. It meant moving ever closer in order to launch a final assault, should one be necessary. Despite the damage inflicted—a quarter of Tokyo was destroyed by incendiary bombs on March 10—the B-29 raids, taking off from the Marianas since late November 1944, were not enough to overcome Japan's desperate resistance. For this reason, Nimitz, after taking Iwo Jima in the Bonin Islands in February and March, was forced to capture Okinawa in the Ryukyu archipelago, after a bloody battle (May–June). To oblige Japan to surrender unconditionally, Operation Downfall was developed, to be carried out in two stages: Operation Olympic, beginning in November, against the south of Kyushu, and then Operation Coronet, in March 1946, which called on the combined forces of MacArthur and Nimitz, reinforced by troops fresh from their victory over Nazi Germany in Europe, to land close to Tokyo.[7] But this plan would never be put into effect, for President Truman decided to put an end to the war as quickly as possible by using the new atomic bomb. On August 6, 1945, the B-29 *Enola Gay* dropped the bomb which destroyed Hiroshima, and since there was no reaction from the Japanese government, Nagasaki was hit in turn three days later. Finally, on the 15th, Emperor Hirohito ordered the unconditional surrender of all his military forces. The United States, in destroying the Japanese aggressor, had achieved the objective it had set itself after Pearl Harbor. It also was now without a rival in the Pacific, which had become an "American lake."

From Defeat to Victory

Would this have come about without the war? Certainly not to the same extent. Actually, if peace had reigned, as it did during the 1920s, the United States would never have needed to develop its fleet and, without naval supremacy, could not have succeeded in dominating the ocean so easily. Nevertheless, within the framework of the kind of liberal international order envisaged by Cordell Hull, its relative economic power would have been sufficient for it to have emerged as the economic focus of the Pacific Rim countries. At the time, Japan possessed neither the industrial resources, financial means, nor capacity for innovation, needed to challenge American supremacy. As for the British, despite the extent of their possessions, their power in the region was already in decline. So it is surprising that Japan attempted to challenge American power, and even more surprising that it was able to put it on the defensive for several months, between December 1941 and May 1942, and to resist its onslaught until almost the end of 1943.

It would be unfair to ascribe the early defeats entirely to mistakes committed by the American civil and military authorities. The Japanese were able to put the limited means they had at their disposal to remarkably effective use. They had equipped themselves with a disciplined Army and Navy, hardened by demanding exercises, extremely well-trained, and equipped with the most modern war materials. In 1941, the pilots of the Japanese naval aviation had hundreds of hours of training under their belts, their dive-bombers and torpedo planes were capable of much better performances than those of their American adversaries, while the Zero fighter was considered the best in the world—certainly the most maneuverable, if not the best armored. The Navy was equipped with giant battleships which, with their 460 millimeter guns, were supposed to guarantee its mastery of the seas. Almost all the Japanese Navy men were convinced that this type of vessel would fight the final battle, which would decide the outcome of the war. Yet the admirals did not neglect the aircraft carrier. In 1922 they launched the first carrier to be constructed as such, and on the eve of Pearl Harbor they possessed ten of them, against three for the United States Navy, which was obliged to keep its four others stationed in the Atlantic. Yamamoto already saw the aircraft carrier as the paramount naval weapon of the future, while most of the admirals saw them as no more than support vessels for the fleet of battleships. All Japanese sailors were trained in nighttime combat on cruisers and destroyers which were unequaled. As for the Army, it had mobile divisions whose soldiers displayed a spirit of self-sacrifice in the purest samurai tradition. In general, the Japanese were convinced that "spirit" was more important than equipment, and that their warlike mentality was so superior to that of the Americans that it would easily compensate for their inferior productive capacity. To win battles, it is not enough to be prepared, for a share of luck is also necessary. But luck, during the early months of the war in the Pacific, was mostly on the side of the Japanese. It was with them in the attack on Pearl Harbor. However great the care with which the raid was prepared, Vice Admiral Nagumo's fleet of six aircraft carriers might have been spotted and incurred much greater losses than the 130 men and twenty-nine planes which were the cost of the operation. But there was also an element of bad luck, for if Japanese bombs totally destroyed three battleships and 162 planes, inflicted serious damage on thirteen ships, and lesser damage on five others, killing 2,403 and wounding 1,178, they left the facilities almost intact and had the disappointment of finding that the three American aircraft carriers were out of port on that Sunday morning of December 7, 1941.[8]

What was good luck for the Japanese was bad luck for the Americans. In an ideal world where information could be communicated instantaneously, the disaster of Pearl Harbor might have been avoided. Roosevelt's detractors have gone to great lengths to prove that given all the information they had, thanks to the success in deciphering the Japanese codes, the leaders in Washington should have been expecting an attack on the fleet. Some even read into the failure to do so a deliberate intention on the part of a Machiavellian president to tempt the enemy into just such a misstep so that he could involve his reluctant country in the World War. But this conspiracy theory, so typical of American historiography, lacks any serious foundation. Aware that the struggle against Germany was a priority, the American leadership was attempting, in December 1941, to delay the outbreak of hostilities in the Pacific for as long as possible, while not giving in to all the Japanese demands. They had nothing to gain from a conflagration in the Pacific. The commissions of inquiry, on the other hand, have established an impressive list of failures in the chain of command—a lack of precision in the orders, and the priority given to controlling the population of Japanese origin on Hawaii instead of preparing to meet aggression from without. These were all minor details, but they resulted in the engine of war misfiring too frequently, just when it really mattered. However, there was no reason to suspect that the Japanese were about to attack Pearl Harbor. Their offensive was expected to take place in Southeast Asia. It was underestimating them to think they were incapable of such a complex operation, even if such a hypothesis had been put forward now and again. In this respect, the Americans were victims of their racism.[9] Curiously, the military authorities on Hawaii are blamed for not having taken the necessary measures while MacArthur, in the Philippines, came out heaped with honors, although he allowed his air power to be destroyed long *after* the attack on Pearl Harbor.[10]

After such an unpromising start, it is surprising that the United States was able to remedy the situation relatively quickly, first by stabilizing the positions in the summer of 1942, and then by launching major offensives, beginning in late 1943.

The Pacific war was fought between a single country and an alliance. It was impossible for Japan to communicate with its Axis allies in Europe, and in its "co-prosperity sphere" it could only rely on the support of the territories it had occupied. The puppet governments it installed in China, the Philippines, and Burma hardly provided it with any useful assistance, in addition to which it lost too much time in establishing an effective administration in the former European colonies. It quickly lost control of the shipping routes within its own empire, for in December 7, 1941, the Americans, abandoning their earlier policy, launched an all-out submarine war against Japanese commercial shipping, which was rarely organized into convoys. The Japanese, on the other hand, contrary to the Germans, opted to use their submarines in naval engagements rather than in raids on shipping.[11] By November 1942, in spite of a major shipbuilding program, the Japanese fleet had been reduced to a lower tonnage than at the outbreak of hostilities. In 1944, American submarines sank almost 2.5 million tons of merchant shipping.[12] As a result, the Japanese war industry was being slowly strangled for lack of raw materials.

In their war against the Japanese, the Americans were supported by a broad coalition in which they were, indeed, the key player, but which provided them with room for maneuver that Japan did not have. Toward Latin America, their rear was secure. Only Chile, where the strong German influence and pro-Nazi sympathies on the political right favored neutrality, showed, at best, considerable reluctance to provide support. Like Argentina, Chile took advantage of the decision by the Rio conference in January 1942, which went no further than to "recommend" that diplomatic relations be broken off with the Axis powers, to drag its heels while making ever more demands, such as first being supplied with war materials to enable it to resist any Japanese attack (a most unlikely eventuality, given the great distance), and guarantees for their copper and nitrate exports after the war. Finally, after a year of arduous discussions, economic and political pressures finally overcame Chile's reluctance. In January 1943, its government broke off relations with the Axis powers after the United States extended its protection to Chile—which had hardly been possible to do earlier. Chile, however, did not declare war on the Axis until February 1945. The essential thing for the Americans was to obtain the strategic war materials they needed, and this kind of support was in fact always given quite readily.[13] On the western side of the ocean the most disappointing ally turned out to be China. Chiang Kai-shek feared the Communists more than he did the Japanese, and his fragile power depended on the favors owed him by certain of his generals. He therefore had no intention of committing all his forces to the struggle against the Japanese aggressor and was reluctant to develop a genuinely professional Army as the American Stilwell urged. Instead, he called for the use of aerial power, following the recommendations of another American, Claire Chennault, the commander of the Flying Tigers, and complained endlessly of a shortage of supplies and armaments. He finally wore out the patience of his strongest supporters in Washington, particularly when his troops collapsed in southern and central China in 1944. The American high command could no longer count on Chiang Kai-shek, who was granted the status of a great power without having done anything to deserve it. China had just one advantage to offer: it tied down a number of Japanese divisions which would have been more usefully employed elsewhere in the Pacific. The Americans would have liked the Soviet Union to intervene in the Far East once Germany was beaten, believing that this would hasten the Japanese collapse. Stalin confirmed his agreement to this at the Potsdam Conference in July 1945, before the atom bomb was ready. The successful testing of the new weapon very soon made the Soviet intervention almost superfluous, but it had been planned for several months and it was by then too late to cancel it, so that on August 8 Stalin's troops entered Manchuria, annihilating any remaining Japanese hopes that he might act as mediator between them and their Anglo-Saxon adversary. We must not underestimate the contribution made by the British Empire all through the war in the Pacific. Australia and New Zealand provided excellent bases in the rear, as did New Caledonia and the New Hebrides, in French Polynesia. The Americans could station their soldiers there safe from Japanese attack, and construct storage facilities, ports and air fields from where counteroffensives could be launched. Until 1943, MacArthur had more Australian than American troops under his command.[14] Their control of the South Pacific provided the United States with the strategic depth it needed to destroy the Japanese war machine.

Without any wish to belittle the part played by other allies—the British, for example, dispatched a task force of aircraft carriers to help in the invasion of Okinawa—we can say with certainty that, in reality, the war in the Pacific became increasingly an American affair. The United States owed this predominant role to its crushing material superiority. Its industry was safe from any attack from without; it possessed the necessary raw materials (synthetic rubber replaced the latex from the rubber plantations of Southeast Asia), abundant sources of energy, and enough productive capacity not to have to sacrifice the needs of its civilian population to the same extent as the other belligerent nations. Not only, therefore, were any material losses quickly replaced, but its armed forces received ever-increasing quantities of more and more sophisticated armaments. One after the other, bottlenecks were freed. An impressive logistical system was put in place to supply the fighting units out across the ocean. Despite some waste, the war in the Pacific was won by a rational, organized, efficient America, whose wealth was unrivaled at the time. The more farsighted among the Japanese soon saw what folly it had been for their country to attack such a power. Japan lacked the necessary domestic industrial resources to conduct a lengthy war. It had equipped itself for a brief, total war, not for prolonged resistance. It was able to mobilize all its forces only by imposing terrible privations on its citizens. Japan itself, unlike mainland America, was vulnerable to aerial attack. On April 18, 1942, sixteen B-25s, under orders from General Doolittle, taking off from Vice Admiral Halsey's aircraft carriers, carried out a raid on Tokyo which, though ineffective, was disquieting enough to cause the Japanese high command to make their attack on Midway seven weeks later. Starting in November 1944, bombing raids carried out by B-29s reduced the major Japanese industrial cities to rubble. All that remained to the Japanese to enable them to confront such a hail of metal was their nihilistic and suicidal heroism. The most famous instance of this is the kamikaze pilots who crashed with their planes on the decks of Allied ships, in order to cause maximum destruction. This "divine wind" was supposed to save the empire, like the typhoon which had swept back the Mongol invasion in the thirteenth century. And, indeed, starting on October 25, 1944, when they first appeared in the naval battle off Samar, in the Philippines, they caused heavy losses. In one month, off Mindoro and Luzon, they sank twenty-four ships and caused greater or lesser damage to sixty-seven others. In the engagements around Okinawa, between April 6 and June 22, 1945, the kamikazes launched ten major attacks, destroying or damaging almost 200 American vessels and four of the five British aircraft carriers, at a cost of 1,900 pilots—a high price to pay for the results achieved.

For the United States to have lost the Pacific war it would have been necessary for the Japanese not to commit a single error, and for the Americans to make a number of mistakes—a highly improbable scenario. The Japanese high command, apart from the poor decisions mentioned above (the inefficient use of submarines, the failure to organize convoys, and the delay in exploiting the resources of the conquered territories), dispersed its forces over too many islands. The American tactic was, therefore, not to attack all of these directly but only those which would be useful as bases, isolating the rest by destroying their supply lines. In addition, the Japanese neglected information-gathering, whereas the Americans were able to decipher the Japanese secret codes, giving them a considerable advantage in the Battle of Midway, despite their numerical inferiority. Until September 1944 the imperial troops were instructed to defend the islands on the beaches. They made ferocious charges, to shouts of *banzai!*, which invariably ended in a bloody massacre. Too late, on Peleliu, in the Philippines, at Iwo Jima, and, most effectively, at Okinawa, they opted for an in-depth defense with well-prepared positions which were very difficult to take, inflicting losses the American military considered excessively high—almost 10,000 men killed or wounded on Peleliu, 6,000 dead on Iwo Jima, and 12,300 on Okinawa. If it had been put into effect earlier, this tactic would have slowed down the attackers, who lacked reserves, since at the time most of their efforts were concentrated on Europe.

The American generals and admirals have sometimes been accused of not having been as concerned as they should have been for the lives of their men.[15] Indeed, if the conflict in the Pacific was of short duration, it was often exceptionally bloody because of the desperate resistance put up by the Japanese. In total, the Americans lost 100,000 dead and almost 200,000 wounded in this theater of operations, while the Japanese Army and Navy lost 1,750,000 dead in eight years of fighting, between 1937 and 1945, with an additional 400,000 civilians killed by bombing.[16] The disproportion between the two countries seems enormous, but American public opinion was much affected by the increasing intensity of the fighting: 53 percent of the GIs who lost their lives did so in the final year, between July 1944 and August 1945. Nevertheless, these figures show what good use the Americans were able to make of their military power. From year to year they improved their logistics, discovered more efficient landing techniques, and improved the coordination of their different armed forces. By the end of 1942 they had taken the initiative and never lost it again, leaving the Japanese to wonder where the next blow would fall.

The American victory depended on their mastery of the seas and on the new factor which made this possible, the aircraft carrier.[17] The Pacific war sealed the fate of the battleship, which until then had been the most important ship in naval battles.[18] Henceforth, enemy fleets would fight major engagements out of sight of one another, for it was the air power they carried which would decide the outcome. There was an element of chance in this. Everything depended on the skill, anticipation, and reflexes of the officers, and on the tenacity, but also on the luck of the pilots. In the Pacific, in 1942 the war took on the appearance of a gigantic chess match between the Japanese and American admirals. After his success at Pearl Harbor, Yamamoto made the mistake of dispersing his three groups of carriers throughout the ocean, instead of attempting to destroy the enemy's smaller number of carriers. The standoff in the Coral Sea (May 6–8) gave him a short-term advantage. But a month later he squandered this in the Battle of Midway where, setting out with eight carriers (five of them heavy carriers) against three, he ended up losing four heavy carriers, with only one American carrier being sunk. In the six naval battles near Guadalcanal between August and November 1942, Nimitz saw his forces cut to tatters, particularly after the engagement off the Santa Cruz Islands (October 26–27). This left him only one remaining carrier to call on, and even it had suffered damage.[19] Fortunately for him, Admiral Kondo had also suffered serious damage, with only two light carriers left, and had lost many more first-rate pilots, who would turn out to be irreplaceable. This was why the two adversaries avoided any further naval confrontations in 1943, while they nursed their wounds. In the meantime, the American shipyards were turning out new ships, so that in 1944 the Pacific Fleet had seven carriers of the *Essex* class (27,000 tons) and eight of the *Independence* class (11,000 tons). Nimitz organized his ships into heavily protected "task forces," with the result that not one large carrier was sunk, while the Japanese Admiralty continued to suffer loss after loss. In the Battle of the Philippines Sea (June 19–20, 1944), three carriers were sunk and four damaged, while 400 planes were shot down in "The Great Marianas Turkey Shoot." The sea war reached its climax in Leyte Gulf (October 23–25, 1944), the greatest naval battle in history, involving 282 ships, 187,000 sailors, and hundreds of planes. The Japanese made a final effort to disrupt MacArthur's landing in the Philippines, but despite a carefully thought-out strategy, their combined fleet was totally destroyed. The Americans were now masters of the Pacific thanks to their squadrons of carriers, which allowed them to strike at will, with minimal risk, under the command of such experienced officers as F. J. Fletcher, R. Spruance, W. Halsey, and M. Mitscher.

In the American victory, the point which has given rise to the most debate is the use of the atom bomb to end the conflict.[20] Was it moral? Did it really serve a useful purpose? Unconditional opponents of nuclear warfare condemn President Truman's decision. He justified it by his desire to save thousands of American lives. If Japan itself had had to be taken at the same kind of cost as the Pacific islands—Tarawa, Peleliu, Iwo Jima, and Okinawa—the invaders would have suffered heavy losses. Though these would certainly not have amounted to the million or half-million sometimes spoken of, they would certainly have been enough to disgust an American public which was longing for peace. The president had to take into account the growing intensity of the fighting and the fanaticism of the Japanese soldiers, who were prepared to fight to the last man rather than surrender. A very small number of Japanese prisoners was taken. The argument that this was partly due to the Allies' reluctance to take prisoners does not seem to hold much water, although there were undeniably massacres (but how typical were they?). The collective suicide of the civilians who threw themselves off the cliffs at Saipan gave a foretaste of the costly guerrilla war which would be fought on Japanese soil. As supreme commander, Truman had the responsibility to save his soldiers' lives, even if this meant confronting the Japanese with the prospect of total annihilation if they did not surrender. We cannot be at all certain that the Tokyo government would have accepted surrender without the bomb. Even after Hiroshima and Nagasaki the Supreme Council for the Direction of the War was equally divided between those who wanted peace and those who wanted to fight to the bitter end. It was necessary for Prime Minister Suzuki to make the extraordinary move of calling on the emperor to intervene to break the deadlock, and for War Minister Anami, who had been a hard-liner until then, to accept his decision and refuse to join the coup attempted by fanatical officers on August 14–15, 1945. Without the bomb there was a good likelihood (but there can be no total certainty on this point) that the government would never have dared to surrender, fearing assassination by ultranationalists. "Operation Downfall" would have been followed for the invasion of Japan, and there would certainly have been many more dead—even among civilians alone—than in the two bombed cities. Even if Truman did not possess all the information, he nevertheless had a good grasp of the situation, and from his viewpoint, the United States had no rational choice other than to use the bomb. It alone could deliver the salutary shock which would lead to peace.

Seen in this light, the other arguments put forward by revisionist historians seem to me to carry little weight. If, it is suggested, the Allies had replaced their demand for unconditional surrender with more flexible terms, implying that the imperial institution would not be in any danger—which was in any case the American administration's intention—Japan would have capitulated before any invasion became necessary. Here again, it is far from certain that the ultranationalist officers would have accepted such a humiliation or that the emperor himself would have intervened to oblige them to cease hostilities, for this argument fails to take account of the samurai mentality and of the belief that nothing is lost until defeat is final. The terrain of the Japanese archipelago, like that of Okinawa, is well-suited to guerrilla warfare and to developing a multitude of lines of defense. Another argument dear to revisionists is that the bomb was intended as a warning, less to the Japanese than to the Soviets. Washington is supposed to have already seen that the USSR would become its principal enemy in the future, and wanted to forestall the troops it was preparing to send to Manchuria and Korea from gaining a foothold in Japan, so that the bomb was mostly a warning to Stalin to proceed with caution. But even if Truman distrusted the Communist dictator's ambitions, it is unlikely that this argument weighed heavily in his decision. The facts speak for themselves. On August 6 and 9, Japan was still resisting, and it had simply become necessary to put an end to a war that had already lasted too long, and which threatened, if it continued, to cost the lives of too many American boys. But a final question remains: If the Americans had been dealing with white Nazis rather than yellow fanatics, would they have resorted to such an extreme solution? This brings us to the cultural dimension of the war in the Pacific.

CULTURAL ASPECTS OF THE PACIFIC WAR

If the Pacific conflict seems to have been a "war without mercy," to borrow the title of a book by the historian John Dower, this is because it was not simply motivated by economic and political factors, but was also a confrontation between two different worlds, each with an ideology strongly grounded in the concept of race.[21] It is characteristic that the Americans spoke of their enemy in Europe as Nazi Germany, with the emphasis on "Nazi," thus implying a distinction between the German people, which had been temporarily led astray, and its evil leadership. However, this did not prevent sometimes resorting to blanket bombing, as on Dresden, on February 13, 1945—a foretaste of what awaited Tokyo a month later. The Japanese, on the contrary, were viewed as a single entity, with the same traits ascribed to both the people and its leaders.[22] They were often depicted as subhuman, whose bestial cruelty was attested to by the massacres carried out by the imperial troops in China, in Bataan, and in Manila. In studying the Japanese national character, even anthropologists open to a more objective analysis, such as Ruth Benedict, emphasized their differences rather than their similarities.[23] Even when they were victorious, the Japanese character was basically different from the American. There was therefore no need to have as many scruples in dealing with them as with European enemies. The best evidence of this attitude was provided by the United States itself, where the 120,000 persons of Japanese origin, whether United States citizens or not, were deported from the Pacific Coast states to detention camps in the interior, on President Roosevelt's orders (February 19, 1942).[24] This evacuation did not apply to the Japanese-Americans in Hawaii, who were better integrated into a multiracial society, and watched over by the local military authorities. In California, in Oregon, and in Washington State, pressure exerted by the press, by patriotic organizations, and by envious economic competitors, resulted in a speedy evacuation of the coastal zone as a security measure to avoid any possibility of a fifth column being formed. It is true that some residents of Japanese origin had not renounced all their bonds of loyalty to the Japanese emperor, but the vast majority were loyal to the United States, and many young *nisei* covered themselves with glory on the battlefields of Europe, or played an invaluable role in information gathering. Today, nobody denies that the internment was an injustice and against American law, even if such a hysterical reaction was quite understandable in early 1942, when it seemed that nothing could halt the Japanese offensive. The facts that it was judged constitutional by the Supreme Court several years after the event, and that it was not until 1988 that the victims of this discrimination were granted compensation, show to what extent the Japanese were not thought of as ordinary people at the time, contrary to the German-Americans and the Italian-Americans, who were not subjected to a similar expulsion from the East Coast states.

The Japanese had an equally Manichean view of the world. Believing themselves to be racially pure, a peerless nation, superior to all others, they viewed the war as a mission to destroy Anglo-Saxon influence in the western Pacific. The Japan Yearbook for 1943–1944, in studying the causes of the war, makes several references to the unique Oriental character and to the notion of race:

> The first cause of the Greater East Asia War lies in the rise and expansion of Nippon and the United States in the Pacific Area somewhat similar to the cause of the First World War which was the conflict of Germany and Britain. It is entirely a new war. Many foresighted thinkers and statesmen on both sides of the Pacific had been apprehending a possible clash of the two Pacific powers and had endeavored to avoid it by organizing several international associations for better understanding between the two great nations and making joint investigations on the Pacific problems. But all their efforts failed to attain their peaceful purpose and the war broke out in 1941, because the destiny or the natural trend of racial expansion of these two nations had been such as needed an adjustment in terms of superiority or fitness of either nation for dominating the Pacific Area with its real power. It is a matter of racial movement. In older days, racial movement on the globe had been easier and more natural than in the present century in which there was created a seemingly fixed boundary or international system in regard to the sphere of influence of each nation.[25]

Ultranationalist propaganda depicted the Americans as white devils and materialists, lacking in all the qualities which made for the beauty of Japanese spirituality. The bombing of cities was seen as demonstrating a lack of humanity in contradiction with the principles proclaimed by Anglo-Saxon civilization in other areas. Such crimes were condemned, quite forgetting that the Japanese Army was the first to have massacred women and children in Nanjing in 1937 and to have displayed its contempt for the Geneva Convention, as a punishment for which several hundred officers would be condemned to death and executed when the war was over.

But if the images of each nation promoted by the media of the other were loaded with negative connotations all throughout the hostilities, members of both governing elites had a much more lucid perception of how much the two countries had in common. The memory of the 1920s, when the internationalism dear to Wilson's heart had seemed to take precedence over the self-interest of nations, remained in people's minds as the best proof of the fact that dialogue is not impossible between two cultures who, despite their differences, nevertheless share many common interests and values. The historian Akira Iriye has shown very well the similarity of attitudes in each country, making conflict between them seem all the more absurd, and partially explaining why a reconciliation between the two countries could be brought about so rapidly after 1945.[26] The war could not be conducted entirely as a conflict between two races. Japan had made an alliance with Hitler's Germany, the ultimate champion of the white race. The United States maintained close—if not entirely confident—relations with Chiang Kai-shek's China, but the exclusion laws were considered offensive by an ally who had lost millions of men resisting the Japanese invader. This is why Roosevelt, in 1943, managed to obtain—though with some difficulty—the abolition of these measures, dating from 1882. The Chinese were allowed an annual quota of 105 immigrants and were to be allowed to take out United States citizenship. It seemed a small concession, but it had considerable symbolic value, for it eliminated the notion of race from American immigration laws.[27] The Pacific was not to be the stage for a titanic confrontation between the white and the yellow races, as the nativists had once dreamed.

The war created an American presence in areas of the Pacific where scarcely any had been found since the end of whaling. In the South Pacific, in particular, hundreds of thousands of GIs were stationed at the bases used as staging areas for the counterattack. In Australia and New Zealand they found themselves in countries which spoke their own language, and which were happy to have their protection against a possible Japanese invasion, at a time when Britain was in no position to provide effective assistance. More than the New Zealanders, who remained very attached to their British roots, the Australians felt a strong affinity with the Americans. Naturally, as is always the case when foreign soldiers are present in great numbers, some tensions did emerge, with young Australians criticizing the GIs for having too much money, for enjoying an excess of material goods, and for winning the attention of too many of their young women. Rioting between the two groups broke out in Brisbane in 1943, but, basically, relations were good. Ties across the ocean, which until then had been almost nonexistent, were established when thousands of young Australian women married Americans, and United States culture made considerable inroads in Australia.[28]

Nearer to the equator, the Polynesian and Melanesian archipelagoes had many charms to offer, particularly the former with their accommodating young women. The novelist James Michener, in his *Tales of the South Pacific,* has given a vivid account of the fascination exerted by these voluptuous islands on the puritan American soldiers. The nurse, Nellie Forbush, finally finds happiness with the French planter Émile de Becque, de Becque's racially mixed daughters manage to win the affections of some GIs on their way through, and Lt. Joseph Cable carries on an impossible love affair with Liat, a Polynesian girl of Vietnamese origin.[29] In 1949, a number of these stories, pieced together by Oscar Hammerstein and set to music by Richard Rodgers, provided the plot for the musical comedy *South Pacific,* with the well-known songs "Some Enchanted Evening" and "Younger than Springtime."[30] The Americans left many souvenirs of their stay behind them. For the first time, the natives came in contact with whites who made no attempt to subject them, or who, even better, reduced the privileges of the European planters and introduced a certain equality into their relations, awakening their desire for emancipation. The sight of black American soldiers bearing arms made a strong impression on the Melanesians.[31] This friendly invasion by an Army so rich in provisions, and abandoning enormous stocks in its wake, gave new life to the already ancient cargo cults, which received a fresh impetus and became linked to the hope for decolonization—for why work for the Europeans, when such riches arrive from the sea?[32] The Americans left many delayed-action bombs behind them which, in the years to come, would change the mentalities and the geopolitics of the Pacific.

The Pacific war was not simply the extension of an existing trend, but marked a new departure where the United States' role in the ocean was concerned. In September 1945, when the tumult of war finally died down after raging for 45 months, the Americans exercised complete mastery over the entire ocean. No one but they had the capacity to deploy a fleet from one coastline to the other, encountering no obstacles and no adversaries. Their frontier had been advanced to the very shores of the Asian continent. Never in history had any power achieved the kind of supremacy over the ocean which the United States had now attained. The Pacific completely lost its status as a somewhat marginal region, far removed from the world's center of gravity. For diplomats and strategists in Washington the problem was now to decide how to keep control over their lines of communication so as to ensure the Pacific remained an American lake.

TEN

Stormy Waters

    In 1945 the United States was the undisputed master of the Pacific. Its principal adversary, Japan, lay in ruins: it no longer had a fleet, its cities had been leveled by incendiary and atom bombs, its industry had suffered terribly, it was occupied by American troops, and the Supreme Commander for the Allied Powers, General MacArthur, was playing the role of a shogun. The European colonial powers—Britain, France, and the Netherlands—had seen their prestige melt away with the Japanese victories and now would have to reconquer their empires over the opposition of native nationalists extremely unwilling to tolerate a return to the former state of affairs. The British dominions in the South Pacific felt betrayed by the mother country and sought shelter under the wings of the American eagle. The Philippines were preparing to attain independence and willing to grant military bases to the former ruling power. Not only did the United States control the entire ocean, and particularly the former Japanese possessions in Micronesia, but it also had troops stationed on the Asian mainland in a South Korea now liberated from the Japanese colonial yoke, and in China, where its ally, General Chiang Kai-shek (one of the Big Five in accordance with Roosevelt's wishes), controlled most of the country, nominally at least, except for a few regions in the north and Manchuria which remained loyal to Mao Zedong's communists. Only the USSR was in any position to threaten this overwhelming supremacy, but its Far Eastern territories were too far removed from its center of gravity to serve as a springboard for any attack, and Stalin, who distrusted Mao, preferred to make his country's defenses secure by consolidating the Soviet position in Eastern Europe. For him, the Pacific was not a strategic priority, even if he could be expected to jump at any opportunity which presented itself. At the dawn of the Cold War, it seemed as if the American position in the Pacific would be safe from any challenge for many years to come.

THE FREE WORLD VS. THE COMMUNIST BLOC (1945–1972)

However, the United States' supremacy encountered two often indistinguishable obstacles: communism and nationalism. The former was generally perceived as a vast attempt to subvert the basic institutions of the so-called "free world," based upon the American values of free enterprise, the free circulation of ideas, political democracy, and individualism. Nationalism was the response of subjected peoples to colonial domination or imperialist exploitation. It was the seed from which the nation-state would grow, along with the modernization required for it to emerge from underdevelopment. When existing structures were resistant to this desire for change, nationalism often aligned with communism to free itself from the shackles of the past. Once victory had been achieved, the interests of the different social classes reemerged, so that wherever the middle class or an indigenous land-owning oligarchy retained control of the process, a break would ultimately develop between communism and nationalism, with the stronger of the two coming out on top. The America of the 1950s, preoccupied by the Cold War with the Soviet Union, tended to ignore the subtle relationship between these two revolutionary forces, viewing all nationalists as cryptocommunists and a Trojan horse for the Kremlin, or, at best, as naive individuals led astray by the enemy's propaganda.[1]

The Western Pacific Line of Defense

To avoid another unpleasant surprise like the attack on Pearl Harbor in 1941, American strategy in the Pacific consisted in establishing a first line of defense as far as possible to the west. In its confrontation with the mainland of Asia, Washington's policy was to exert total control over the Pacific Ocean in order to make its lines of communication between the two shores as secure as possible. To achieve this objective, it was necessary to establish strong positions on islands which could be used as naval or air force bases—as "unsinkable aircraft carriers"—or at least keep them out of the Communist enemy's hands.

As had been the case during World War II, mastery of the seas depended on fleets of warships with groups of aircraft carriers as their nucleus. The United States stationed two such fleets in the Pacific: the Third Fleet in the east and the Seventh Fleet in the western half. Marine units were kept in a state of alert, while the Air Force constructed runways which would put the enemy within easy range should a "hot" war break out. Pearl Harbor, the nerve center, was no longer *the* forward base, for thousands of miles farther on a string of military installations stretched out along a defensive, or even offensive, perimeter, enclosing a private American domain consisting of the entire Pacific Ocean.

To the north, the Aleutian Islands, extending outward from Alaska, housed bases such as Adak, capable of neutralizing Soviet defenses in the Far East and keeping the stormy waters of the North Pacific under surveillance.[2] Japan, lying to the south of Kamchatka and the Kuriles, which Stalin had annexed for the Soviet Union, had now become a bastion of the free world in the Far East. Before the San Francisco Peace Treaty (September 8, 1951) the Americans were almost alone as military occupiers of the islands which make up Japan. They took over the former naval bases at Yokosuka and Sasebo, and created a new Air Force base at Misawa, within range of Vladivostok and Manchuria.[3] On the very day that the state of war came to an end, with most generous conditions being offered to Japan (it retained its unrestricted sovereignty, was not obliged to pay any reparations, and was subject to no restrictions on trade, fishing, or shipbuilding), a bilateral security treaty negotiated by John Foster Dulles recognized that, by virtue of the individual and collective right to self-defense, the recently defeated nation, which had been left without an Army by Article 9 of its constitution, could, for the time being, call on the United States to discourage any aggression against it, without this implying any formal commitment on its part to intervene.[4] This allowed the United States to keep troops in Japan and nearby, and even to use them, at the request of the Japanese government, to help suppress any riots or domestic unrest caused at the instigation of, or as a consequence of the intervention of, one or more foreign powers.[5] This treaty between

America and Japan was a substitute for the proposed Pacific regional pact inspired by the Atlantic Pact, which had never come into being because of opposition from the British, their dominions, and countries which had suffered under Japanese occupation.[6] In Washington, the Senate was also reluctant to make a former enemy commitments comparable to those which, in a departure from its isolationist tradition, it had just made to its Western European allies. There would never be anything like the Atlantic Alliance in the Pacific, where there was no similar sense of belonging to a common civilization. To obtain ratification of this security pact by the Senate, the American administration made a complementary administrative agreement (February 18, 1952) applying to military bases in Japan. The right to extraterritoriality, which the Pentagon asked for, could be suspended in certain cases, on a request from the Japanese. Furthermore, Prime Minister Yoshida, in a letter dated December 24, 1951, but published on the following January 16, undertook not to sign any bilateral treaties with the People's Republic of China and to make peace with Chiang Kai-shek's Nationalist government on Taiwan, paving the way to the establishment of diplomatic relations, which took place on April 28, 1952.[7] Japan's reacceptance into the international community was thus achieved at the cost of its close dependency on the United States. In 1954, 210,000 American troops were stationed at 300 bases and other installations in Japan, and the national defense force, created during the Korean War, was rearmed under Washington's aegis.

In the 1950s, these treaties provoked growing opposition in Japan. The nationalist right complained of their unilateral and unequal nature, and tried to come to terms with Moscow. The Socialists favored neutrality. As for the Communists, they adhered faithfully to the Kremlin's anti-American stance. On the other hand, Prime Minister Kishi (1957–1960), who still bore a heavy burden of guilt for having been a member of Tojo's war cabinet, could not conceive Japan's future role in Asia other than in partnership with the dominant power, the United States. For this to become possible, the Security Pact of 1951 had to be revised to establish a more equal relationship. Since the Americans were in agreement, there was little difficulty with the new principles. If the preparation of the final text took some time, this was only because of the extremely complicated Japanese domestic politics and the factional rivalries within the Liberal Democratic Party. Finally, on January 19, 1960, a new U.S.–Japan Treaty of Mutual Cooperation and Security established a framework for reciprocal obligations. Article V stated: "Each Party recognizes that an armed attack against either Party in the territories under the administration of Japan would be dangerous to its own peace and safety, and declares that it would act to meet the common danger in accordance with its constitutional provisions and processes."[8] To contribute to its security and to the maintenance of peace in the Far East, Japan granted the United States use of facilities and sites for the use of its forces. An exchange of notes clarified the principle that any "major" change in the deployment and equipment of these forces, or any use of bases for operations outside Japan, would be subject to prior consultation. All these provisions were unacceptable to the Socialists, who came under pressure from the major trade union and who were more hostile to Kishi than they were to the Americans. The Communists and leftist students, on the other hand, were anti-American rather than anti-Kishi. Blunders committed by the premier aggravated the tension. Violent demonstrations and highly emotional mass movements,[9] fueled by the international crisis, were sparked by a refusal to hold new elections and by the way the vote approving the new pact was forced through the House of Representatives on May 20, so as to ensure its ratification by June 19 at the latest, this being the date when President Eisenhower was supposed to arrive in Tokyo. Kishi was humiliatingly obliged to request a postponement of the visit, and had to resign shortly after. It is amazing, in the circumstances, that after surviving a few anxious moments during the Vietnam War, particularly during the visit of the nuclear aircraft carrier *Enterprise* to Sasebo in January 1968, the pact should have survived in relative peace since then, and still be in effect. The 1960 riots revealed a Japan suffering from an inferiority complex toward the United States, against whom it could let off steam in this way without running much of a risk. These events had no impact on the Americans, whose military protection enabled Japan to launch a feverish economic expansion unhampered by the cost of its own security.

South of the Japanese archipelago, the Ryukyu Islands, which had been taken at great cost in 1945, provided an ideal strategic base facing the mainland of Asia. The San Francisco Treaty had made them a United Nations trust territory administered by the United States, while maintaining Japan's "residual sovereignty"—a compromise between the annexation advocated by the Pentagon strategists and their eventual return to Japan, the solution preferred by the Department of State.[10] The

American military administration transformed Okinawa into a powerful nuclear base and an indispensable telecommunications center—the key to the western Pacific. The United States had no intention of losing control over this. But this semicolonial status was a cause for deep resentment in Japan and in the Ryukyus. Was it not preferable to maintain good relations with the Japanese people, who accepted the American presence, rather than impose an intolerable domination which in the long term would weaken the strategic utility of the bases? This question, first asked by pro-American Prime Minister Sato in 1964, came up again after 1966, when Okinawa no longer appeared so essential, and at a time when intercontinental ballistic missiles were becoming the major weapon of deterrence. But two problems remained. Could a Japan suffering from "nuclear allergy" accept that the Americans continued to stockpile atomic weapons there after the islands were handed back? Would the decision to use weapons be subject to the provisions of the 1960 pact? After much dithering, Sato adopted the view which prevailed among the public, and proposed that Okinawa should be returned denuclearized, with any external use of American forces being subject to prior consultations between the two allies. But Henry Kissinger, chief of the National Security Council, played his cards astutely in order to safeguard America's vital interests. The Japanese-American communiqué signed by Sato and Nixon on November 21, 1969, is full of ambiguities. It was indeed provided that in 1972 Okinawa would be definitively returned to Japan, "subject to the conclusion of . . . specific arrangements with the necessary legislative support,"[11] and that, as provided by the 1960 pact, the United States would retain the sites and armaments necessary to maintain the security of both countries. On the question of nuclear weapons, the Japanese government believed it had obtained an agreement from its ally to remove all stockpiles prior to the agreed date for transfer of power, although it is far from certain that the prohibition has not been occasionally infringed by American forces since that time. The requirement for prior consultations led to differing interpretations. Did it mean acceptance of a Japanese veto? Or was there nothing to prevent the United States from carrying out its international obligations to defend countries in the Far East? Lastly, in spite of the explicit references to South Korea and Taiwan, both important keys to Japanese security, Japan, true to the policy it had followed since 1951 but against its ally's wishes, did not allow itself to be drawn into a regional system which might require it to intervene militarily outside its own territory. At the expense of some ambiguity, the return of Okinawa did nothing to weaken the close ties between Washington and Tokyo.

On Taiwan the Americans had a more aggressive ally than Japan. After the collapse of his regime in mainland China, Chiang Kai-shek had taken refuge on Taiwan in 1949, where he awaited a Communist assault. The Truman administration, disappointed by his incompetence, would gladly have abandoned him to this fate if the Korean War had not broken out on June 25, 1950.[12] In Washington's eyes, Chiang Kai-shek, like Syngman Rhee, even though his regime allowed no freedom of expression, suddenly became a buttress of the free world, with his fierce anticommunism, endowing him with new attractions as the Cold War began to intensify. On June 27, at the same time as he ordered his air and naval forces to assist the South Korean troops to push back the Communist aggression, Truman observed that the occupation of Taiwan would pose a direct threat to security in the Pacific zone. Consequently, to prevent any invasion by Mao Zedong's troops, he stationed the Seventh Fleet in the strait separating Taiwan from the mainland and, to give an appearance of evenhandedness, he called on Chiang Kai-shek to halt any military operations aiming at the retaking of China—an improbable eventuality, given the overwhelming strength of his adversary. Under Eisenhower, Taiwan and its dependency, the Pescadores Islands, became an offensive base or fortress which had to be defended at all costs, as was demonstrated by the signing of a mutual defense pact on December 2, 1954, allowing the United States to station ground, air, and naval forces on the island. Communist shelling of the coastal islands of Quemoy and Matsu provoked two serious crises. After the first of these, between September 1954 and April 1955, Eisenhower was given carte blanche by Congress to take whatever measures he considered necessary for the defense of Taiwan and the Pescadores. During the second crisis, which was shorter (August 23–October 6, 1958), he threatened to use nuclear weapons, following the "brinkmanship" policy favored by Dulles.[13] Such a to-do about two small islands without any real strategic value seemed absurd, as long as there was no chance of launching any assault to "liberate" Communist China, as the Guomindang put it. A return to relative calm did not prevent Washington from giving unwavering support to its ally which, as a consequence, was allowed to continue to occupy China's seat on the United Nations Security Council.

Completing the crescent of islands along the fringe of the Asian mainland, the Philippines, which had been independent since July 4, 1946, provided secure bases for the former mother country.[14] This had been accepted in principle in 1944 by President Quezon, in exile in the United States at the time, and confirmed by his successors Osmeña and Roxas (the latter a former collaborator with the Japanese, instantly rehabilitated by MacArthur). However, it was some time before the agreement was signed, for the Pentagon wanted to base ground troops in Manila, at the risk of creating friction with the local population, and was also asking for extraterritorial rights. For its part, the Filipino government tied the granting of bases to economic aid, even if this meant making an additional concession respecting "parity"—i.e., for equal treatment of American and Filipino investors.[15] Under the compromise finally agreed to, signed on March 14, 1947, and ratified without any opposition by the Senate of the new republic, the United States dropped its request for Army bases, since in any case it lacked the troops needed to man them. The United States was content with Air Force and Navy bases (Clark and Subic Bay, respectively), for which a 99-year lease was granted (reduced to 25 years in 1959), which it was free to use as it saw fit. The Philippines did make a concession where extraterritoriality was concerned, though in Europe the American forces enjoyed no such privilege.[16] The local elite, despite its penchant for corruption, was assured of military assistance in putting down any revolt, such as the one fomented by the Communist Huks among the downtrodden peasants in central Luzon. The close alliance between the elite and the United States thus continued, and was indeed further strengthened by a reciprocal defense pact (August 30, 1951). The Philippines, in the eyes of the Democratic administration, was an essential link in the chain of islands making up the United States' vital defense perimeter in the western Pacific[17]—a point of view shared by the Republicans, who were to find Ramón Magsaysay (1953–1957) an ideal president, extremely cooperative and capable of crushing the Huks, thus sparing the United States the necessity of supplying any substantial amount of aid or of making major concessions on the everlasting question of the bases.

Behind this first line of defense, the Micronesian archipelagoes captured from Japan, the Marshalls, Carolines, and Marianas, were spread out across the North Pacific over 35° of longitude. The United States held on to Guam and was granted trusteeship over the remaining islands by the United Nations. The Navy would have preferred an out-right annexation by right of conquest in order to further reinforce its positions along the lines of communication between the Far East and the South Pacific, but the Department of State felt this would be con-trary to the objectives of war as they had been formulated during the conflict with the Axis powers, and that it was out of the question to use one hand to undermine European colonial domination while grasping at new territorial gains with the other. A compromise was fi-nally reached in 1947. Micronesia and ten other colonies were made trust territories of the United Nations (which had inherited the League of Nations mandates), but with a unique status as "strategic area pro-tectorates." This represented a considerable advantage for the United States. Its administration was answerable only to the Security Council, where it had a veto, and in return for advancing the economic and social development of the native peoples it was allowed to fortify the islands, and to exclude any United Nations missions from them, in the name of defending its own security, and peace in general. No change in this status could be made without its consent. The system of strategic protectorates thus gave the United States exceptional privi-leges. It was completely free to exclude hostile powers from the islands, thus reinforcing its supremacy over the ocean.[18] In fact, Micronesia was not to play any major role during the Cold War years, for it was too far removed from the "front" along which United States forces were sta-tioned. It was principally used as a site for atomic tests in 1946 and early in the following decade. Even before the United Nations, with Russian consent, had approved the trusteeship, Bikini Atoll in the Marshalls had been used for this purpose. Eniwetok was used as a site for testing the hydrogen bomb, while the Pentagon developed a rocket-testing facility on Kwajalein. Strategic forces were stationed only on Guam, from where B-52s took off to pound North Vietnam. Around 1970 the American position in Micronesia seemed under no threat, despite a growing desire for autonomy, or even independence, among the native peoples.[19]

South of the equator, the American defensive perimeter was far-ther from the Asian mainland. Since the war against Japan, the United States had two loyal allies in the South Pacific, in Australia and New Zealand, though the latter felt less affinity with it because of its deep at-tachment to Britain. But the two Anglo-Saxon nations now found they could no longer count on the Royal Navy to protect them from the

Japanese, whom they still mistrusted, or against communist aggression, particularly after the Franco-British debacle at Suez in 1957.[20] Despite some displays of irritation at the relative indifference shown them by Washington, which was quite aware that the region was far removed from the points of tension, Australia and New Zealand set out to court the United States, sending troops to Korea, as they would do later to Vietnam. So, a week earlier, on September 1, 1951, to calm the anxiety caused them by the generous terms of the peace treaty it was about to sign with Japan, the United States agreed to sign the tripartite security pact which gave birth to ANZUS. But despite the fine words about unity and the need for collective regional security, the United States refused to make any binding commitment to intervene against armed aggression, agreeing only to consultations which might lead to whatever action was allowed "in accordance with its constitutional processes." ANZUS did not establish any military organization on the model of NATO, though periodic shared military exercises did reinforce good relations between the three Anglo-Saxon nations. All in all, the benefits seemed equally distributed. Australia felt reassured vis-à-vis an overpopulated Asia, and particularly Indonesia with its imperialist designs on New Guinea and Malaya. As for the Americans, after the 1960s they possessed the telecommunications and surveillance installations they needed to keep watch on Soviet movements, and which were located in a part of the world from which the USSR was excluded.[21] This was enough to make the South Pacific their private preserve.

But there was a gap in the defensive-offensive alignment stretching from the Aleutians to New Zealand, in the shape of Indonesia. There, indeed, Washington's policy came up against an extremely active nationalism. In Japan or the Philippines, however, this factor reared its head only sporadically, while in Taiwan and in the two Anglo-Saxon countries of the South Pacific it did not take the form of anti-Americanism.

The Fate of European Colonialism

In tune with his vision of a postwar world open to international trade, which he saw as the surest path to peace and prosperity, Roosevelt hoped to see an end to the European colonialism which had dominated Southeast Asia before the Japanese offensive. American anticolonialism was a mix of idealism and rational self-interest (for its objective was equal access to raw materials and to markets). It was inspired by the right to self-determination and the doctrine of the open door. From this point of view, the return of Britain to Malaya, of France to Indochina, and of the Netherlands to Indonesia, were viewed as only temporary, intended to allow these countries to achieve independence in the optimum conditions. But very soon Truman, his successor, encountered obstacles to these noble aims, and was obliged to lower his sights. First of all, the colonial powers refused to renounce their claims— for reasons of prestige, but also so that they could keep their hands on sources of raw materials and protected markets, which they considered invaluable at a time when economies starved of hard currencies were attempting to rebuild. Furthermore, these countries were indispensable allies in the Cold War which was developing in Europe. The dominant pro-European faction in the Department of State was concerned not to alienate them. Lastly, the local nationalist parties made no secret of their communist leanings. In Indochina, under Ho Chi Minh's leadership, the Vietminh was a Marxist-inspired movement which intended to transform social structures along Soviet lines. The more tensions grew with Moscow, the more the American leadership was led to moderate its anticolonialism in order to maintain a line of defense in Asia against what it considered a powerful attempt at subversion.[22]

The United States' dilemma can best be appreciated in the case of Indonesia's evolution toward independence.[23] The "Europeanists" in the State Department considered it a priority to support the Netherlands, even if they had no great liking for Dutch colonialism, while the "Asianists" were afraid that their country would be made to appear an ally of imperialism, thus playing into the hands of the USSR. Washington, unable to resolve this debate in favor of one side or the other, remained prudently on the sidelines, allowing the British to occupy the Indonesian archipelago in 1945, and, after the landing by Dutch troops, contenting itself with the Linggajati agreement (November 15, 1946), which established a federal structure combining an actual republic in Java and Sumatra, and semiautonomous territories in Borneo and in eastern Indonesia—an ambiguous document, whose faults soon became evident. When, in July 1947, the Netherlands launched military action against the republican nationalists, the United States adopted a neutral stance—an attitude which in fact favored the colonial power. But when the United Nations came down against the Netherlands, it was to endorse the compromise partitioning of Java, which was signed on board a United States ship, the *Renville*, on January 17, 1948. No durable agreement turned out to be possible between the Dutch federalist proposal and the nationalists' aspiration to create a unified state. In the meantime, throughout 1948, positions were hardening in the Cold War. The "Europeanists" defended the Dutch cause, seeing Holland as an important factor in European reconstruction under the Marshall Plan, while the "Asianists" were afraid that by not supporting the moderate nationalists the United States would help the Communist cause. In September and October, a communist-led revolt, though it was quickly crushed, showed that the danger was real. In addition, Dutch intransigence was becoming more and more intolerable to Washington, and this came into the open when, on December 19, the Netherlands Army launched its second "police action." Without going to the length of cutting off Marshall aid to Holland, the American government brought strong pressure to bear on the Dutch government to put an end to a polarization of whites and colored peoples which was dangerous in the long term, and which in the short term worked in favor of the USSR. The cease-fire agreement on August 1, 1949, opened the way to the round-table conference at The Hague, during which the American diplomat Cochran succeeded in finding a compromise acceptable to both sides. However, the newly independent Indonesia did not side with the free world, but opted instead for neutrality, a position Dulles considered immoral. In 1955, at Bandung, Indonesia hosted a conference of the leaders of a Third World, which had very little liking for the United States. Relations soon began to deteriorate. The CIA supported rebel troops in Sumatra and Sulawesi with the intention of making life difficult for Sukarno, who was aligning himself ever more closely with the Indonesian Communist Party. If Kennedy contributed to the handing back of West Irian, the last remaining subject of disagreement between the Netherlands and their former colony, relations took a turn for the worse when the authoritarian head of Indonesian "Guided Democracy" became involved in a "confrontation" with Malaya. When American aid was cut off, Sukarno responded by saying: "Go to hell with your aid!" So it was with some relief that in 1965 Washington viewed the failure of an attempted Communist coup d'état, which opened the way for power to pass to General Suharto, a fanatical anticommunist who would make the Indonesian archipelago safe from communist expansion.

On the Asian mainland, the Moslem principalities in Malaya and the kingdom of Thailand were also relatively reliable barriers to

communism. On the other hand, the departure of the French from Indochina seemed to open the door for Moscow to exert greater influence in Southeast Asia. In spite of Ho Chi Minh's overtures in 1945,
and its dislike of French colonialism, the leadership in Washington was
unable to bring as much pressure to bear on Paris as on The Hague.
France was too vital an ally in Western Europe for the United States to
force it into decisions contrary to its wishes. It had a veto on the Security Council, so it was impossible for the United Nations to take away
its authority over one of its colonies.[24]

This meant that until 1950 the United States had to restrict itself to
observing the difficulties encountered by the French Army, without
offering any assistance. The limited independence granted Vietnam on
March 8, 1949, gave little cause for satisfaction. But everything changed
with Mao Zedong's victory in China. Ho Chi Minh suddenly began to
look like a pawn in some vast communist plan to gain control of Southeast Asia.[25] The time had come to enforce the policy of containment
which seemed to have met with such success in Europe. On May 15,
1950, Truman granted France $15 million in initial military aid to fight
the Vietminh. The Korean War broke out a month and a half later, enabling Paris to claim, without renouncing its dream of empire, that its
soldiers were fighting in Indochina to defend the same cause, namely,
the defense of the free world against communist aggression. The dollars
flowed—$133 million in late 1950, $500 million in 1953, and as much as
$1.1 billion the following year, which was to be the last year of the In-
dochina War. The United States paid one-third of the cost of this war in
1951–1953, and 78 percent in 1954. Yet this massive financial aid was
unable to prevent the disaster of Dienbienphu. To oblige General Giap
to lift the siege on the French garrison, American Chief of Staff Admiral Arthur Radford, and Vice President Nixon, advocated the use of
B-29 bombers and possibly even nuclear weapons. General Ridgeway
was strongly opposed, as was President Eisenhower. At a meeting of the
National Security Council at the end of April 1954, where the question
was raised, Ike became irritated with his advisers and protested that it
would be sheer folly to use the atom bomb against Asians for a second
time in less than ten years. Dulles, in spite of all his efforts, failed to persuade the British to take part in a "joint action." There was, therefore, no
intervention by American forces, neither on the ground nor in the
air.[26] The Geneva Accords (July 21, 1954), sealed the fate of French Indochina.[27] Independent Vietnam, temporarily partitioned along the 17th

parallel, was to have been united in 1956, after free elections. But the United States refused to sign the accords, as did the new prime minister of South Vietnam, the anticommunist nationalist Ngo Dinh Diem. It was bent on preventing the fall of any more "dominoes," in the form of the neighboring kingdoms of Laos and Cambodia. This decision was to sow the seeds of the second war in Indochina—an American war, which took over where the French had left off, but one which would last as long, and the outcome of which would be no more positive.[28]

The Sino-Soviet Monolith

Revisionist American historians have often reproached Washington with failing to see that Ho Chi Minh was a nationalist rather than a communist. In fact, the Vietminh leader was both a nationalist *and* a communist. The error was rather not to have seen that attachment to a common ideology (with sometimes considerable local variants) does not necessarily mean sharing the same strategic and political positions. The "Cold War culture" destroyed any awareness of such subtle distinctions.[29] The specter of a communist bloc acting collectively under Moscow's leadership in order to achieve universal subversion left no room to take advantage of the divisions, which were nevertheless quite evident. America found itself obliged to protect its Asian marches, not only in the nearby archipelagoes, but also on those parts of the mainland which still escaped domination by international communism.[30]

Yet, in 1945, immediately after World War II, the situation still did not seem serious. For almost a half-century, China had been considered an ally, having earned the sympathy of many Americans, mostly Protestants. Chiang Kai-shek's conversion to Christianity encouraged high hopes. Instead of Japan, once an enemy and now humbled in defeat, Washington wanted to make China the major power in the region, capable of frustrating any designs Stalin might have had on the Far East. Unfortunately, American foreign policy was backing the wrong horse. This was not for lack of good advice, for General Stilwell and a number of well-informed advisers, such as the "three Johns"—John Stuart Service, John Carter Vincent, and John Paton Davies—as well as Owen Lattimore had shown it would be impossible to use Chiang Kai-shek, with his excessively corrupt, inefficient, and even fascistic regime, as the basis for a Chinese policy. The "Generalissimo's" only concern was to keep his faithful followers around him, at the expense of the peasants and even of a section of the new indigenous middle class. His trump card was, however, the fact that there was only one alternative—Mao Zedong's Communist party. But the Washington administration was more than reluctant to see the communists gain power, in spite of the advice from some experts who were able to see that Mao would never be Stalin's puppet.[31] In 1945, Roosevelt's personal envoy to Chiang Kai-shek, Patrick Hurley, adopted a firm anticommunist stance. The following year, his successor, the more impartial General Marshall, made a vain attempt to bring about a reconciliation between Chiang Kai-shek and Mao, but the truce was short-lived. When he became secretary of state in 1947, Marshall dispatched General Wedemeyer. His report, which was kept secret, though it was very critical of the nationalist regime, recommended saving it by sending massive economic aid and 10,000 American military advisers—an effort which the Democratic administration, being careful to reserve most of its limited resources for Western Europe, rightly considered ineffective and impracticable. At this point, nothing stood in the way of a communist victory in China.[32] Some government circles actually came to hope for such a result, without saying so openly.[33] Up to the outbreak of the Korean War, Acheson refused to commit the United States to the defense of Taiwan, for fear of losing prestige in defending a lost cause. He preferred to wait for the island to fall, so that the United States could recognize the new regime in Peking and explore any possible weaknesses in the recent Sino-Soviet alliance,[34] for Mao might turn out to be another Tito—one could never be sure. This moderate view was not universally shared by the Washington bureaucracy, nor by the Defense Department in particular. In addition, from 1950 on, the Republicans, eager to regain power and reverse the consequences of the New Deal, made the theme of the "loss of China" a major reproach to level at their Democrat adversaries. A powerful "China lobby," comprising influential senators, rich businessmen like Alfred Kohlberg, and noted journalists like Henry Luce and his wife, Claire Booth, employed systematic denunciation and intellectual terrorism to stifle any views which failed to idolize Chiang Kai-shek's regime.[35] These tactics succeeded in destroying the Institute for Pacific Relations, and the reputation of many established experts—in short, silencing any sources of lucid analysis, and, simultaneously, depriving American foreign policy of all subtlety or flexibility. This was at a point when the Korean War was providing a reason to adopt a more forceful attitude in the cause of maintaining its credibility—a Western version of the fear of "losing face" which is supposed to be a cause of so much humiliation for Asians.

Whatever revisionist historians may say, Washington was quite correct in thinking that China had deliberately taken sides with the Soviets following Mao's visit to Moscow, culminating in a pact of friendship, alliance, and mutual aid, signed on February 14, 1950, and that it had been providing aid to the Vietminh since January of the same year to help extend the territory held by communists. For his part, Mao was overfearful of the supposedly aggressive intentions of the United States, and this prevented any defusing of the tensions.[36] Korea became the stage for the military conflict which began in November 1950. Henceforth, the American leadership would consider the People's Republic of China an implacable enemy. They would block its entry to the United Nations, subject it to a trade embargo, and ensure Chiang Kai-shek survived on Taiwan, thus prolonging the standoff between two hostile Chinas. Should they have been able to detect early signs of the future Sino-Soviet conflict at this time? In all fairness, there was no reason to hope for any such development as long as the USSR and China were collaborating so closely, and Mao was adhering to the Stalinist model of development.[37] Certain historians see American policy under Eisenhower as an attempt to open up a fissure in the monolith by raising the economic and military stakes, particularly during the Quemoy and Matsu crises, so as to show Mao he should not expect a great deal from the USSR[38]—a stance which can just as easily be interpreted, it seems to me, as a desire to thwart Peking's revolutionary ambitions and throw a sop to the Republicans of the China lobby.

During two decades—in the 1950s and 1960s—the Asian seaboard of the Pacific was, together with the Iron Curtain in Europe and, to a lesser extent, the Middle East, one of the primary battlefields of the Cold War. There, the winds of détente were hardly felt. Officially, the United States' intention was to apply a policy of containment, or even to regain lost ground, wherever possible. The "domino theory" meant defending each and every domino, even if it had little strategic or economic value, for if one fell, all the others might well be toppled, one after the other, in a chain reaction. In the long term, not only would all Asia become communist, but the archipelagoes which were part of America's first line of defense would be lost in turn. In such a nightmare scenario, Mao's China was the primary agent of international communism, whether it was seen as a satellite of the Kremlin or as an even more dangerous opponent than Khrushchev's and Brezhnev's USSR because of its revolutionary messianism.

Far from the public stage, behind the walls of the Department of State, there was no shortage of thinking about how to bring about a crack in the monolith.[39] From 1955 on, when hopes that a Maoist form of Titoism might develop came to nothing, détente with the USSR seemed the more hopeful possibility. It seems Kennedy even briefly sounded out the Kremlin to see how it would react to the destruction of the Chinese nuclear test site, but to no effect. The Peking government became Public Enemy No. 1 for American foreign policy. This made it impossible to take advantage of the rift between China and the USSR which became visible in 1958 and official in 1960, for the Vietnam War reinforced the notion that in Asia it was Peking rather than Moscow which represented a threat to United States interests and which was fostering subversion. The Lyndon Johnson administration was further hamstrung by the accusation of having "lost China" which was still being leveled at the Democrats. Under these conditions, it was only a Republican—and, what is more, one who was considered to be strongly conservative—Richard Nixon, who was in a position to question the wisdom of attitudes which had been deeply ingrained by the Cold War.

"Hot" Wars: Korea and Indochina

If the Cold War took the form of a "long peace" between the two great powers who possessed nuclear arms, on occasions it turned into a "hot" war on the shores of the Pacific. For the United States, as for the USSR, Europe was the principal arena in which the future of the two rival systems would be decided, and where any military action could rapidly turn into a third world war. Each therefore restricted itself to its own sphere of influence, taking care not to provoke its adversary excessively. But the Asian seaboard of the Pacific lacked this central character—an indication that a "frontier" still existed, even if it was becoming ever more intangible.

In both Korea and Indochina the United States intervened militarily to halt what it considered to be invasions by the communist bloc. In each case, however, a civil war was already taking place. On the one side were revolutionary forces attempting to create a new society along Soviet lines (collectivization of the means of production, central planning, a single party, and suppression of fundamental liberties), and on the other the more or less conservative elements of society, who wanted to maintain private property but who were seldom genuine liberal democrats.[40] The United States bears part of the responsibility for partitioning countries which had originally been unified. In 1945, the United States and the USSR had agreed to divide Korea along the 38th parallel, similar to what happened in Germany. After bloody internal conflicts both occupied zones became transformed into authoritarian regimes ruled over by charismatic leaders, Kim Il-sung in the North and Syngman Rhee in the South. In Vietnam, the elections promised in the Geneva Accords were never held because of the opposition of Ngo Dinh Diem, backed by the Americans, and Washington's policy was to divide the country permanently along the 17th parallel. When this was done, any attempt by northern forces to restore national unity by the use of force could be interpreted as a foreign aggression rather than as a continuation of the earlier civil war.

In Korea, there was no doubt that the North was the aggressor. Skillfully playing his Chinese and Soviet protectors against one another, and to some degree with their consent, or at least with their military assistance, on June 25, 1950, Kim Il-sung launched his well-equipped divisions in an attack on the ill-prepared South Korean Army, which was only saved from a total rout by Truman's snap decision to come to its assistance under the UN flag.[41] In South Vietnam, on the other hand, it looked as if an internal revolutionary movement was responsible, namely the Vietcong, when it took up the struggle against the increasingly dictatorial and corrupt Ngo Dinh Diem in 1960. In fact, however, the Vietcong was merely the southern arm of the Communist party in the North. It had no policies of its own, and its hit-and-run strategy had been developed in Hanoi. Ho Chi Minh carried on his campaign throughout all of Indochina, and it was impossible for Laos to remain neutral, because the trails used to transport war materials to the communist fighters in the South passed through it. The American government therefore felt justified, under the Eisenhower administration, and even more so under Kennedy, and later Johnson, in sending military advisers, and, beginning in 1965, troops, to help roll back armed subversion from without.

Despite the voluntarily limited nature of their military involvement, the United States fought two of the major wars in its history in Korea and Indochina. The losses sustained were far from negligible. In Korea, 34,000 were killed, and 103,000 wounded in the space of three

294 THE "AMERICAN LAKE"

years, between June 25, 1950, and July 27, 1953. In Vietnam, 47,000
were killed and 153,000 wounded in eight and a half years, between
August 4, 1964, and January 27, 1973.[42] In financial terms, these two
wars accounted for almost the same percentage (14 percent to 15 percent) of the annual GNP at the time[43]—a much lower share than
during the two world wars or the Civil War. Actually, neither ever degenerated into total war, for, all in all, the Washington leadership
showed restraint. In Korea, if it allowed the success of the Inchon landing to go to its head and permitted MacArthur to advance as far as the
Chinese frontier on the Yalu—leading to the intervention by Mao's
Army and the disastrous retreat across the 38th parallel—it refused to
grant its daredevil general freedom to conduct operations as he pleased,
including even the bombing of the Manchurian sanctuary. He was
eventually relieved of his command in April 1951, to avoid the United
States becoming involved in "the wrong war at the wrong place at the
wrong time and with the wrong enemy," for the highest priority was
still to remain on guard against the Red Army in Europe.[44] In Vietnam
also, if the American Air Force bombed villages and dikes in the north
on several occasions, over varying periods of time, and pounded the
trails in Laos and Cambodia, it never attacked China, through which
all the materiel provided by the Vietcong's communist allies had to
come, and the Army never crossed the 17th parallel. The Washington
administration was unwilling to risk a more widespread conflagration
in a region where it had no vital interests at stake.

The Korean and Vietnam wars have numerous traits in common,
but the outcomes appear to have been very different. The former,
whatever the feeling at the time, was in the final analysis a success for
America, for the North Korean offensive was halted and the 1953
cease-fire line was almost identical with the previous border.[45] The
United States had given itself limited objectives, which the enemy had
been forced to concede. In Vietnam, on the other hand, even though
they were not defeated on the ground, the Americans lost the very
thing they had sent as many as 540,000 men to save in 1968, namely, a
noncommunist independent state. As was France's fate in Algeria, military victory was transformed into political defeat. For there to have
been any chance of total victory, as Colonel Summers remarked, President Johnson would have had to declare war on North Vietnam in
order to arouse American public opinion, or, if that was impossible because of international reactions, the Army would have had to restrict

itself to halting Northern infiltration by establishing control over the demilitarized zone along the 17th parallel, from the sea to the Mekong River—i.e., to the border with Laos—instead of dissipating its energy in battling internal subversion, a task which would have been better left to the South Vietnamese.[46] The Vietnamization policy adopted by Nixon in 1969 failed to put any such policy into effect. The progressive withdrawal of American troops, completed after the Paris Accords were signed on January 27, 1973, entrusted the country's defense to a South Vietnamese Army which was inadequately prepared, poorly commanded, and quite unable to resist the onslaught of the Northern divisions in 1975.[47] Within a month and a half, the eight years of American efforts were brought to nothing. As Ho Chi Minh had foreseen, the American people were not prepared to tolerate such a long ordeal with no tangible results to show for it. The Vietnam War affected them deeply, making them reluctant in future to join in any wars which did not have precise objectives, and resulted in an undeniable loss of prestige.

The Strategic Triangle (1972–1988)

The Soviet Threat in the Pacific

Toward 1970 it was becoming clear that the United States was beginning to lose some of its dominant position in international affairs. The USSR was gradually making up ground in the nuclear field, and seemed likely to catch up in the not-too-distant future. While the United States was bogged down in Vietnam and undergoing a moral crisis under which all its institutions were subject to scrutiny, the leadership in the Kremlin appeared self-assured: within their sphere of influence they were reestablishing the order which had momentarily been disrupted by the "Prague Spring," and in 1969 their troops were confronting the Chinese army in the Far East. In short, the USSR now had to be seen as the rising, and hence the most threatening power in the Pacific zone.

Up to this point, the Soviets had based their strength on their continental land mass, and had not aspired to any mastery of the seas, which meant that they were denying themselves the ability to push far beyond their own frontiers in order to defend their own interests, or those of communism in general. Conscious of this weakness, the Soviet Navy, after the fall of Khrushchev, was busy building up a fleet able to attack enemy lines of communication, using the submarines based at Vladivostok and Petropavlovsk as its mainstay. From these ports they could fan out all over the western Pacific. Previously, their lack of support stations and friendly ports had prevented their venturing into tropical waters. The North Vietnamese victory in 1975 enabled them to take over the former American base at Cam Ranh, offering a direct threat to the American military installations in the Philippines.[48] Fishing and merchant vessels were used for espionage, and, to facilitate refueling and repairs, the USSR signed treaties with those archipelagoes which had won their independence from the European colonial yoke. Fishing rights were granted in exchange for economic aid. When early approaches to Tonga in 1976 failed because of Australian objections, the USSR gained a foothold in the Cook Islands (1978) and Kiribati (1985). The Sandinista revolution in Nicaragua in 1979 gave it the opportunity to establish its influence in Central America. By the time Gorbachev came into power, the USSR seemed to have broken out of its maritime encirclement to venture into the Pacific.

Analysts at the time insisted on the increased power of the Soviet fleet in the Pacific—the largest of the four fleets Moscow had stationed on various seas or oceans. Yet it was in no position to confront the American Navy, or to offer it as serious a threat as Japan had done around 1940. Its range of vessels was much smaller, and included no attack aircraft carriers. In winter it was contained within the Seas of Okhotsk and of Japan, which are normally icebound in that season, and from which in any case it could only exit through passages which were easy to keep under surveillance. If it was to become a great naval power in the Pacific, the USSR would have had to undertake an extremely expensive shipbuilding program, and this the increasingly desperate state of its economy would not allow. This was why, on July 28, 1986, Gorbachev broke with his predecessors' policy. In a speech delivered at Vladivostok, dealing basically with the imminent Soviet withdrawal from Afghanistan, the reduction of tension on the frontier with China, and nuclear weapons control, he proposed a long-term cooperation with Japan for the use of the seas, suggesting that Vladivostok, until then a closed port, be transformed into "our window opened widely on the East." Moscow would support the efforts of the countries of the South Pacific to create a nuclear-free zone, and suggested holding negotiations to reduce naval activity in the Pacific—"in particular, nuclear-armed ships."[49] It was a clever speech which, although it was of course an effort to deny the United States an important element in its superiority, did display a desire for de-escalation, and introduced a change in the political climate which was soon to bear fruit in progress toward disarmament. After 1987, the Soviet threat in the Pacific faded away. This was one consequence of American strategic renewal under President Reagan, and of the new triangular international relationship which came into being in 1972.

The China Card

After two decades of ostracizing Communist China, the United States had indeed made a complete about-face. Despite the spasms of the cultural revolution agitating the radical Maoists, the new Republican administration abandoned its earlier way of thinking, which had ruled out any kind of rapprochement. President Nixon and his adviser, Henry Kissinger, conscious of the relative decline of the United States vis-à-vis the USSR, rejected ideological posturing in favor of taking the difficult steps required by the national interest. This meant that any enemy of the Soviets became a potential friend of the Americans, whatever the political or economic regime in power. It so happened that Sino-Soviet antagonism was growing constantly, fueled by ideological, territorial, economic, and strategic conflicts. By this time, the rupture seemed complete. So why not conclude at least an informal alliance with Maoist China? The United States had no real bone to pick with China, while the policy of Vietnamization and the Guam doctrine meant that there was no more than the remotest possibility of a confrontation between the two countries. By making allies of the swarming Chinese masses, the Americans would tie down a considerable number of Red Army divisions, and oblige Moscow to devote more attention to the situation on its frontiers instead of expanding its power in the Pacific.[50] If Washington was playing the China card, the Peking government, on the other hand, needed American support. With equal realism, it toned down its diatribes against Yankee imperialism in order to take better aim at the "social-imperialism" of its former Soviet partner. Mao knew perfectly well, in spite of his posturing, that China was relatively weak in comparison with the superpower with which it shared a long frontier. So he, too, was tempted to play the

American card against Moscow. This opened the way to a reversal of the situation as spectacular as the one brought about by the German-Soviet Pact of 1939.

After a phase of secret diplomacy, the situation evolved rapidly in 1971. In April, Peking invited an American table tennis team touring Japan to come to China. Kissinger made two trips in July and in October, the latter just as Red China became a member of the United Nations, occupying the seat previously held by Taiwan, over American objections. An invitation was issued for President Nixon to visit between February 21 and 28. At no time did the United States consult its Japanese ally on this about-face, and Japan, with little reluctance, was obliged to reverse its own policy. In this new strategic alignment, however, China and the United States remained far apart on a number of important points. The "Shanghai Communiqué," signed on February 28, 1972—i.e., eleven months before the cease-fire treaty in Vietnam—made no attempt to paper over these differences.[51] After a unilateral exposition of its positions by each party, the Chinese and the Americans recognized the five principles dear to Zhou En-lai which were to provide a basis for their mutual relations, and condemned any attempt to achieve supremacy (this was aimed at the USSR, without naming it). But a substantial obstacle remained, in the form of Taiwan. Peking strongly affirmed that there was only one China, and that Taiwan was a province only temporarily removed from the mother country's authority. For his part, Nixon, who was still an ally of Chiang Kai-shek, "acknowledged" that all Chinese, whether on the mainland or on Taiwan, rejected the concept of "two Chinas," and allowed that the matter would have to be peacefully resolved by the Chinese themselves. As a gesture of goodwill, he announced that American military forces would begin a withdrawal. The verb "acknowledge" is ambiguous, but it did represent accurately the positions taken by both sides, for Peking could read it as meaning that the United States was showing some sympathy for the Chinese interpretation, favoring reunification, while Washington was actually taking a neutral position.[52] But because of such ambiguities, the ten years which followed the Shanghai Communiqué were not marked by any displays of genuine amity.

The new Sino-American relationship was dominated by strategic considerations. The two countries had the same enemy, in the USSR, but lacked even one fundamental value in common. It was a marriage of convenience, rather than a passionate relationship.[53] Neither of the two partners was willing to make much of a commitment to the other, for the new triangular situation introduced an element of instability, with each nation retaining its freedom to act without being able to exercise any real control over the other. The possibility of deception kept the basic distrust alive. The United States did not want to give any firm assurances to China, to avoid becoming involved in any escalation of the confrontation between Moscow and Peking, but neither could it exclude the possibility of a Sino-Soviet rapprochement which would be disadvantageous to it, for, between the end of the Maoist era and the accession to power of Deng Xiaoping, the political situation in China seemed extremely unstable. For its part, the Peking government was unable to place complete trust in a partner who had been conducting such an irresolute foreign policy ever since its withdrawal from Vietnam, and even more so after the Watergate scandal, which had weakened the presidency. The Chinese were nervous that in the 1970s Soviet concessions might tempt the Americans to accept a degree of détente. On the other hand, President Reagan's Cold War rhetoric, early in the following decade, caused them some uneasiness, for fear they might be drawn willy-nilly into a confrontation without their having very much say in the matter. Such an atmosphere was not conducive to a long-term relationship, particularly since the Chinese Communist party preferred to limit severely any economic, cultural, or military interchanges in order to avoid domestic unrest.

Nevertheless, an important obstacle was removed when, on January 1, 1979, under the Carter presidency, the United States and China established diplomatic relations and exchanged ambassadors. In addition, the Americans recognized that the Peking government was the only legal Chinese government, and that there was only one China, of which Taiwan was part. The mutual security treaty of 1954 was revoked, so that relations with Taiwan would henceforth be conducted on an unofficial basis. Congress, more distrustful than the president of China's ultimate intentions, insisted on a firmer commitment to the former ally, and the law of April 10, 1979, dealing with relations with Taiwan, specified that any attempt to resolve the fate of the island by other than pacific means, including a boycott or an embargo, would be considered a threat to the peace and security of the western Pacific and to American interests. Consequently, the United States would supply Taiwan with defensive armaments in sufficient quantities, and consider responding to any attack by communist troops. The new relations with the "Taiwanese people" (no longer called "Chinese") were entrusted to an American Institute in Taiwan (AIT), a nongovernmental agency staffed by employees seconded from the State Department—an embassy in fact, but not in name. Taiwan, for its part, was represented in the United States by a Coordinating Council for North-American affairs.[54] The question of arms sales, and of warplanes in particular, remained very sensitive. It was of little significance that the United States stated their preference for "one China" and rejected any solution along the lines of "one China, one Taiwan," for the Peking government feared that once the old generation of the Guomindang had died off, those on the island who favored independence would have the wind in their sails, with the result that any improvement in their military strength would encourage them in their aspirations. In 1982, when Reagan's rearmament policy was beginning to take effect by weakening the Soviet threat, Peking began to feel there was room for it to protest against any sale of modern FX fighters to Taiwan. The American administration, fearing the loss of a valuable advantage, was obliged to make significant concessions in the communiqué of August 17. It agreed that American sales of armaments to Taiwan would not be increased qualitatively or quantitatively beyond the level prevailing since 1979, and even announced the intention to reduce them gradually in the future, as long as China adhered to a policy of peaceful reunification.

This laid the foundation for a marked improvement in Sino-American relations for the remainder of the decade. The strategic alignment based on an essentially negative objective—to prevent Soviet hegemony—gave way to a much more positive vision, encouraged by changes in the USSR under Gorbachev. The United States declared its willingness to cooperate in the modernization of China, while Deng Xiaoping pressed on with economic reform, for which foreign aid—especially American—was required. The newfound friendship between the two countries went through a kind of honeymoon period of a few years, during which commercial and cultural exchanges became more numerous and the United States helped to modernize China's defensive armaments. However, things did not go as far as the creation of a formal alliance,[55] for neither of the two partners wanted to have their hands completely tied. The differences between the two economic and political regimes were still substantial, giving rise to an insurmountable conflict between the American desire to see respect for human rights, and the old communists' fear they would lose control if they relaxed their dictatorship. The fragility of Sino-American relations became evident in 1989 after the bloody repression of the democratic movement of the "Peking Spring."

American Reactions

The China card was played by the United States as a partial response to the Soviet intrusion into the Pacific. The United States' options were limited by its national interests. Since 1970, the region had been playing an increasing role in United States economic activity, rising from a position of secondary importance into one of prominence, finally surpassing even Western Europe. It was, therefore, impossible for the Americans to disengage by falling back on the eastern part of the ocean. Their failure in Indochina made them unwilling to consider any further military interventions on the Asian mainland, but never caused them to question the need for mastery of the ocean itself. However, in the 1970s, some pessimistic analyses suggested that the American Navy had lost its supremacy, and indeed, during the Vietnam War and later, under the Ford and Carter presidencies, funding for naval construction was substantially reduced. In 1972, the Navy had in all 448 large warships (i.e., more than 250 ft. in length), which was fewer than in 1960, when it had 472. By 1978, the number had fallen to 289, and was at the lowest point since 1941, even though the USSR maintained a much greater number. It is true that, under Admiral Rickover, the United States had a crushing advantage in nuclear propulsion, but the reduction in the number of ships in service made it impossible to conduct a war effectively in several oceans at the same time. In spite of this, however, in 1974 the Pacific was the only ocean where the American fleet had an advantage over the Russians in the number of ship-days spent outside their home waters.[56] To make the best use of his slender resources, Carter endorsed an essentially defensive strategy of "swinging" the Pacific Fleet over to the Atlantic in order to protect shipping routes, for the defense of Europe was still considered the top priority. The Reagan administration abandoned this way of thinking in 1981. New Navy Secretary John Lehman laid down what was then termed "*the* Maritime strategy," based on close cooperation between the four armed services (Navy, Army, Air Force, and Marines). The slogan "a 600-ship Navy" was intended to encourage efforts leading to the introduction of a much broader policy in which warships would not only be used to protect lines of communication, but also to attack an enemy. The American fleet would be organized around fifteen groups of aircraft carriers. In the Pacific, the Seventh Fleet would always have two of these available, without any need to transfer support to the Indian Ocean or to the Persian Gulf, as had been the case under Carter. The "maritime strategy" finally signaled the end of the Pacific frontier.

Mastery of the seas implies total freedom of movement. However, nuclear weapons were a source of great anxiety for many Pacific countries, concerned about the possibility of radioactive contamination. In 1984, the Labour government in New Zealand refused to allow any warships carrying them into its ports. The United States was irritated by such an uncooperative attitude on the part of an ally, and since, in the interests of maintaining the secrecy indispensable to its global strategy, its policy was neither to confirm or deny (NCND) that its ships were carrying the contentious weapons, it suspended the guarantees provided New Zealand under ANZUS. The same fear of nuclear weapons led the South Pacific nations to sign the Treaty of Rarotonga, on August 6, 1985, establishing a nuclear-free zone. Each signatory committed itself to preventing the stationing of any nuclear explosive device on its territory, but remained at liberty to decide whether to allow visits by foreign ships and planes.[57] The United States was invited, together with the four other permanent members of the UN Security Council, to ratify the additional protocols, but refused to do so, in order to preserve its freedom of maneuver in the South Pacific.

This desire for independence on the part of New Zealand and the member nations of the South Pacific Forum was, when all was said and done, no more than a minor inconvenience for American strategic interests, for the region was situated well back from the front lines of the Cold War. Nevertheless, the United States was determined to hold on to its assets in Micronesia—the rocket launching site on Kwajalein, the training center for jungle warfare and amphibious landings on Palau, as well as any potential bases it might need should it be obliged to abandon those it had in Japan, Taiwan, or the Philippines. The essential concern was to prevent an enemy from getting their hands on them. This explains why decolonization advanced at such a snail's pace. The Americans kept their grip on the former Japanese mandates by proposing various solutions to the inhabitants.[58] Commonwealth status, as applied in Puerto Rico, would have been preferred by the administrations of the trust territories. The twenty thousand islanders of the Northern

Marianas (Rota, Tinian, and Saipan), which were inhabited by Catholic Chamorros, with ties to Guam and with every intention of continuing to benefit from American financial largesse, chose commonwealth status in a referendum held in 1975. This made them American citizens, while leaving them a large measure of local autonomy. The hundred thousand islanders of the Federal States of Micronesia, created in 1979, opted for an agreement entering into free association with the United States. This was ratified by a plebiscite in June 1986.[59] Washington took the opportunity to terminate unilaterally the trusteeship it administered on behalf of the United Nations. This free association agreement allowed the Americans to take care of the diplomacy and defense of the new federation for at least fifteen years. The same arrangement was made for the forty thousand Marshall islanders, which split off from the rest of Micronesia in 1979, although they chose the same status, with special security provisions applying to Kwajalein for thirty years. Only the fourteen thousand inhabitants of the Republic of Palau remained under a direct mandate. However, there was a contradiction between their Constitution, which prohibited nuclear weapons unless 75 percent of the population approved, and the Compact of Free Association ratified by Congress in October 1986. The United States was determined to avail itself of all the resources necessary to defend the archipelago. It was therefore left to the islanders to resolve the contradiction—apparently a delicate matter, since a dozen referendums failed to provide a solution, while allowing them to continue to enjoy the financial benefits of Washington's trusteeship. The American style of decolonization, by virtue of the close links maintained with the new nations—the Federal States of Micronesia and the Marshalls—united friendly countries in regional groupings such as the South Pacific Forum, with the result that, indirectly, the United States could continue to have a say in their affairs.[60]

Despite the rearmament programs embarked on by Reagan, and the consolidation of their rear, it was no longer possible for the Americans single-handedly to protect the Pacific from communist aggression with the 180,000 soldiers and sailors permanently stationed there.[61] Under the Guam doctrine formulated by Nixon in 1969, Washington called on its allies to share the burden, urging them to make more effective use of their human resources to ensure their own security, and halting all military aid to the better off among them. The end of the Vietnam War deprived the Pacific zone of the priority accorded it (in addition to the

Middle East) in American aid policy throughout two decades (cf. Appendices, Table 14).[62] Taiwan was cut off from gifts or special loans after the rapprochement with Red China, but the Taipei regime no longer needed these, as was also the case for South Korea after 1987. These two countries now had enough wealth to buy whatever armaments they required, and were numbered among the best customers of American arms manufacturers, along with Australia, Thailand, Singapore, and Japan. The Washington administration tried to pressure its Japanese ally to cease sheltering behind Article 9 of the Constitution imposed on it in 1947, which forbade it to maintain an Army. The Self-Defense Forces which had taken the place of an Army since the Korean War were limited in scope, since the defense budget was not allowed to surpass 1 percent of the GNP, whereas the United States devoted 5.2 percent of its budget to this purpose in 1980, and as much as 6.5 percent five years later. Japan gave the impression it was cutting costs in the area of security, enabling it to devote all its energies to economic expansion, at the expense of the nation acting as its protector. In its great majority, the Japanese population supported a policy ruling out any intervention abroad, even if there was still some nostalgia for the imperial era. As long as détente with the USSR continued, the Japanese preferred that their government keep its distance from the United States. But when the Cold War flared up again in 1979, they supported maintaining a special relationship with Washington, for only the United States could guarantee the security of the archipelago against its powerful Soviet or Chinese neighbors. The Liberal-Democrat Prime Minister Nakasone, a straightforward nationalist, then advocated strengthening the Japanese defense forces to allow them to play a role in protecting shipping routes communication up to over a thousand miles from their coast, and so bear its share of the burden. In 1975, Japan still stood only eleventh in the world in military expenditure. It was sixth in 1987, catching up on France, Britain, and West Germany. But despite this financial effort, the Americans still felt Japan was not doing enough in proportion to its wealth, and that it was not accepting enough responsibility for the western Pacific. However, this view was not shared by countries which still had negative memories of Japanese imperialism, most notably China, nor indeed by some American analysts who feared that in the long term a resurgence of the warlike Japanese mentality might add to the already strong tensions in the area of economic relations, as it had done before 1941. A giant in production and in trade, Japan lacked either the will or the means to emerge entirely from behind the American military shield, even if this meant it had to acquiesce to abrupt changes in direction over which it had very little say.[63]

THE END OF THE COLD WAR

The end of the Cold War was one of these unpredictable changes. In 1989, it became obvious that Gorbachev had given up on confrontation with the West and that a new era in international relations was beginning, based on agreement between the two great nuclear powers. Even more unforeseeable was the USSR's collapse under the weight of an economic crisis, and its breakup in December 1991, when an eruption of nationalist passion revealed the artificiality of the union which had until then been held together by an iron-fisted dictatorship. Russia inherited the immense Siberian territories, with their Pacific seacoast, but it was no longer a threat, not only because it needed Western cooperation, but also because its badly maintained and demoralized Navy was prevented for the foreseeable future from pursuing the expansion into the Pacific it had begun two decades earlier. Russia, with its arsenal of nuclear missiles, could still make a nuisance of itself, but it no longer presented any real threat to American supremacy. The United States was now the only superpower in the Pacific, and the only one capable of deploying its fleet across its entire expanse. It really did seem that the ocean was an American lake once again.

And yet, the end of the Cold War has not ensured American hegemony. Actually, the greater the international tension, the easier it has been for the United States to maintain its leadership, for its allies need it too much to develop an independent policy. However, now that the Soviet threat no longer exists, they see America's relative strength as an obstacle to their own objectives. The mortar holding the alliances together has tended to crumble as the need for them has diminished. In addition, the leaders in Washington, the Republican George Bush and, even more so, the Democrat Bill Clinton, have been unwilling to take all the world's problems onto their shoulders. The American people have expected them to give their foremost attention to finding a solution to economic and social problems, and support intervention abroad only when the country's vital interests are at stake—as in the 1991 Gulf War against Iraq. It has wanted to cash in the "dividends of peace," the first signs of which have been a decline in military expenditure, a reduction in personnel, the decommissioning of numerous warships, and the end of the naval strategy so dear to Ronald Reagan.

All around the Pacific, feelings toward the Americans have remained as ambiguous as ever. On the one hand, there is resentment, and sometimes irritation, at the extent of their power, which it is feared they will use in their own exclusive interest. On the other hand, whenever they pull out, or even threaten to do so, anxiety grows at the thought of losing a policeman whose presence has been essential in maintaining a semblance of order. The United States is the only power capable of playing the role of a regional counterbalance, for it makes no territorial claims, nor does it aspire to hegemony. Its strategic objectives are modest: to maintain the *status quo*. It is therefore viewed positively—not, as Robert Scalapino has remarked, because it is the most loved, but the least feared. The western Pacific is indeed a region rich in potential conflicts, even more so than Europe. Territorial disagreements exist between most countries. Japan is determined to recover the four southerly Kurile Islands (the "Northern Territories"), taken from her by the USSR in 1945, but Boris Yeltsin, because of the strong nationalist feelings in the former Soviet empire, has been unable to yield any ground on this point. Nor has Russia reached a final agreement with Peking regarding its far eastern frontiers. In the South China Sea, the small islands, supposedly rich in potential oil deposits, or considered to have strategic value (the Paracel and Spratley Islands), are the object of conflicting demands by China, Taiwan, Vietnam, the Philippines, and Malaysia. The Sino-Vietnamese antagonism which resulted in a brief confrontation between the two countries in February 1979, after Hanoi's troops invaded Cambodia to get rid of their Khmer Rouge adversaries in Phnom Penh, has been simmering constantly. Even more dangerous is the situation in Korea. Despite all efforts, reunification of the country seems as far off as ever. Communist leader Kim Il-sung, who held dictatorial power in the North until his death in July 1994, was unable to accept democratic elections as a step toward reunification, for his regime would have been totally obliterated. He attempted to transfer his power to members of his family. As the North Korean economy was in ruins, he counted on possession of nuclear arms to give him some weight in the international arena, thereby running the risk of creating a profound instability in the whole region.

At the beginning of the twenty-first century, then, there is no shortage of risks in the western Pacific, apart from North Korea's atomic policy. Many fear that Communist China will attempt to become a dominant regional power once again. People point to its efforts to strengthen the People's Liberation Army, providing it with modern armaments and the means to intervene outside the Asian mainland.[64] A conflict with Taiwan, where, since the liberalization of the regime, apologists for independence have been able to speak freely, is a distinct possibility. In the circumstances, an American presence is strongly desirable. By virtue of their naval power, only the United States can play the role of peace brokers to prevent one of the countries in the region from threatening the independence of its neighbors. In opposing Vietnamese imperialism in Cambodia, the Americans have been able to call on support from China and the other countries of Southeast Asia. Their joint pressure was successful in bringing about a withdrawal of the invading troops in 1989, but it was not until the final collapse of the Khmer Rouge nine years later that something resembling peace was restored to the former kingdom, where hostilities had been ongoing ever since Washington included it in the sphere of hostilities in 1970, during the Vietnam War. Similarly, the restoration of diplomatic relations with the communist government in Hanoi did not occur until two decades after the departure of the last American troops. The problem of U.S. soldiers still listed as "missing in action" (MIA) postponed the lifting of the embargo for some time. Finally, in 1994, the Clinton administration put an end to a somewhat inglorious episode, the memory of which could not be allowed to delay readjustment to the new strategic order for much longer.

The end of the Cold War might have made the whole system of forward military bases established after 1945 redundant. In the Philippines, the Americans would have liked to hold onto the naval base at Subic Bay and Clark Air Force Base. However, when the eruption of Mount Pinatubo made the latter unserviceable, the 1991 negotiations only concerned Subic Bay. A treaty which would have allowed the Americans to use it for 10 further years in return for a rent of $203 million per annum, and substantial economic aid, was rejected by a majority of the Philippines Senate on September 16, despite appeals from President Corazón Aquino. In other times, this might have been considered a major setback, but it was not viewed as such by the Americans, since the Russians were no longer using their base at Cam Ranh in Vietnam, and the Philippines no longer possessed unique strategic value.[65] To compensate, Singapore, a Chinese city which felt somewhat isolated, with Malaysia surrounding it, offered facilities to the American Navy. In 1991, almost 45,000 troops were stationed in Japan and 40,000 in South Korea. In 1977, President Carter, as a goodwill gesture to help further détente, had thought of withdrawing all United States troops from the Korean peninsula within the next five years, much to the distress of the Seoul regime. A more realistic assessment of the situation led him to revise this decision in August 1979, when the Cold War reintensified. The presence of American soldiers in South Korea and their participation in military maneuvers ("Operation Team Spirit") played an important part in containing the threat posed by the motorized divisions of the Pyongyang regime, which was still suspected of wanting to reunify the country by force, as it had attempted to do in 1950.[66] The presence of American troops also provided a bargaining chip to help persuade the Communist dictatorship to accept international inspection of its nuclear installations. In the face of the threat from North Korea, Japan could not do without American protection, unless it was itself prepared to undertake a nuclear armament program quite incompatible with its constitution, and which was likely to cause anxiety among all its neighbors, as well as profoundly destabilize the region. The United States was welcome for the purpose of maintaining peace, less by playing an active role than by simply being there.[67] Should it ever lose its bases in Japan, it could still retain a forward defense position on Guam and on the new island states, which still enjoy United States protection after four decades. On the other side of the Pacific, the Panama Canal Zone was still controlled by American forces, with 10,000 men stationed there permanently. However, under the treaties ratified in June 1978, total sovereignty over the Canal Zone was restored to Panama in the year 2000. This, however, does not pose any real threat to the United States strategic interests in the Pacific.[68]

The principal problem for the Americans in the Pacific is no longer a military one. With the end of the Cold War, economic power has come to play a more important role than military might. More meaningful requirements for leadership are a prosperous business environment, innovative research facilities, and an educated populace, rather than armored divisions, warships, and fighter planes. The United States is the only military superpower in the Pacific and in the world, but it makes little use of this strength apart from bringing reckless dictators like Saddam Hussein to their senses. In any case, it shows no indication of wanting to use it to expand its hegemony. At most it exerts a certain amount of blackmail by suggesting that in the future, unless it is granted trade concessions, it may pay less attention to security in the western Pacific. But it is doubtful that this threat is taken seriously by the Pacific Rim countries, for American economic interests are too important in the region for the Washington government to give up its active presence and fall back on the "western hemisphere" alone. Despite the fact that the waters of the "American lake" can be turbulent at times, the United States must continue to patrol them, for if it did not, a major storm might well blow up.[69]

ELEVEN

The Great Economic Transformation

In economic terms, if the Pacific long remained a frontier for the United States, this was because trade relations across it were unbalanced. On one side of the ocean there was an increasingly powerful country, thanks to its demographic growth, its high standard of living, and level of industrialization, producing capital goods and manufactured mass-market consumer goods, and, on the other, still developing states and colonies producing raw materials or tropical commodities. This was a valid description even for Japan, the only country capable of catching up to the leading pack of industrial nations, but which was content to sell the Americans partly finished goods and to supply world markets with cheap merchandise of mediocre quality. However, this relationship began to change in the 1960s. On the economic plane, the following decades would witness the extraordinary rise of several Far Eastern countries. Japan very soon joined the club of great modern powers, and even challenged American leadership in certain respects, as did the Western Europeans. South Korea, Taiwan, Singapore, and Hong Kong took the same path a little later, as more recently did Communist China. The Pacific's integration into the world economic space on an equal footing with the Atlantic—some spoke of its supremacy, or even of its future hegemony—consummated the disappearance of the frontier.

Growing Trade and a Worsening Trade Balance

Since the end of World War II, commercial exchanges between the United States and the Pacific area, measured in current dollars, have grown at a dizzying rate. Between 1948–1949 and the end of the twentieth century, imports have risen from around $1.5 billion to $200 billion, and exports from $2.3 billion to $130 billion (cf. Appendices, Table 15). If we discount the effects of the heavy inflation which has prevailed during the past half-century, resulting in a fivefold increase in prices, we still find a much more impressive boom in trade relations than anything seen during the two earlier chronological periods.[1] Between 1950 and 1990 imports grew twenty-seven times, and exports fourteen times, corresponding to an annual average growth rate of 8.6 percent and 6.9 percent, respectively.[2] This accelerating trade in goods and services transformed the Pacific into an economic zone no less vital to United States interests than the Atlantic. By the end of the 1970s there was no longer a frontier in the sense in which we have been using the term. The growing significance of the Pacific can be seen even more clearly by its steadily increasing share of American external trade. Between 1947–1949 and 1989–1991 its share of American exports increased from 18 percent to 30 percent, and its share of imports from 23 percent to 40 percent. Recall that in the 1930s the percentages were 21 percent and 32 percent, respectively. So if World War II and the years immediately following it saw a relative decline in the flow of trade with the Pacific zone, more noticeable in imports than in exports, there was subsequently a return to the earlier volume, starting in 1970 in sales to, and in 1982 in imports from, the region, with both showing continuous growth. But, over the long term, the predominant place occupied by the Pacific in American external trade has been a recent phenomenon, for a good number of the postwar years were actually spent recovering to the high levels of the past. This makes the boom even more spectacular, and seems to justify the speculation that the twenty-first century will be the "Pacific century." This could indeed turn out to be the case if the trends of the past four decades continue in a straight line!

Traditionally, the United States has always run a trade deficit with the Pacific region. In this respect also, World War II brought about a profound change. Until 1967 the Americans sold more than they bought in the region, except for five years, three of which were during the Korean conflict.[3] Beginning in 1968, on the other hand, not only did the trade deficits return, but they became ever larger, contrary to the nineteenth- and early twentieth-century trend of an improving trade ratio. After the alert sounded in 1971–1972, this was generally reduced to below 70 percent by 1976, and even to below 50 percent in 1985–1987. The emergency resulting from the first oil crisis in 1973 demonstrated how much the American economic position had deteriorated relative to a region with which trade was increasing.[4] Under the Reagan presidency, the trade ratio stood at about the level it had been at the end of the nineteenth century, before the large-scale export of manufactured goods began. The trade deficits with the Pacific were so substantial that the surpluses with other trading zones were not enough to make up the difference, with the result that American foreign debt began to increase very rapidly, turning the major world power into a net debtor instead of creditor.

These trends are so marked that coincident fluctuations are of little significance. Imports, which depended on American economic prosperity, suffered slightly from the recessions under the Eisenhower presidency in the 1950s, and rather more severely as a result of the two oil crises, but this did not prevent them from resuming their extraordinarily rapid growth under Reagan. As for exports, they were sustained by a high demand in the developing countries, who were in the process of industrialization. Exports tended to grow more rapidly after the end of the 1960s, with slightly more marked fluctuations, caused either by the economic slowdowns due to the oil crises, or between 1981 and 1985 by the inflated value of the dollar against other currencies (cf. Appendices, Graph 4). From long-term point of view, these were only temporary circumstances which have little effect on the overall impression of ultrarapid and relatively continuous growth.

The commercial geography of the Pacific also underwent profound changes. Most important was a marked decline in trade with the Latin American countries of the eastern Pacific seaboard. There is nothing surprising in the fact that by far the greater proportion of imports came from these countries during World War II, for their location at a great distance from the theater of operations made them a secure source of raw materials and tropical produce to supply the "arsenal of democracy." They held on to this supremacy until 1958 (cf. Appendices, Table 15), even accounting for half of all imports from the Pacific in 1954, and until the same year they were also the principal market for goods manufactured in the United States. But then, within a single decade, their role became marginal, so that in 1990 they accounted for only 5 percent of the imports and 8 percent of the exports. This huge relative decline justifies the fact that present-day analysts, when they speak of the Pacific, only consider the western half of the ocean, and usually omit Latin America. However, we must not forget that forty years earlier things were very different. It was the Far East which made up the ground lost by Latin America. Oceania—essentially Australia and New Zealand—remained of minor significance because of its small population, while it offered the advantage of allowing the United States to build up a regular trade surplus with it after 1954.[5] As the end of the twentieth century approached, Japan, Southeast Asia, and the "Four Little Dragons" (South Korea and the three Chinese islands of Taiwan, Hong Kong, and Singapore) accounted for a preponderant share of United States trade with the Pacific. In 1949, on the eve of the Korean War, they accounted for 57 percent of exports and 45 percent of imports, and in 1990–1991 for 80 percent and 83 percent, respectively. They were responsible for the deterioration in America's balance of trade. By 1950, Southeast Asia had recovered its traditional surplus position (except during the Vietnam War, between 1966 and 1973), but in relative terms its market share shrank considerably from the 1950s on, to the advantage of the countries which had begun the process of industrialization earlier. Beginning in 1965, in every year Japan sold the United States more than it purchased from it. In spite of a rapid increase in Japanese imports from the United States, these were increasingly outstripped by exports, so eager had American consumers become for Japanese products. Between 1984 and 1989 the trade ratio finally fell below 50 percent and even dropped to 33 percent between 1985 and 1987—accounting for the increased tension over trade between the two countries arising from the enormous trade deficits, which in some years came to around $50 billion, and as much as $60 billion in 1993.[6] By the early 1960s Japan had recovered the special position at the center of trade between America and the Pacific which it had first achieved at the end of the nineteenth century. After 1965 approximately half of United States imports came from Japan, and almost 40 percent of imports were destined for it. In the wake of Japan, though some way behind, came the "Four Little Dragons," who in the 1970s, and even more so during the following decade, turned out to be trading partners of primary importance. The trade deficit with them, even though there was one, was less disproportionate than with Japan. Relative to the size of their internal markets they bought more from the United States—though it must be said that they were more susceptible to pressure, particularly in the case of South Korea and Taiwan, whose national security, in the final analysis, depended on American military power.

These changes in the ranking of countries can be explained by the profound changes in the way production and, consequently, commercial activity were organized. Japan, followed by the "Chinese islands" and South Korea, were the first to set off down the path of the contemporary industrial revolution, upon which the rest of Asia and certain Latin American countries such as Chile have set out more recently. At the beginning of the 1960s, the organization of trade between the United States and the Pacific countries differed very little from what it had been early in the century (cf. Appendices, Table 16). Japan clearly emerged from the pack, not only because of the value of its trade, but also because of the way imports and exports were distributed between types of products, reflecting a level of development which was already exceptional for the region. The diversification of goods, the volume of imports of raw materials (cotton and scrap metal) to supply its textile factories and steel foundries, and the preponderant share of manufactured articles in its exports (radios, steel sheeting, and tubing, for example) were all traits which made it a unique trading partner and source of imports to the United States, with one hundred times the weight of South Korea and twenty-five times that of Taiwan.[7] As before, the other countries sold a limited range of light manufacturing products—rubber, copper, tin, sugar, coffee, meat, and bananas (cf. Appendices, Table 17). Within a quarter of a century the situation had changed completely, involving all the newly industrialized countries of the Asia-Pacific. By the mid-1980s, American imports from the Pacific consisted mainly of manufactured articles, with Indonesia, a major oil producer, the exception (cf. Appendices, Table 18). All the imports from Hong Kong, Taiwan, and South Korea were commodities of this type, but even developing countries, such as Malaysia, Thailand, the Philippines, and China, were now mainly exporting goods produced in their factories and workshops, which took on a relatively greater importance compared with commodities originating in their fields, plantations, or mines.[8] Some countries produced mass-market consumer goods (inexpensive electrical appliances, clothing, shoes, cameras, toys, jewelry, musical instruments, and watches). Others, such as Singapore, and its neighbor Malaysia, instead offered a range of "machinery," i.e., electrical equipment (electronic components and telecommunications apparatus), computers, office supplies, medical and optical instruments, pumps, etc. Conversely, the United States sold proportionately fewer manufactured goods to the "Four Little Dragons," and there was a noticeable reduction in the percentage of such goods being sold to a number of developing countries. On occasions even the large proportion of foodstuffs (cereals and soybeans) and raw materials (cotton, wood, and skins) being exported might create the—mistaken—impression that the United States was less developed than certain of its Asian competitors. Partly finished articles and consumer goods now accounted for only a marginal share, and the strengths were still in chemical and aeronautical products, as well as in a vast range of electrical and other "machinery," including capital goods for heavy industry. With the emergence of newly industrialized countries, trade with the United States increased, often consisting of articles of similar types, but becoming increasingly specialized and complementary. More and more alliances were created between companies in these countries, pushing nations who still had not chosen the path of modernization to the fringes of what was now a global economy.[9] In Southeast Asia, the transition was rapid. In 1970, for example, machinery represented 12 percent of Taiwan's exports to the United States, and consumer goods 60 percent, with 34 percent being made up of clothing and 14 percent of consumer electronics. Likewise, in South Korea, transistors made up 7 percent of sales, and consumer goods 62 percent, including 35 percent in garments. The great transformation took place during the years of the Vietnam War. Since then Taiwan's potential has been developed, on the one hand by a diversification of production, and on the other by an increase in the volume of trade, with more and more substantial movements of merchandise starting in the 1970s.[10]

From Positive to Negative: Direct Investment

Merchandise is more mobile than capital. There is always a hiatus between trading activity and the flow of capital. The net balance of direct investment—i.e., the difference between the amount invested by Americans in the Pacific and investments in the United States by inhabitants of Pacific countries—demonstrates the comparative strength of the various economies and of their relative levels of development.[11] Prior to the 1960s, Washington paid little attention to the influx of long-term capital.[12] In 1980, the country's net investment balance with the other Pacific nations was still very positive, at around $25 billion (cf. Appendices, Tables 19 and 20). This was the case even for Japan, with which it stood at $2 billion. Then, within a few years, particularly from 1987 on, the situation deteriorated rapidly, to the extent that in 1992 there was a negative net balance of $18 billion—entirely attributable, it is true, to the abrupt transformation of the United States' economic and financial standing vis-à-vis its Japanese competitor, with whom it was now running a negative balance of $70 billion.

This spectacular change should not make us lose sight of an essential fact: the Pacific still lagged far behind Europe in attracting American investment, and accounted for only about 20 percent of the total in 1992—exactly the same figure as in 1929—compared to 50 percent in Europe (16 percent in Britain) and 14 percent in Canada alone. At the end of the twentieth century the Atlantic region had preserved its predominance in this respect, contrary to the evolving trend in trade relations.[13] In investments, as in trade, Latin America has suffered a sharp decline, of which Asia and Oceania have been the beneficiaries. In 1960, 53 percent of direct American investment in the Pacific region still resided there, while by 1990 its share had fallen to 19 percent, mostly in Panama, a financial center and tax haven. This huge decline can be explained first of all by the nationalist policies adopted by Peru and Chile as they attempted to regain control over their natural resources. Foreign capital was blamed for economic underdevelopment, and both governments hoped that by nationalizing American firms they would be able to put the profits to work to modernize their countries. So, in Peru, the Army, which seized power after a coup d'état in October 1968, nationalized the International Petroleum Company, a subsidiary of Standard Oil, on the pretext that it owed $690 million in back taxes. In 1973 it was the turn of the Cerro de Pasco mines. President Garcia (1985–1990) pursued the same policy by revoking the contracts of three foreign companies and nationalizing one of them. In Chile, in 1968 the Christian Democrat Eduardo Frei launched a program to "Chileanize" copper, under which the Santiago government purchased shares of subsidiaries and increased its participation in the mining of the ore which was the country's chief export. His successor, Socialist Salvador Allende, abandoned what his supporters in the Popular Union considered to be only a half-measure, and in September 1971 the Chilean Congress voted unanimously to completely expropriate two American companies, Anaconda and Kennecott, without paying any compensation, since the profits they had accumulated since 1965 were considered "excessive," amounting to more than 10 percent annual return on their investment—a measure the dictator Pinochet made no move to repeal, despite the high esteem in which he professed to hold free enterprise. However, even in the absence of nationalism or Marxism, Latin America would have seen its share fall considerably because the importance of mining and extractive industries was being reduced as the industrial and service sectors boomed. But only certain Asian societies and Australia were ready for industrialization. American investments were indeed attracted to low-wage countries, but there was a preference for those where the productivity of the labor force was relatively high (as was the case on the "Chinese islands"), but above all for countries able to provide a sizable, credit-worthy market (Japan and Australia).[14]

Such countries were themselves a potential source of capital investment in the United States. The falling value of the dollar after 1971 made it less expensive to acquire American companies, and to establish or expand subsidiaries in the United States. It made exports more difficult, and stimulated domestic production, particularly since manufacturing costs were rising less rapidly in the United States. The two oil crises accelerated oil and coal production in countries which imposed no restrictions on foreign investment in natural resources, and where political stability made for a high level of confidence. So, from 1978 on, the Japanese and the Australians to a much smaller extent invested more and more in America. With enormous surpluses in its current account, Japan was able to establish a net positive balance in direct investments by 1981, and by 1992 it had become the primary foreign investor, ahead of Britain, whose predominance had until then been unchallenged. Even more than its incursions in the industrial field—particularly in the automotive industry, in which eight of its companies installed factories right across the country, from California to Tennessee—spectacular Japanese purchases of prime real estate led people to think, in the 1980s, that that country's irresistible financial power would enable it to acquire an ever-increasing share of the American economy. But this failed to take account of coincident fluctuations, which have little respect for exponential projections.[15]

THE ECONOMIC ORGANIZATION OF THE PACIFIC

There are three possible frameworks for economic relations between the United States and the Pacific, namely, bilateralism, regionalism, or multilateralism. Under the first, American trade with each country is regulated on an individual basis—a modus operandi typical of the initial phase of the frontier. Subsequently, as trade grows, it becomes necessary to choose between a regional or a multilateral type of organization. Multilateralism seemed the ideal framework in the eyes of the principal architects of American foreign policy after World War II. United under the umbrella of an institution intended to be universal, or under the aegis of a dominant power, countries would agree to create a vast global market within which tariffs would be minimized, so that the comparative advantages analyzed by Ricardo would be given free rein, without specialization having negative consequences for development. Failing this, if too many tensions should arise, it was still possible for the United States to help create a number of more or less open trading blocs, each constituting a free-trade zone for its members, and excluding nonmember countries. This type of regional organization, like a multilateral one, signifies the frontier's disappearance, for in both cases the integration of formerly peripheral, or even marginal, economies leads to a blurring of differences, and even their ultimate elimination.

The Quest for Multilateralism

In 1945, the major economic problem facing the Americans in the Pacific was to find a way to integrate the victorious nations and the vanquished, together with their colonies, into a comprehensive ensemble within which the former exclusionary barriers would be replaced by the free movement of goods and capital. This was to lay the economic foundations for future peace. Yet the initial political decisions did not point in that direction. The occupation authorities in Japan wished to dismantle the structures responsible for bolstering the country's imperialism. They imposed a kind of New Deal, with the objective of bringing about a profound transformation of Japanese society. The *zaibatsu*, or economic conglomerates, blamed for having encouraged military expansionism, were to be dissolved. Some Americans even envisaged limiting Japanese industry to light manufacturing, as a way of curbing any future imperialist impulses. Deprived of its colonial markets, or in a state of forced subjection, Japan, with its 80 million inhabitants, would probably have remained dependent upon United States for many years to come, for it lacked sufficient food supplies to feed its people adequately, as well as most of the essential raw materials needed to supply its manufacturing industry. A low rate of growth would only give rise to popular discontent and create a following for the Socialists and Communists. If, as a result, Japan became drawn into the Soviet orbit, this would represent a major setback for America's position in the western Pacific. It was for these reasons that Washington made an about-turn in 1947. Japan's economic recovery, rather than reform, became the priority. The engines of production had to be brought back to life, for in 1948 the manufacturing index stood at only a quarter of its 1944 level. Just like in Western Europe at the time, there was no shortage of qualified workers eager for employment. It only remained to find a way to kick start industrial activity.

A priori, the best solution would have been for the Americans to open their borders to Japanese goods. With the dollars earned, the Japanese could have bought whatever they needed to import. But this type of bilateral trade turned out to be impossible. First, Japanese-manufactured articles were generally unsuited to the American market because of their inferior quality, and whenever they were marketable, the corresponding American sectors felt they were the victims of unfair competition arising from excessively low wages. They, therefore, brought pressure to bear on Congress to impose quotas, or to maintain, or even increase, the high tariffs. Japan's ongoing trade deficit with the dollar zone led the Americans responsible for establishing a program of economic reconstruction, and Joseph Dodge, in particular, to reinvolve the former enemy in a system of multilateral trade. Japan would develop its commercial activity with the Pacific countries of the sterling zone, buying raw materials from them and selling them manufactured goods in return, possibly accumulating a surplus in the process. Unfortunately it would not be a very useful one until the conversion of assets into foreign currency was allowed, or, better yet, total convertibility was introduced. So it was that Southeast Asia assumed an ever greater importance in American plans for the Pacific, not only because of its strategic function in opposing communist expansion, but also because of its crucial role in helping reestablish the Japanese economy, thereby ensuring that country would opt to remain in the Western camp. However, this plan encountered numerous obstacles. The peoples of Southeast Asia were too poor to purchase great quantities of manufactured goods, however cheap they were, and their leaders could not accept being reduced to merely supplying raw materials, thereby forgoing the benefits of industrialization in their own development. Furthermore, the European countries who were still clinging to their colonial possessions relied on these markets to help them rebuild their own economies, and took a dim view of any incursions by the Japanese, who had so recently been their enemy. By the time the Korean War began, the Americans, despite all their efforts, had made hardly any progress toward multilateralism in the Pacific. In Japan, Joseph Dodge was reduced to imposing an austerity program to combat inflation. He decreased costs, lowered the standard of living, and channeled a growing portion of industrial production into the export market, all without any assurance in the medium term that this would help to balance the current account.[16]

The Korean War suddenly provided a solution. Japan became the closest support base for the powerful American war machine, with so-called "off-shore orders" for war materials being added to expenditures by the troops themselves. This was a considerable help in remedying the shortage of dollars and making up the trade deficit. Idle production capacity was progressively brought into service, so that by 1955 the manufacturing index had returned to its 1944 level. It was not by opening markets in Southeast Asia, but rather thanks to its need for military supplies, that America reprimed the pump of trade in the Pacific region. From then on, growth would depend on technology transfers—in naval shipbuilding, for example—and an ever greater volume of exports. The United States, which accounted for 18 percent of Japanese foreign sales in 1952–1953, did so for around 30 percent between 1965 and 1972 (cf. Appendices, Graph 5). Textiles, steel products, and shipbuilding opened the way for Japan's increasing economic power.[17] This increase in trade with the United States did not take place at the expense of the other Pacific countries, for the absolute value of Japanese foreign trade increased by leaps and bounds, with the result that Japan imported increasing amounts from Australia and Southeast Asia, while the share of its imports coming from the United States declined from 35 percent between 1952 and 1954 to 28 percent between 1965 and 1969. The triumph of multilateralism was complete when Japan made its entry, with America's blessing, into the major international bodies, GATT and OECD. In 1966 the founding of the Asian Development

Bank paved the way for active cooperation between the nonsocialist countries with interests in the Western Pacific.

As long as the dollar remained strong, it was not difficult for the United States to accept the headway made by Japan. There seemed to be little concern that Japan was taking an inadequate share in its own defense, enabling it to concentrate most of its resources on economic expansion. In opening the United States market to Japan, the aim had been to divert its attention from Communist China, which, in any case, would have had neither the means nor the desire to import large quantities of Japanese manufactured goods. Indeed, the Tokyo government skillfully exploited this appearance of dependency, continually invoking the weakness of their country as an excuse not to follow Washington in its military adventures, and to obtain an increasingly less deserved preferential treatment in the economic domain. At times the United States was driven to react by lobbies worried by the Japanese penetration of the American market. The alliance was seriously shaken in 1969–1971 when, just as the Okinawa question was being quietly settled, the problem of imports of synthetic textiles and woolens reared its head. In his presidential campaign, Nixon had made a commitment to restrict imports of these items, and pressure groups from the relevant industries exerted a considerable influence on the discussions, even though the textile sector was only a minor player in the total field of American manufacturing. An accumulation of gaffes and misunderstandings fueled a dispute which it took two years to resolve, particularly since Japanese Prime Minister Sato had great difficulty in persuading the MITI bureaucracy and the textile interests to accept the compromise solution which he personally preferred.[18]

The pattern established by Japan also applied, with a time lag, to most of the other countries in the Pacific region. The opening up of the American market provided an outlet indispensable for any development strategy based, not on import substitution (generally destined to fail, as seen in Latin America), but on exports, and hence on a certain amount of specialization. This was the path successfully taken by the "Four Dragons," and more recently by others. By becoming part of the network of multilateral trade, these countries finally built up trade surpluses with the United States, Hong Kong from 1959 on, Taiwan from 1970, South Korea from 1976, and Singapore and Thailand from 1984. Technology transfers and liberal advice on productivity contributed to their success. But what was still lacking was a readiness by Asian societies to welcome, and put into effect, a far-reaching economic transformation. Their transition was initially based on low labor costs, by a still hard-working labor force, and on the undervaluation of other national currencies. Then, as wages increased, a high savings rate and remarkable growth in education began to play a part, with the result that less intensive industries were abandoned for products incorporating a high degree of added value and of research and development. The textile age was replaced by one of consumer electronics, and, in the case of Korea, by the automotive industry.[19] Overall, the American policy of multilateralism met with success. The Pacific finally became fully integrated into the industrialized world, on an equal footing with Western Europe. These fresh sources of goods also had the advantage of being new customers, even if the trade deficits the Americans built up with them were a cause of concern.

The Regionalist Temptation

During the economic reconstruction of the world following World War II, the United States treated the Pacific and Atlantic zones quite differently.[20] In the latter case, it implemented the Marshall Plan and encouraged the Western European nations to join together in a European Community. There was no initiative of this kind in the Pacific— no actual coordinated program, no encouragement to unite in order to bring about a common external tariff and similarly oriented policies. There were good reasons for this. The European countries belonged to the same cultural tradition, had achieved relatively comparable levels of development, and formed a compact geographic zone within which trading exchanges had long been customary. The Pacific, on the other hand, even if only its western portion is taken into account, was immense, contained a number of extremely varied cultural traditions, and had never experienced intensive trading activity. In addition, colonialism and imperialism had broken up the region into a number of markets oriented toward centers outside the Pacific rather than to the countries closer by, despite the undeniable signs of regional cohesion which had already begun to emerge before World War I.[21] In addition, whereas Western Europe possessed the human resources and complementary economic characteristics required to ensure the success of a coherent and temporarily massive aid program, the Pacific countries, except for Japan, had no diversified industrial base. Their principal exports, whether raw materials or tropical produce, were in competition with one another. The mostly peasant population lived in such poverty that it was out of the question for them to buy manufactured goods on a regular basis, so that without a credit-worthy market, entrepreneurs were unable to take advantage of the economies of scale which would have made their factories competitive. The technological, scientific, and economic elites were still too few in numbers to help their nations begin rapid progress. The Point Four Program formulated by Truman in 1949 was intended to remedy some of these weaknesses, but in any case time would have been needed to obtain tangible results, and Congress preferred to spend money on programs whose objectives could be achieved in no more than a few years. In fact, a four-year Marshall Plan for the Pacific could never have achieved the same success as in Western Europe. In spite of its relative wealth, the United States did not have infinite resources to call on after World War II, so the wiser path was to give priority to Europe, the major stake in the conflict between the two blocs, rather than to a Pacific where bilateral relations would be enough to safeguard what was most essential, i.e., Japan and, to a lesser degree, Australia and the countries of Southeast Asia. This indicates that around 1950 the Pacific still retained some of the characteristics of a frontier.

The idea of a Pacific community appeared late in the day.[22] Far from being of American origin, it was put forward by certain Japanese academics in reaction to what they perceived as the construction of a European bloc. In 1965, Professor Kojima put forward a proposal for a Pacific free-trade zone. Three years later Foreign Minister Miki called a conference to discuss the proposal, but the United States was cool to it, for its interests were global rather than regional. Yet the idea did not die. Professor Hugh Patrick of Yale and the Australian Peter Drysdale appeared before committees of Congress to put the case for an organization for trade and development in the Pacific, on the model of the OECD, which would include the five developed countries bordering the ocean (the United States, Canada, Japan, Australia, and New Zealand), those of the Association of Southeast Asian Nations (ASEAN), founded in 1967, and the nonsocialist countries of northeast Asia (South Korea, Taiwan, and Hong Kong)—a proposal espoused by Japanese Prime Minister Ohira. However, this was to no avail, for there was too much fear that such an organization would be dominated by Japan, with its recently restored economic power or, worse still, by a joint

American-Japanese hegemony over the underdeveloped countries. The factors making for division were stronger than those encouraging the Pacific countries to join in a cohesive effort.[23] ASEAN hoped to turn Southeast Asia into a zone of peace, liberty, and neutrality—which meant freeing it from American control. In 1992 it set itself the objective of creating within fifteen years a customs and economic union along the lines of the European Community—a very challenging undertaking because of the disparities among its members, and one which hampered attempts to create a broader multilateral or regional organization.[24] In Latin America also, regional groupings ran counter to American objectives, whether it was the Central American Common Market, created in 1960, or the Andean Pact of 1969, both pale copies of the EEC, and which soon came to nothing.[25] The strengthening of European cohesion, the creation of a single market in 1993, and the projects for European Union, raised the specter of an enclave from which the rest of the world would be excluded, with the result that in the 1980s numerous attempts were made to create a regional organization. The United States made North America its priority, forming a free-trade zone (NAFTA) with Canada and Mexico, which was approved by Congress in November 1993. To prevent its isolation between these two large customs unions, Australia made the case for a specific organization for Asian-Pacific economic cooperation (APEC), including the fourteen countries previously suggested by Patrick and Drysdale, with the addition of Communist China.[26] The most daring dreamed of endowing it with a structure and objectives comparable to those of the EEC or of NAFTA, but interests were too diverse for such a community to come rapidly into existence. In fact, APEC took a similar form to the OECD—a body for the exchange of information and for discussion, intended to encourage cooperation and interdependence, but with no common policies or tariffs, nor even a unified market. Since 1989, annual meetings have fleshed out this framework for an open regional agreement. After speaking of a future "new Pacific community" in Tokyo in July 1993, President Clinton called a meeting of the fifteen APEC countries, though his objective was to exert pressure on Europe in the GATT negotiations rather than to lay the groundwork for a real free-trade zone in the Pacific. However, for the first time, the fact that this happened showed that the Pacific would be placed on an equal footing with the Atlantic in the future economic organization of the world.[27]

The United States' economic relations with the Pacific countries were even more difficult than with Western Europe, particularly when the balance of payments to American interests were regularly in a deficit position in the 1960s. Disputes about the law of the sea, or the positive trade balances accumulated by America's Asian trading partners, fueled an atmosphere of tension which was incompatible with a genuine sense of community.

The Shrinking of the High Seas

Traditionally, international law had recognized a limit of three nautical miles for territorial waters. In 1945, however, Truman took the unilateral decision to extend the territorial limit to the extreme edge of the continental shelf, reserving exclusive sovereign rights for the United States. This example was not lost on the developing countries. The Latin American nations bordering the Pacific complained that they were deprived of the advantage of a continental shelf, since their coastline fell off sharply into a deep trench. In 1947, Chile, Peru, and Ecuador extended their jurisdiction to 200 nautical miles, rather than the twelve-mile limit agreed on by the Geneva conference of 1958. They thus appropriated the fish-rich waters of the Humboldt current, to which they shortly added further claims. In 1970, the seabed with its mineral riches was included. These steps led to conflicts with Californian fishermen whose boats were seized, an act to which Washington reacted by imposing an embargo on fish from the Latin American countries. Such actions were an indication that the area of the high seas was being nibbled away. It is true that free passage for ships through national waters was recognized by the third UN Law of the Sea Conference in 1982, but the principle of Exclusive Economic Zones (EEZ) extending 200 miles out from the coastline granted coastal countries absolute jurisdiction over natural resources. In the Pacific, the areas now included within territorial limits removed a considerable amount of the high seas. The seabed was declared the common patrimony of all humanity, and placed under the control of an international authority whose status was ill-defined, for it was uncertain whether it had been given considerable power and exclusive rights over the exploitation of the mineral-rich seabed, as the underdeveloped countries desired, or was merely to be a coordinating body, in line with what the United States and other industrialized nations wanted. This question has still not been settled.[28]

Periodically, tensions have arisen over fishing rights. The Pacific, because of its size, contains more fish than any of the other oceans. It is particularly rich in tuna, which Americans consume in great quantities. To satisfy this demand, up to around 1980 a large fleet based in California and Hawaii supplied the canning factories operated by multinational food companies such as Heinz and Ralston Purina. These companies owned their own fishing vessels, and bought any additional fish they needed from independent fishermen under contracts guaranteeing prices and quantities. Their main fishing ground at the time was the eastern Pacific. But overfishing of the tuna stocks obliged them to shift their activity more and more toward the central and southern Pacific, closer to the Exclusive Economic Zones established by the recently independent archipelagoes. For the multinational companies themselves, this shift was not much of a threat. They based their ships in Pago Pago in American Samoa, and, when operating costs became too heavy, they sold their boats, buying what they needed from foreign fishermen. The independent American tuna fishermen did not have this option, and rebelled against having to add to their already heavy expenses by paying for fishing rights. In their opinion the archipelagoes had no proprietary rights over tuna, a fish which migrates over long distances. The response of the island governments was to proclaim that they and they alone were entitled to sell fishing rights, and that any offending vessels would be seized. For taking such action, Papua New Guinea, in 1982, and the Solomon Islands, in 1984, exposed themselves to American reprisals. Amid this tension, Kiribati, and later Vanuatu, signed fishing agreements with the Soviet Union, whose Navy was beginning to pose an increasing threat. In 1987, therefore, the Reagan administration opted for conciliation, and brought the confrontation to an end. A multilateral agreement was signed with the fisheries agency of the sixteen nations of the South Pacific Forum, with the Americans agreeing to pay 9 percent of the value of the catch in return for access to the Exclusive Economic Zones. The major portion was paid out of the federal budget, since the association of American tuna fishers refused to contribute more than 2 percent. Since then, relations between the United States and the islands of the South Pacific have improved considerably. The contracts with the USSR were not renewed on their expiry, and the islanders' resentment was turned against Korean, Taiwanese, and Japanese fishermen, who provided inaccurate information on the size of their catch. Contrary to the United States, Japan refused to sign a multilateral treaty and threatened to cut off development aid if its tuna fishers were refused access to waters over which the independent archipelagoes claimed exclusive sovereignty.[29]

American-Japanese Discord

The financial stakes were much greater in trade disputes between the United States and Japan. Since 1965, trade deficits built up year by year to a level that Washington considered alarming. The difficulty was to decide whether they were a reflection of the two countries' relative economic strength, or of a failure by the Japanese trading partner to respect the rules of the game. If it was the former, the Americans had only themselves to blame and would have to take the necessary steps to rectify their shortcomings and make up the lost ground. If it was the second, Japan would have to be punished for failing to respect the code of ethics governing peaceful international relations. The debate seemed all the more confused in that the arguments were constantly shifting, and did not provide a very useful basis for the discussions which have continued throughout the past quarter-century.[30]

Some Americans attributed their trade deficits to unfair practices by the Japanese. Japanese wages, they said, were too low, making labor costs much lower than for the highly unionized American industries, making it easier for Japanese goods to penetrate the U.S. market because of their competitive pricing. This argument was partially valid during the 1960s, even though labor costs made up only a small part of the value added in a number of sectors. However, it no longer holds for more recent years, since in 1992 the average hourly cost in dollars was identical in the United States and in Japan.[31] The stratified Japanese labor market, where large firms guaranteeing lifelong employment existed alongside a multitude of small and medium companies whose workers did not enjoy the same kind of pay or opportunities for advancement, had its equivalent across the ocean, where unionized employees enjoyed incomes and social benefits much superior to those employers granted their nonunionized workers, particularly in states recognizing the "right to work"—i.e., the right to be hired without joining a union. The growth of American trade deficits coincided with the convergence of wages in the two countries. The currency exchange argument equally fails to hold water. It is true that in 1949 the establishment of a rate of 360 yen to the dollar did undervalue the Japanese currency, allowing their companies to gain a share of the market which would have been out of reach had the rate of exchange been based on purchasing power parity. The rate still in effect in 1970 certainly gave Japan an undeserved advantage. However, the following year saw the abandonment of fixed rates of exchange, the cornerstone of the Bretton Woods international monetary system. Ever since then, the dollar has tended to decline substantially against the yen (cf. Appendices, Graph 7).[32] In 1993, with the exchange at 110 yen to the dollar, considerably raising the cost of Japanese exports and—in theory at least—reducing the cost of imports to Japan, the deficit nevertheless reached $60 billion. The United States also complained for a long time that the Japanese markets were protected by high customs duties and quotas. This was indeed the case until around 1970, allowing the automotive and electronic industries to develop, in the same way as textiles, steel, and shipbuilding had done earlier, shielded from foreign, and particularly American, competition. But since then Japan has dismantled its protectionist barriers. Import quotas have disappeared and customs duties are on average among the lowest in the world. Neither low wages, nor an undervaluation of the yen, nor customs protectionism, can explain Japan's recent huge trade surpluses, and a picture which would have been appropriate for the Japan of the 1960s bears no relation to the reality of the 1980s. There remains the answer of "structural obstacles." The Japanese economy and mentality are said to possess characteristics making them resistant to any temptation to buy foreign goods—a kind of consumer ultranationalism which supposedly erects an invisible, but impenetrable, barrier against goods manufactured abroad.[33] This is blamed on the traditional values still prevailing in a country which has been poor for many years, and was shut off from the rest of the world for two and a half centuries during the Tokugawa regime. Others speak of the sense of vulnerability in a nation concerned for its future in the twenty-first century, when the large numbers of postwar baby boomers will have to be supported in their old age by a much smaller generation of workers, making a very high savings rate necessary to provide the capital to generate pensions in the future. Above all, foreign trade appears to be controlled by a few large companies, the *sogoshosha*, linked to the principal Japanese conglomerates that have systematically preferred markets abroad at the expense of local customers, whom they condemn to pay high prices. The state itself, through the Ministry for International Trade and Industry (MITI), is allied with these powerful private interests. The close links between the public bureaucracy and business leadership creates a wall against foreign imports and the entry of capital from abroad. It is claimed, in short, that Japan does not play according to the rules of free trade, which it only invokes when it wants to protect its own sales.

The Japanese responded to these "Japan-bashing" arguments, repeated ad nauseam in the press, in periodicals, and in books, often in an aggressive tone, by pointing to the undeniable efforts they have made to open their country to trade, as is shown by the strong growth of imports from the United States, and to their considerable contribution to solving the problem of the balance of payments through massive purchases of federal public debt securities and increases in their direct and portfolio investments. In their sometimes virulent ripostes, the Japanese attributed their trade surpluses to failures of American productivity, which they attributed to a labor force that has lost its willingness to work hard, and to a management more concerned with short-term profits than with long-term growth and excessively preoccupied with the balance sheet because of a fear of hostile takeovers, resulting in inadequate investment in research and development. The claim was made that the proportion of the federal budget devoted to military research resulted in fewer and fewer positive spin-offs for the public at large, costing the United States its lead in high technology, as is shown by the number of new patents issued. The American automobile industry, for instance, stagnated during the 1980s, at a time when the Japanese firms of Toyota, Nissan, Honda, and others were winning an increasing share of the market with their imports and local assembly plants. Did some economists not go so far as to declare that the organization of industrial work had evolved from "Fordism" to "Toyotaism"? It was beginning to look as if American corporations had turned into dinosaurs, incapable of making the changes in their management style that were essential if they were to maintain their position in what has become a global world economy. In addition, criticism was leveled at the state of American society, with its neglect of primary and secondary education, leading to a high percentage of illiteracy among Americans, while Japanese youth have a much greater enthusiasm for education. Racial tensions and inner-city crime were seen as symptoms of serious social malfunction. In general, it was claimed, consumption was given too much importance in the United States at the expense of savings and investment, causing deficits both in the federal budget and in the balance of payments, which had gotten out of control under the Reagan presidency. In short, Japan was depicted as the ant whose success was earned at the expense of the American grasshopper.

There is much truth in all these analyses. The Americans themselves were indeed largely responsible for their economic weakness vis-à-vis Japan. Nevertheless, it does seem that if free trade had been working perfectly, trade balances should not have remained so disproportionate over such an extended period.[34] Although it has become the second economic power in the world, Japan still does not seem prepared to accept the responsibility it bears to ensure the viability of the international trading system. In the nineteenth century, Britain generally had a negative trade balance, but surpluses in its current account balance still allowed it to invest abroad.[35] The Japan of the end of the twentieth century, while building up trade surpluses and a positive balance in net investment, still failed to play the driving role which justifiably could have been expected of it. It was accused of being a mere machine for production, "Japan Inc.," living parasitically off an America which, in contrast, was prepared to accept its global responsibilities, both militarily and economically. The "Four Dragons," who copy the Japanese strategy for development, do not deserve the same criticism. First of all, individually, their importance is much less than that of Japan. In addition, their trade with the United States is more balanced—in 1992 they purchased as much as Japan from America and sold her a third less.[36] From the American point of view, there is indeed a problem with Japan.

The Americans do not have an unlimited range of solutions to call on to remedy this situation, which is undermining their leadership in the Pacific. Theoretically, they could refuse to continue to guarantee Japanese security and force Japan, situated as it is in the unstable region of northeast Asia, to devote a much higher proportion of its GNP than 1 percent to national defense. This would certainly diminish the efficiency of the Japanese economic machine. But as long as the Cold War lasted there was no likelihood that such a solution would be adopted, for the Japanese archipelago provided an invaluable forward base in the confrontation with the Soviet Union, and was of major strategic importance to American global interests. Since 1989 the threat of withdrawal might have been more credible, but for the United States this would have meant abandoning its responsibilities in the western Pacific, where its presence was still necessary to prevent tensions from turning into open conflict. How could it possibly withdraw voluntarily from a region where its commercial dealings were the largest in absolute value, and where its investments were growing substantially?[37] In these circumstances, the Americans have been led to seek bilateral accommodations with Japan by using, whenever necessary, reprisals or threats against an ally extremely sensitive to anything which seems likely to affect its markets. Periodically, the United States has exhorted Japan to pursue an active budgetary policy in order to stimulate internal demand and reduce the share of its GNP devoted to exports. Voluntary limits on exports offered by Japanese industry did slow down the influx of Japanese-made products, while still appearing to respect the freedom of trade. First applied to textiles in the 1960s, limits were imposed on television sets in the 1970s, and on automobiles in the following decade. But this had perverse consequences. The restrictions applied to the number of units thus encouraged Japanese producers to export more expensive models, bringing them even higher profits and encouraging further modernization of their factories, thus increasing their long-term competitiveness against American companies.

It would be much healthier if a more even balance of trade could be achieved, not by decreasing imports from Japan, but by increasing American exports to that country. Washington's negotiators have attempted to open up a market which they considered too closed, if possible without giving the Japanese time to attain a level of productivity comparable to that of America, in the tried and true protectionist tradition. The history of these negotiations is a long litany of promises made by the Tokyo government, but rarely kept to any acceptable extent.[38] It has been easy for the Japanese to point out American inconsistency in preaching free exchange on the one hand, while at the same time attempting to tamper with its workings as soon as the United States seems to be on the losing end. Trade "administered" by the state is in contradiction with the principle of free enterprise, for it is companies which buy and sell, not national authorities. Even if they wanted to, the Tokyo government could not oblige Japanese merchants to import American goods for which there is no demand among Japanese consumers. Nevertheless, on several occasions Washington has tried to get its trading partner to put hard numbers to import objectives, which they would have to respect, or face sanctions. It was believed that quantitative criteria, based on market share, would indicate the extent of Japan's openness. In 1986, for example, the Reagan administration was able to obtain an agreement that the share of the Japanese market in semiconductors taken by American exporters would be increased from 10 percent to 20 percent by 1991. When this objective was not achieved by the date set, surtaxes applied in reprisal obliged Japan to partly live up to its commitment. In the 1993–1994 negotiations the Clinton administration based its demands on the low foreign penetration of the Japanese market, which amounted to only 3 percent in the case of automobiles, whereas Japanese manufacturers held 29 percent of the American market, and 5 percent in telecommunications, an area in which United States imports from Japan came to 28 percent. Certain Japanese sectors were almost entirely closed to foreign competition, insurance being an example.[39] After the conclusion of the Uruguay Round in December 1993, American rice producers were able to hope that a door which had previously been jealously kept shut to them, at an exorbitant cost to Japanese consumers, might have been opened by even a crack. Similarly, contracting firms gained greater access to bidding on Japanese projects. But even if Japan had completely opened up its markets it would have taken some time for equilibrium to be restored, for in certain sectors the United States has completely abandoned production to other Asian trading partners. It has been a long-term undertaking to put the American economy back into shape, after the far-reaching effects of the Reagan administration's neglect. And the challenge is not just to transform the mind-set of the Japanese and make them into consumers rather than savers, for it will also require an educational effort whose results will certainly not be felt right away. There is no chance that transpacific trade disagreements will disappear overnight, as if by magic. The United States and Japan will have to continue to coexist under these sometimes uncomfortable trade conditions, for there can be no question of war as a solution, and the close ties based on the reciprocal interpenetration of their economies are much stronger than the media make them seem at times of tension between them.

The Reemergence of China

Between 1950 and 1971, the American embargo had put an end to all commercial relations with Communist China. The objective of the new policy initiated by Nixon was essentially strategic, aiming at readjusting the balance of power in the Far East in order to resist Soviet expansionism. The mirage of the vast Chinese market, which had attracted the leadership in Washington on several occasions in the past, played no part in this initiative, which aimed only at bringing Maoist China back into the international arena. Indeed, until the official resumption of diplomatic relations in 1979, Sino-American trade was insignificant, at less than $1 billion, accounting for only 1 percent of trade between the United States and the Pacific (cf. Appendices, Table 21). China was still very much influenced by Maoist ideology, and by an almost fanatical desire for self-sufficiency. It had no articles of marketable quality, and had no currency reserves which could allow it to make massive purchases abroad.

All this changed when Deng Xiaoping took power. The new brand of communism still rejected political freedom, but, to provide an outlet for the people's impatience, it committed the country to a far-ranging campaign of modernization which, after a few setbacks, was able to reestablish some forms of unrestricted capitalism, at least in the coastal regions. This new China was open to foreigners, and investments were welcome. For the United States, the fabulous Chinese market it had dreamed of for so long seemed to be becoming a reality at last. After 1980, trade grew at an extraordinary rate, but not without some disappointments. Even more than Japan, China is an exporting country. From 1986 on, the trade ratio with China fell below parity, with deficits growing from year to year. By 1992, with a trade surplus of $18 billion, China was in second place after Japan ($49 billion), but ahead of Taiwan ($9 billion) and Canada ($8 billion). The extremely low cost of labor attracted investment from Far Eastern countries now involved in more sophisticated forms of production. For the manufacture of textiles, toys, and inexpensive electronic devices for export, the coastal areas close to Canton and Shanghai offered ideal conditions in the short term. For the United States, if these imports put pressure on domestic prices and diminished the risk of inflation—for American consumers snapped up the cheap goods—China nevertheless posed a problem no easier to resolve than that of Japan.

Deng Xiaoping's China felt less tied to the United States than did Japan. It was self-sufficient in matters of defense, and had no further need to fear a Soviet threat after the collapse of the communist regime in Moscow. Its desire for independence came into conflict with some of the guiding principles of Washington's foreign policy. It was criticized for selling arms and missiles to unstable regimes such as Pakistan and Iran, of failing to respect human rights by jailing dissidents, or condemning them to the firing squad. Its commercial practices were reprehensible, whether the export of articles made by prisoners at a ridiculously low cost, the excessive undervaluation of the national currency, the piracy of intellectual property, or the way it got around the quotas it had agreed to by routing exports through complicit third countries.[40] The Sino-American trade dispute became more and more serious. The Tienanmen massacre in June 1989 revealed the true nature of the Peking regime. But should America make outlaws of the communist leadership and punish all the Chinese people, or was there any hope that encouraging trade might help lay the economic and social groundwork for future liberalization? This was the dilemma confronting the Washington administration. It had one effective weapon: the annual renewal, or the suspension, of the most favored nation (MFN) status, which allowed China to enjoy the same tariffs as other countries. Peking would have liked to be granted this status definitively, thus removing the sword of Damocles suspended over its head. Washington, on the other hand, preferred it be approved by Congress on a yearly basis, making it possible to bring some pressure to bear on behalf of political prisoners, and encouraging greater circumspection in arms sales as well as in less questionable trade practices. China, at its stage of development, could not afford to lose the American market, which supplied it with enormous dollar surpluses. It therefore made temporary concessions, while refusing to give in on the essentials, knowing very well that the American presidents, from Reagan to Clinton, had no desire to break off relations with a rapidly expanding potential market of 1.2 billion people, or with a major nuclear power with a veto on the UN Security Council. In 1994, under pressure from American industrialists involved in exporting to China, Clinton separated the question of most favored nation status from the cause of human rights. The disagreements between the United States and China seem destined to continue until the leadership in Peking adopts a more responsible attitude, although they are not an obstacle to a relatively willing cooperation between the two countries.[41]

The Pacific, at the end of the twentieth century, became the most important region for United States foreign trade. It has still not achieved the same importance in the area of direct investments, but from now on it will weigh as much as the Atlantic in the balance of the country's overall interests.[42] For the first time in history, the increasing interdependence of economies has resulted in the area being able to overcome its fragmentation and create a closer relationship between its different regions. This has not always proved to be a good thing, as is shown by the example of the drug trade. Narcotics reach the United States, where there is a steady demand, from the heroin-producing Golden Triangle on the frontiers of Burma, Thailand, China, and Indochina and along the cocaine routes passing through Bolivia, Peru, and Colombia, the homes of the Cali and Medellín cartels. The enormous profits made by the traffickers are laundered in other sectors, weaving a complex network of dealings along the coastlines and across the breadth of the ocean.[43] Nevertheless, basic economic relations conform to a more legitimate pattern. In less than three decades, the frontier, after inheriting several very disparate levels of development rooted in its long history, has disappeared. Does this not demonstrate that, if capital is mobile, it enables mentalities and cultures to evolve rapidly, freeing them from the straitjacket of inertia which had prevented the development of individual countries?[44]

TWELVE

Toward Cultural Symbiosis?

‒ If it sometimes seems so difficult to resolve conflicts between opposing interests, this is because there is an underlying confrontation of
value systems. It is easier to reach a compromise if nothing but material interests are at stake, for even when the initial positions seem very
far apart there is always the possibility of drawing them closer together.
A difference in values, on the contrary—even when these are not intangible ones—may mean the gap is unbridgeable. Values define the
culture and identity of a civilization. Nevertheless, as economic activity
increases, the various cultures of the Pacific are brought into much
closer contact than ever before. Does this mean that they will also increase their borrowings from one another? And if so, as their knowledge of one another deepens, will this not provoke forms of resistance
which emphasize the differences? What holds out the best promise of
future peace—a heterogeneity more favorable to diversity but which
may also lead to a rejection of the other, or a degree of homogeneity
which will be less interesting, but more likely to lead to a reduction in
tension?

THE AMERICANIZATION OF THE PACIFIC

The transformation of the Pacific into an "American lake" after World
War II made it ripe for Americanization, which is to say for the spread

of American culture values, customs, and tastes. In the 1950s and 1960s theoreticians of social change saw this as a stage underdeveloped countries had to pass through if they wished to free themselves from the shackles of tradition. Societies, they noted, develop in stages, following a certain path of evolution. As the United States was the most advanced nation at the time, the fastest track to modernization necessarily meant taking on its economic, social, and political characteristics. Technology transfers lead to this kind of modernization, for if they are to be successful, any functionally incompatible structural and cultural factors must first be overcome. The sociologist Talcott Parsons describes how change depends on the acquisition of four increasingly widespread "evolutionary universals," namely, bureaucratic organization, money and the market complex, a universalistic legal system, and democratic association.[1] The various versions of this theory of modernization all insist on the idea that "modern" values and institutions always follow the lines of those found in the industrialized Western world, and particularly in the United States, so that development is largely a matter of imitating these models, while underdevelopment results from tradition, which therefore must be rejected if modernity is to be achieved. According to this view, Americans cannot hope to modernize underdeveloped countries on their initiative, for the effort must come primarily from the society itself. However, the spread of American norms spurs individuals dissatisfied with traditional structures to reject them, thereby assisting their countries to be eventually numbered among the most progressive. Modernization points the way to an inevitable convergence of societies.

Agents of Americanization

The Americanization of the Pacific has been brought about by many different agents. As in the nineteenth century, missionaries remained active, though less visible. With the communist victory, China was "lost," not only in a political but also in a religious sense, for the Protestant churches were obliged to abandon their universities, schools, and hospitals in that country. The Seventh-Day Adventists found new souls to convert in the underdeveloped countries, particularly in Latin America. In Guatemala they made inroads into the position of the Catholic Church by offering spiritual comfort to the most deprived sectors of the population, while at the same time diverting their attention from revolution and collective action. Such sects promote conservative policies which suit the interests of certain American business elements. This is also true of the Mormon Church, whose first mission in Polynesia dates from 1844. Around 1980 it had approximately 700,000 members in the different foreign countries bordering the ocean (not counting Canada and Mexico). It was quite successful in expanding outward from its base in Hawaii, particularly to countries where a strongly right-wing government held power, as in Pinochet's Chile. Temples were built in New Zealand in 1958, in Tokyo in 1980, in Tonga, Papeete, Western Samoa, and Santiago de Chile in 1983, and the following year in Sydney, Manila, Taiwan, and Guatemala.[2]

From the United States came educators and technicians whose mission it was to help modernize developing countries. The Alliance for Progress, launched by Kennedy, set out to contribute the knowledge required to break out of the vicious cycle of poverty. The Peace Corps enlisted young Americans in a noble cause, much as the Protestant churches had done between 1890 and 1930. These steps expanded an initiative begun under Truman with the Point Four Program he outlined in his inaugural address in 1949, and with the Fulbright scholarships. In addition, many students from the Pacific region attended American universities, where they received the most up-to-date training in science, technology, medicine, economics, and business management, so that when they returned home—or at least those among them who wanted to do so, for many chose to stay in the United States, which became the beneficiary of a brain-drain—they would form an elite prepared to begin the process of breaking free from a constricting traditionalism.[3] There is no doubt that after experiencing the freedom of life on American campuses the Asians and Latin Americans had, for the most part, sufficiently absorbed Western values to make them wish to help their own societies evolve.

Apart from its military and economic power, America enjoyed considerable cultural prestige. Many foreigners saw American civilization as the most advanced and one which, in the second half of the twentieth century, best corresponded to the aspirations of young people in particular. This image was propagated by the media, in films, television series, records, cassettes, and advertising, encouraged by the Washington government. People showed how up-to-date they were by wearing jeans, drinking Coca-Cola, and listening to rock-and-roll. American companies profited handsomely from this infatuation with things American, which, thanks to more and more efficient telecommunications, was able to reach rural areas just as easily as it did the large cities. In Japan, baseball became the national sport, but in a way remained faithful in spirit to the Japanese mind-set with its rejection of individualism.[4] The United States also had the advantage of speaking the only truly international language (though not the one with most native speakers). The British colonial inheritance made it easier for American culture to penetrate the English-speaking countries (Australia, New Zealand), and wherever English continued to play a major role alongside the national language, whether this was Malay (Malaysia), Chinese (Singapore and Hong Kong), Fijian, or some other Polynesian, Melanesian, or Micronesian tongue. In the Philippines, English replaced Spanish and remained the primary language of communication even after independence, ahead of Tagalog. In Japan, Taiwan, South Korea, and even in Communist China, a good proportion of the scientific literature was written directly in English. Neither Japanese, nor Chinese, nor Korean became major international languages, despite the economic success of the countries where they are spoken. News about Far East countries is essentially carried by English-language newspapers and magazines. This leads to a strong imbalance between the United States and the rest of the Pacific, for relatively few Americans read or speak a Far Eastern language, or even Spanish (except for the "Hispanics," of course), while the Pacific elites are compelled to acquire a minimal knowledge of English if they wish to escape the limitations of their nationality and open themselves to international intellectual currents. This results in a lack of balance in the availability of news, since Japanese newspapers, for example, carry many more articles on events in the United States and on cultural developments on the other side of the Pacific than American newspapers and magazines do on Japanese culture, economics, or politics. The same is true for Australia or Chile. Consequently, the elites, and even the ordinary people, have a better knowledge of America than Americans do of the various foreign countries, with the exception of specialists (as competent as they are numerous) working in universities and research institutes.

A New Deal for Japan

The principal attempt to instill American values and transform a society took place in Japan after its defeat in 1945. To destroy any possibility of a return to aggressive militarism, the Supreme Command of the Allied Powers (SCAP)—in effect, General MacArthur and his advisers—resolved to eliminate the social factors and the institutions which had allowed an authoritarian regime to achieve power in the 1930s. The New Deal, Roosevelt's experiment in reform, served as a model. It also supplied administrators convinced that social engineering could profoundly modify behavior. Of course, it was not necessary to reject all the Japanese tradition, but if Japan was to enter the modern world, minds would first have to be freed from the dead weight of its feudal past. The promising thing was that the Japanese appeared ready for change. Their behavior was most amenable from the moment the occupation began, as if they were bewildered by the experience of the first invasion their country had ever suffered. Faithful to their ethic, which is based on adapting to situations rather than on supposedly universal norms, they were quite willing to cooperate as long as they were well treated. Furthermore, MacArthur, once victory was assured, made no attempt to punish his former enemy, instead making it his mission to reform the Japanese people—a task made much easier for him by the fact that he enjoyed the support of the conservatives in Congress. In the space of two years, he supervised an "induced revolution,"[5] neither totally spontaneous nor entirely imposed from above. The trick was to encourage the Japanese government to transform the empire into a democratic nation, even if this meant forcing its hand when things did not move quickly enough or when the old order dug in its heels. MacArthur had the advantage of having to deal with a country which was already developed and which had previously been acquainted with parliamentary government. This explains his success in Japan, while in China America's ally, Chiang Kai-shek, turned out to be incapable of introducing the essential reforms.

The Americanization of Japan was never fully accomplished. The emperor, after renouncing his claim to divinity—i.e., his role as high priest of the Shinto state religion—retained his position as head of state under the new Constitution, although he no longer held the reins of power.[6] The Americans introduced the sovereignty of the people, the basis of any liberal democracy. If the bicameral parliamentary system was closer to the British than to the American presidential model, a number of institutions bore traces of American influence. Rights, broadly defined, were written into the basic text, with a fully independent judiciary to protect them and to monitor the law. In the courts, defense and prosecution were placed on a strictly even footing, and both lawyers and judges were allowed to cross-examine witnesses. At the local level an extensive decentralization modeled on the American federal system was introduced. Provincial governors were elected, and there was provision for referenda, for citizen initiatives, and for the right to recall elected officials. The national police was abolished and replaced by local forces in the towns and cities.

The spirit of the New Deal was particularly evident in economic reforms. SCAP took on the *zaibatsus*, the conglomerates blamed for having encouraged militarism.[7] American specialists in antitrust legislation advised that holding companies, considered to be feudal family institutions, be dissolved and that a law similar to the Sherman and Clayton Acts be introduced, forbidding any agreements which attempted to restrict trade. In December 1947, SCAP forced the Japanese Diet to adopt Law FEC-230, allowing companies to be broken up, meaning that any company of monopoly proportions—a concept kept deliberately vague—could be dissolved or reorganized by the Liquidation Commission. As a counterweight to business interests, the occupying authorities granted complete freedom to the labor unions, who considerably increased their membership and reinforced the democratic camp. Farmers, who until then had been exploited by the landowners, benefited from drastic agrarian reforms which enabled them to buy land cheaply, resulting in the creation of a class of small landowners unreceptive to communist propaganda. This economic democratization laid the foundations for Japan's future boom by creating a mass domestic market for consumer goods.[8]

Japanese society emerged from this process partially transformed. The education system, designed to produce devoted servants of the state, was remodeled along progressive American lines. The hitherto rigorous control exerted by the central government was transferred to autonomous elected local boards. Discrimination against girls was ended. Curricula were completely overhauled, indoctrination being replaced by a pedagogy intended to produce good citizens. The number of universities was increased in the hope of making them accessible to the masses. Family relations were also affected. The old Civil Code had given the head of the family complete power and confined married women to an inferior status, but this kind of authoritarianism was abolished, and equality between the married couple was introduced. In 1947 the results of the occupation appeared to have been a "magnificent success," as Kazuo Kawai puts it.[9]

If the idea of reforming Japan along American lines and turning it into a genuine democracy had a weakness, this was not that it lacked ambition, but rather the temptation to move too quickly at times. After two years the initial impetus ran out of steam. The Japanese economy was at a standstill, and the communist threat was growing. Reform was no longer the priority and gave way to the need to rebuild industrial capacity. This "reverse course" made it necessary to water down some of the measures inspired by the New Deal. SCAP no longer supported the union movement, and the policy of breaking up companies, held responsible for the stagnation, was emasculated, so that the list of 1,200 companies originally scheduled to be broken up was reduced to 325, then to 30, then to 19, and finally to 9. The concerns of the American bureaucrats William Draper and Joseph Dodge grew closer to those of the Japanese conservatives. The project of Americanizing Japan had been obliged to yield to strategic and economic necessity.

In no other Pacific country—not even in its trust territories—did the United States undertake a similar attempt at modernization. In any case, it lacked the power to do so. Nevertheless, its influence was felt everywhere, more or less intensely, and after varying amounts of time had gone by. In the end, the Pacific came to share, to different extents, certain values of the Atlantic world, such as free enterprise and democracy.

Changing Images

American attitudes toward the peoples of the Pacific are not determined solely by the extent to which the latter are Americanized. Americans are not always put off by what is different. However, an obvious adherence to the same fundamental values makes amicable relations much easier, as long as there is no fundamental conflict of interests. One can only be struck by the fluctuations in the ways Americans viewed foreigners, as recorded by polls or revealed by an analysis of the press, audiovisual media, novels, and comic strips. No stereotypes last forever—which is to say that, contrary to the anthropological notions put forward during World War II, there is no such thing as national character, for individual differences are much more significant than the traits which are supposed to be typical of a particular people.[10] Perceptions of other peoples vary according to the situation, rather than to any supposed changes in their nature.

Where China was concerned, the Americans, in attempting to establish a special relationship, have been perpetually torn between euphoria and disillusionment. The historian Warren I. Cohen lists five distinct periods in American attitudes toward the Chinese, with an era of deference (1784–1841) giving way to one of contempt (1841–1900), and then of paternalism (1900–1950). The communist period was characterized by a total shift from fear (1950–1971) to respect (1971–1988). The Chinese national character underwent no violent changes when diplomatic relations became closer, and yet within a very short time American perceptions changed dramatically.[11] In January 1968 a Gallup poll measuring the opinions expressed by Americans about 28 countries put Communist China in last place with positive opinions at 5 percent and very negative at 75 percent. At the same time, 72 percent had a positive opinion of Japan, with 19 percent very positive and only 9 percent very negative. Other percentages of positive opinions were: Australia, 89 percent; Chile, 66 percent; South Vietnam, 62 percent; North Vietnam, 7 percent; and Gaullist France, 49 percent.[12] Between May 1972 and April 1973 the percentage of negative opinions of Communist China fell from 71 percent to 43 percent, while the positive opinions rose from 23 percent to 49 percent—without in any way detracting from Japan's reputation, which rose from 64 percent to 70 percent in favorable, and fell from 30 percent to 22 percent in unfavorable opinions.[13] This trend continued as diplomatic and commercial relations were improved. In 1988 a poll found that 70 percent had a positive view of China, with esteem being higher among young people between twenty-four and twenty-nine (76 percent) than among the older generation (56 percent). This score was almost identical with Japan's (72 percent), though in its case there was a much stronger contrast between age groups (81 percent and 55 percent, respectively).[14] So it can be seen that at any given moment public opinion is composed of a number of strata in which memories of the past still linger—memories of World War II and the Korean War for those who lived through them—mingled with a sense of the prevailing climate. Stereotypes do not disappear completely, but to the extent that they do describe an aspect—but no more than an aspect—of reality, they represent a stock of responses to any situation liable to arise. So, the Japanese are seen alternately as cruel warriors, fanatically devoted to their emperor, as sensitive gardeners lovingly tending their chrysanthemums, or as relentless and cunning businessmen, bent on taking over the entire world.[15] The assimilation of the samurai and the businessman is almost natural. Disagreements can quickly revive the perception of a hostile Japan. This is why, between May 1989 and November 1991, the percentage of favorable opinions fell from 69 percent to 48 percent. In the latter year, 43 percent of Americans claimed to have friendly feelings toward Japan (51 percent in 1951), 8 percent unfriendly feelings (25 percent in 1951), while the majority (47 percent) were actually neutral or indifferent (compared to 18 percent in 1951). Almost half saw the Japanese as possible future aggressors.[16] But it only needed economic tension to decline for positive characteristics to once again predominate over the negative stereotypes. Which goes to show that there is an advantage to cultivating multiple images.

RESISTANCE TO AMERICANIZATION

If Americanization is viewed as a benefit by a majority of Americans, there is still a minority which criticizes it as not being in the best interests of the peoples concerned, while abroad there are powerful movements which reject it in the name of tradition or of revolution, or simply in defense of their national culture and identity. Dependency theory, developed in the 1950s and subsequently amplified, attacks the theory of modernization. It views the United States as the center of the developed world with, orbiting around it at a distance, certain countries whose underdevelopment is seen as a result of unfair trading practices. The flow of capital controlled from the system's core leads to an international division of labor which works entirely to the advantage of the most advanced nation-states. The similar concept of world models identifies multinational companies—until recently most of them American—rather than countries, as the driving force. But in all these scenarios the underdeveloped countries are deprived of a large part of their autonomy, for their economic and political elite tend to follow directions from the center and copy its role models, all at the expense of the ordinary people. The resulting tensions lead to a rejection of American values.

Cultural Relativism

However, dependency theory exaggerates the loss of local autonomy. National cultures have strong powers of resistance. We have seen the strength of nationalism in Latin America and in the Far East, where American interests have had to accommodate themselves to it, since they lacked the means to overcome it every time. While Americans, like other Westerners, tend to preach universal values, valid for all times and places, the leaders of Far Eastern countries rise to the defense of a cultural relativism according to which each civilization has its own norms, all of them equally valid. They consider that the rights of the individual as defined and applied in the West are specific to a certain type of society, rather than as a model which is either desirable or applicable everywhere. As a result, the American desire to see these rights respected and defended are seen as no more than a manifestation of their unjustified desire for cultural domination. Furthermore, there is something hypocritical in insisting on respect for human rights in these countries while at the same time showing considerable willingness to overlook violations by others. The United States has urged greater tolerance for freedom of expression on the Chinese leadership, while at the same time lending support to a number of right-wing dictatorships in whose countries their strategic and economic interests have been threatened by movements with communist or progressive leanings. In Vietnam they supported the authoritarian regime of Ngo Dinh Diem after the first Indochina war, despite his persecution of Buddhists and Caodaists. In South Korea they supported Syngman Rhee and then, after his overthrow in 1960, various generals who systematically repressed any opposition. Similarly, in the Philippines, the Ferdinand Marcos dictatorship, between 1972 and 1986, met with hardly any disapproval from America. In Latin America there is a long list of countries where the United States closed its eyes to legal violations, in some cases actually giving these its approval, as in Guatemala in 1954 and Chile in 1973. Indeed, it was only with the Jimmy Carter presidency (1977–1981) that American policy took a more idealistic turn, committing itself to defending human rights as the supreme expression of civilized values and hence as a norm deserving of universal respect. In practice, however, it was out of the question to subordinate all other interests to such an imperative, and American diplomacy was content to save appearances by obtaining concessions of varying significance.

Resistance in Confucianist Societies

The impact of cultural differences can be seen quite clearly in the debates occasioned by the increase in economic power of Japan and the newly industrialized countries of the Far East. The competition they provided was considered unfair because both sides were not playing by the same rules, with the United States as an enthusiastic proponent of free trade on the one hand, and on the other mercantilist nations whose priority was to build up positive current account balances. These choices were expressions of different value systems. Free trade is the product of an individualistic society, centered on the human person, whose rights are protected by the state. Mercantilism, on the other hand, is better attuned to a society rooted in Confucianism.[17] The teachings of Confucianism originally spread from ancient China through Japan, Korea, and Vietnam. The holistic and hierarchical ideology which characterizes it is totally at variance with Western Enlightenment thought. It sees the part as subordinate to the whole, with one of the two principles comprising the whole, the *yang*, being superior to the other, the *yin*.[18] Societies based on Confucian values put the emphasis on family solidarity, respect for those who naturally exercise authority, a taste for study, and a sense of social responsibility. Individualism is rejected, being considered synonymous with egoism, and loyalty is instilled, rather than independent thinking. Since Confucian philosophy considers mankind to be good and perfectible, it sees no need for the separation of powers as a basis for the political order. The paternalism of a benevolent ruler, guided by his moral sense, is supposed to establish harmony among the governed. It is hardly surprising that in Maoist China and in postwar Japan Confucianism was accused of standing in the way of modernization, while attempts were made to replace it with Marxism-Leninism or Western liberalism. But there are different kinds of Confucianism. The sociologist Gilbert Rozman has identified five varieties: an imperial and a reform Confucianism, and three types associated with social classes: the elite, the merchants, and the common people.[19] Looking at one or the other of these traditions, it is possible to find arguments in favor of restoring the prestige of Confucianism and advancing it as a philosophy capable of serving as a basis for the organization of modern society. In Taiwan, Singapore, and South Korea, and even in Japan as the occupation period receded into the past, these principles often seemed more attractive than American liberalism. Cohesive action, group spirit, the subordination of the individual to the superior interests of the collectivity—the family, the company, or society as a whole—are all values fitted to encourage industrialization and the mobilization of human and financial resources without ever really challenging the authority of the government, as long as it ensures the system is functioning properly. Confucianist societies are characterized by their strict discipline. What a contrast with the symptoms of anarchy their apologists observe in the United States—the increase in divorce and illegitimate births, the drug addiction, and the weakening of the work ethic! The limits placed on political freedom, or on the ability of individuals to pursue their own objectives, and on women's rights, are considered acceptable, on balance, when the advantages of social harmony are weighed against them. Lastly, Confucianism is by no means a fatalistic philosophy, for it encourages the pursuit of the highest level of education possible, and the development of will power and courage. For its defenders, then, it possesses all the necessary ingredients to make it a superior cultural dynamo for underdeveloped countries on a mission to catch up with those ahead. And indeed, to the extent that the Japanese, Taiwanese, Korean, and Singaporean societies followed quite different paths from the United States in their quest for ultrarapid growth, it is reasonable to think that their culture did help them make up the lost ground. But we can still ask whether it encourages them to investigate unexplored possibilities through research. It remains to be seen, for even within the large scientific and technological organizations from which most innovation comes there is a need for individualists, even eccentric ones, to explore new paths of discovery. The Confucianist societies, with their leanings toward conformism, are perhaps not the best adapted to embark on this stage of development.

A glorification of the differences between East and West is often used as an excuse to justify undemocratic powers. In the name of a rigid social system those in power gag any opposition, while the so-called consensus is used to defend the dominant interests. Resistance to Americanization is sometimes an ill-disguised front for stubborn conservatism. The example of American democracy is an inspiration for dissidents under authoritarian and even tyrannical regimes who call for greater freedom of expression. Chinese students erected a "Statue of Liberty" in Tienanmen Square during their revolt against the aging communist power structure. An undeniable liberalization was brought about in the recently industrialized countries during the 1980s—in Taiwan after the old guard of the Guomindang faded away, in South Korea, with the Army's return to barracks, and in the Philippines, where the Liberal Corazón Aquino replaced the Marcos dictatorship. As for Japan, it had never pulled back from the liberal democratic regime instituted by MacArthur's Constitution, in spite of the linguistic ambiguities between the English and Japanese texts.[20] The supremacy of the Liberal Democratic Party, which held sole power between 1955 and 1993, makes it a special case among the developed countries, but the lack of an alternative government can be attributed to the political ineptness of the opposition and to the Cold War, rather than to any subtle form of repression. Factional rivalries do not give the impression of harmony and consensus so dear to the Confucianists.

Is Japan then unique, as some of the emperor's subjects, convinced of Japanese racial purity, like to proclaim? On the one hand, of course it is, as are all countries, including the United States, where the concept of exceptionalism has become popular among historians.[21] But, on the other hand, Japan has a great deal in common with other nations which have achieved a similar level of development. The image of an inscrutable Japan hampers the spread of its culture and arouses hostility among publicists willing to exaggerate the threat it poses to America. In the 1980s "Japan-bashing" was very popular in the United States: Japan, since it was so different, could only be dangerous. This was the message proclaimed, for example, in the novels by Eric Von Lustbader, *The Ninja* (1981) and *The Miko* (1984) or in *Rising Sun* by Michael Crichton (1992). In this last work, which is set in Los Angeles, Inspector Peter J. Smith and his superior officer John Connor investigate a crime committed during the official opening of the Nakamoto company's new office building. Apart from its detective plot, the interest of the novel consists in the sometimes extremely didactic way in which the differences between American and Japanese cultures are represented. Dialogues in Japanese reinforces the impression that the book is presenting a serious analysis of the problem. Connor, as an expert, often praises Japan. In the United States, he says, "you think a certain amount of error is normal," and "expect things to go wrong all the time," but "Japan is different. Everything *works* in Japan. In a Tokyo train station, you can stand at a marked spot on the platform and when the train stops, the doors will open right in front of you. Trains are on time. Bags are not lost. Connections are not missed. Deadlines are met. Things happen as planned. The Japanese are educated, prepared, and motivated. They get things done. There's no screwing around."[22] But the constant repetition of such comments creates the image of an antiseptic Japan, where the individual exists only as a member of a group. This is a kind of society which can only repel the average American, in spite of its apparently superior efficiency. In this case, difference leads to rejection.

Images of America

However great their desire to be liked—a desire which faded away somewhat between the 1950s and 1980s—Americans make an ambivalent impression. On the one hand, their economic power, their high standard of living, and their free and easy manner, are appreciated, but on the other hand, their liability to take advantage of their power does create some dislike for them. In Latin America, where the cry of "Yankee go home!" is often heard, it is often followed by the words "And take me with you!"[23] There is a subtle mix of anti- and pro-Americanism, depending on the prevailing conditions and situation. What is certain is that people are familiar with America, and have their feelings about it. In Japan, positive opinions have generally prevailed over negative ones, except in 1973–1974, when they were evenly balanced. But because feelings of dependency exist alongside resentment, the perception of the United States is less stable in Japan than in the European countries.[24] Until 1968, the United States was the country the Japanese professed to like the most, but this percentage was already in decline by the beginning of the 1970s, because of the Vietnam War. For twelve years or so, the popularity of the United States fell below that of Switzerland, which had inspired positive feelings ever since MacArthur proclaimed his intention to make Japan into the "Switzerland of the East." American prestige fell even lower in 1974, when only 15 percent of Japanese expressed a preference for the United States. But after 1980, when the Cold War reintensified, it recovered its lead, and never lost it again, in spite of the ups and downs caused by economic discord. The occupation had left the Japanese with positive memories, and they had no intention of ever again becoming rivals of the United States in the international arena.[25]

In Communist China no polls have been conducted which might allow us to trace the evolution of anti-Americanism in the same way. The ordinary people seem fairly favorable to the United States, despite the intensely vitriolic propaganda campaign against the imperialist enemy waged by the authorities during the Maoist era (and even since then). These official attacks were obviously devoid of objectivity and showed little respect for the facts.

It was rare for the United States to be properly understood, even by specialists. The Communist Chinese "America-watchers" whom David Shambaugh has studied had only a very superficial view of the

"beautiful imperialist."[26] Until the 1980s, Marxists with Stalinist leanings predominated in the universities, the Academy of Social Sciences,
and the party's Central School. They believed that the United States,
with its rampant monopolistic capitalism, was ripe for a socialist revolution led by the working class and that Wall Street was pulling all the
strings in Washington. The ultimate objective of their analyses was to
work out the ties between a given president and some monopolistic
financial group or other—Truman with Morgan, Eisenhower with
Morgan and Rockefeller, Kennedy with Rockefeller, Johnson with the
Texas oil cartel, or Nixon and Reagan with the West Coast defense industries. They believed that the American political and economic
system was driven by a powerful conspiracy, at the expense of the
masses. In the 1980s, rather more subtle commentaries began to be
written by revisionists and non-Marxists who were better informed
about American pluralism, and who had gained a solid foothold in the
government bureaucracy and in the Institute for American Studies in
Peking. But basically the Chinese just found it very difficult to understand America, for their ancestral culture, which had not been substantially changed by a half-century of official Marxism-Leninism, ill-fitted
them to empathize with the essential characteristics of "foreign" cultures. They have been quite willing to borrow Western technology, but
not the democratic and individualist values which go along with it.
Their preference for clearly defined hierarchical relations led them to
see pluralism as the first indication of a disintegrating social order.[27]

OPEN ARMS FOR THE PACIFIC!

Until the mid-twentieth century, the Pacific civilizations had contributed very little to American culture, which until then had originated almost entirely in Europe, with some African elements in the
case of black Americans. The rejection of Asian immigration widened
the gulf. The United States, assured of its superiority, even seemed to
have nothing to learn from such inferior nations. The Chinatowns in
the major cities, mainly San Francisco, New York, and Chicago, were
considered unsavory places, teeming with prostitution and gang warfare, which is how they were portrayed in a Buster Keaton movie. The
works of art from Asia and the islands of the Pacific exhibited in the
museums of Boston or Chicago did little to dispel the prejudices which

were far from limited to the man in the street. But in this respect also everything was to change in the second half of the twentieth century, with the progressive disappearance of the cultural frontier.

The Resumption of Immigration from the Pacific

The 1924 laws imposing immigration quotas had reduced the flow of immigrants from the eastern Pacific to a trickle. All of Asia was entitled to only 1,323 immigrants per annum, and Oceania to 200, out of a total of 153,714. The Immigration and Nationality Act of 1952 did little to improve the situation, allowing for 3,590 and 700, respectively, out of a total of 158,261.[28] The citizens of the Latin American republics, considered to be "special immigrants," were not affected by these restrictions, but at this time little immigration was going to the United States from the countries lying between Guatemala and Chile.[29] Also, the spouses and children of American citizens were exempt, which enabled some Chinese, Filipinos, Australians, New Zealanders, and above all Japanese—women in the great majority—to gain admittance to America.[30] In addition, several thousand refugees came from China and Indonesia in particular. If America had decided in 1924 to put a stop to immigration, for the peoples of the Pacific the door was not completely closed.

The 1965 amendments to the Immigration Act brought about a major change. Quotas favoring Europeans were abolished, and parity was established between the Atlantic and the Pacific. A new selection system was introduced, based on relationship and on professional qualifications. The major groups included in the numerically restricted category were the spouses and dependent children of foreign residents, and the brothers and sisters of American citizens. The spouses, children, and parents of citizens were still exempt from any limitation, as were refugees from Indochina. This meant that immigration replenished itself by an endless chain effect, with immigrants who had taken out citizenship then being able to bring in family members.[31] Immigration figures for the Pacific continued to grow after the end of the 1960s (cf. Appendices, Table 22). With the United States once again welcoming immigration—between 1951–1960 and 1981–1990 the annual immigration rates doubled from 1.5 to 3.1 per thousand of the population, with the Pacific accounting for the largest share (approximately a third), while Europe's share fell gradually from 37 percent in 1961–1970 to 10 percent in 1981–1990.[32]

As in the economic sphere, the share of the Latin American countries of the Pacific seaboard was shrinking, in spite of their closer geographical situation. Its renewed growth in the 1980s can be explained by the bloody political unrest in the Central American republics, principally in El Salvador, but also in Guatemala and Nicaragua. The same explanation applies to Colombia and Peru in the southern continent. The new legislation had the most positive consequences for Asians. There were four dominant groups among the 3.5 million admitted between 1961 and 1990. The Filipinos were the most numerous at 27 percent, with ties to the United States dating back to the colonial period. Next came the Indochinese at 25 percent; the Chinese from Communist China, Taiwan, and Hong Kong at 23 percent; and the Koreans at 18 percent. Japan's share had fallen considerably, to 3.7 percent, since their country's economic boom gave people little cause to emigrate. Elsewhere, underdevelopment and political upheavals drove thousands out in the direction of a more peaceful and prosperous America. Special legislation opened the door to refugees. Out of 1,766,000 refugees admitted to the United States between 1961 and 1990, 48 percent came from the Pacific area, with the great majority originating in the three Indochinese countries, Vietnam, Laos, and Cambodia, from where the "boat people" were fleeing the communist system.[33] The Americans assumed responsibility for the debacle, and, more or less willingly, offered asylum to the people who had supported their war effort in Southeast Asia.

The Pacific Communities in America

This flow of immigrants gradually resulted in an uneven transformation of the ethnic composition of the American population. Recent censuses show that Hispanics from countries bordering the Pacific form a minority compared with the huge numbers from Mexico, Puerto Rico, and Cuba. But if Hispanics are perceived as making up a relatively homogeneous mass, the same cannot be said for the "Asians and Pacific Islanders," who since 1980 have been divided into 49 different groups— an indication that their critical mass had crossed a threshold, requiring a more discriminating method of enumeration (cf. Appendices, Table 23). And indeed, with 7,300,000 individuals, representing 3 percent of the American population in 1990,[34] it is the fastest-growing group, with a 95 percent increase during a decade during which the population of whites grew by only 6 percent, of blacks by 13 percent, of Indians by 38 percent, and of Hispanics by 53 percent. But there were considerable variations depending on origin. The Japanese, who had been the most numerous until 1976, were outnumbered, first by the Chinese and then by the Filipinos (cf. Appendices, Table 23), for once the influx of Japanese immigration had largely come to an end the community only grew by its own, rather weak, natural reproduction, while the other groups were swelled by evermore numerous influxes from abroad. In addition, the Asians and Pacific islanders were spread throughout the country very unevenly, even though they were present everywhere. The strongest relative concentrations were found in the states on the Pacific seaboard. In Hawaii, where 62 percent of the population was neither white nor black in 1990, the Japanese, Filipino, and Polynesian communities made up the majority. In California, where Los Angeles had become the principal entry point for immigration to the United States, the proportion was markedly lower, at 9 percent, but the total number amounted to 42 percent of all the Asians and islanders in the country, with a particularly strong representation of Filipinos, Vietnamese, and Chinese.[35] In the East, the New York City region had the third largest number, due to large concentrations of Chinese and Koreans, with Washington and its suburbs coming next. Among specific cases, it is worth mentioning the attraction of the South for the Vietnamese and the Thais, and of the Midwest for the Hmong, from Laos.

It is surprising that given such ethnic diversity, some observers still speak of a "model minority," as if all the different nationalities from Asia and the Pacific shared the same standards of living and education, and were integrated into American society to the same extent. Average figures do indeed indicate that, taken as a whole, the Asian-Pacific minority has the highest *family* income, ahead of the whites, the Hispanics, and African-Americans. Similarly, their children obtain top grades, to the envy of parents from other ethnic groups. But instead of congratulating themselves on these achievements, the representatives of the "model minority" prefer to put them in context, pointing out that if family incomes are high it is because all family members work (and family income and individual income should not be confused), and that though performances in school and university are exceptional in mathematics and science this is not so in the arts. Also, the persistence of discrimination and of hidden quotas is raised. The fact that they have difficulty finding employment is used to explain the Asians' tendency to fall back on their own initiative to open shops or small businesses. Unable to obtain bank loans, they get together to establish types of co- operatives which enable them to raise the initial funding required to set up a business or a small factory. There is some truth to these complaints, but nowadays discrimination seems to be very limited. The success of many Asians can be attributed to their work ethic, to their desire for edu- cation, and their adaptability—in short, to cultural factors—although we cannot exclude a genetic factor in the case of children of scientists and business executives, who are given priority by the system of prefer- ential criteria for immigration. The influx of Asians has undeniably contributed to the dynamism of the United States economy. Theirs is a genuine success story, as exemplified by Tsung Dao Lee and Cheng Nin Yang, winners of the Nobel Prize for Physics in 1957, by the Korean food stores, and the numbers of students attending the top universities (20 percent of admissions to Harvard in 1990–1991, compared to 5.5 per- cent in 1983), etc. However, the new animation brought to the neigh- borhood of Flushing in New York should not make us forget the deprivation of the Hmong mountain people, who are unable to survive without welfare.[36]

Pacific Contributions to Contemporary American Culture

To an America which glories in its multiculturalism, the peoples coming from the Pacific bring with them a special, and highly visible, quality. This is so in architecture above all. Every major city has its Chinatown, a Chinese neighborhood where the buildings are distin- guished by their pagoda-like style, and "Koreatowns" have been grow- ing up for the past twenty years. On shop fronts, eastern characters stand beside Roman lettering. These Chinatowns attract tourists in search of the exotic. Oriental restaurants were once one of their prin- cipal attractions, but now these are to be found scattered everywhere in downtown areas and in shopping centers, and American palates, in- creasingly attuned to gastronomic diversity, have developed a consid- erable liking for this cuisine—or, rather, these cuisines, for they are extremely varied, with Japanese, Chinese, Vietnamese, and Thai cook- ing all available. In a country so devoted to sports and physical exercise, the Far Eastern martial arts have attracted many devotees, although the spiritual values underlying them in their countries of origin take sec- ond place to physical prowess.

The Orient was long considered by Americans to be a heathen part of the world, crying out for Christian conversion. Because of its diversity, Protestantism was able to satisfy most people's spiritual needs. But, as values began to be questioned in the 1960s, the dominant religion was no longer able to satisfy the needs of large numbers of people. Oriental religions offered an answer for those who wanted to immerse themselves in mysticism, or submit to the will of a strongly united group. Buddhism, for instance, became known in the United States thanks to the World Parliament of Religions, during the Chicago Exhibition of 1893. Subsequently, Japanese missionaries spread its teachings, initially among communities of Asian origin, but more recently among white Americans who were intellectually attracted to a religion very different from Christianity. The Buddhist temples in America belong to the nonmeditative tradition, with its rich liturgy and ritual festivals, and address themselves mainly to Americans of Chinese and Japanese origin. White Americans prefer silent meditation, as taught by Zen Buddhism. In the 1950s, the literary bohemians of the Beat Generation (Jack Kerouac and Allen Ginsberg) claimed—while crudely distorting its message—to have turned Buddhism into a new lifestyle opposed to traditional American values, with the power to save them from their state of world-weariness. A decade later, the apostles of the counterculture, in search of "inner consciousness," after their experiments with psychedelic drugs, often turned to Buddhist meditation techniques to help them realize their "human potential." Christianity lost its monopoly, and in the "spiritual supermarket" the oriental religions became one product among many which were available to satisfy peoples' needs. Today, Zen centers are well integrated into urban society in San Francisco, Los Angeles, and Rochester, so that Buddhism, no longer confined to the arcane underground of the counterculture, is adapting itself to American society.[37]

In a reaction against the secular trend affecting the churches and the growing gulf between the public and private domains, religious sects offer an alternative to all those left rudderless by the disintegration of traditional values. Some reject the social order as corrupt. A charismatic individual can come to personify the path to salvation. Typical of this trend is the Unification Church, founded by a Korean, the Reverend Sun Myung Moon. Arriving in the United States in 1971, Moon claimed to have been chosen by God to complete Jesus' mission. His theology is based on the Bible, but also on the "Divine Principle" developed by

Moon himself, situating the Unification Church on the fringes of Christianity, rather like the Mormons. Moon, the "third Adam," preaches purification as a prelude to perfect matrimonial union. In 1982, at Madison Square Garden, he married 2,075 "blessed" couples, who were supposed to bring about the future unification of the world under the aegis of the church. The South Korean origin of the sect is very evident in its virulent anticommunism. Moon expects America to play the role of an archangel which will crush the satanic forces of communism, but before this can happen, war must first be declared on the decay of moral values and America must be made "one nation under God" once again. Moon's teaching does contain some totalitarian elements. The tentacles of the Unification Church reach into every area of society—politics, diplomacy, the media, and business. With the *Washington Times* it owns the second-largest daily newspaper published in the nation's capital. Nevertheless, its administration does not seem to be above suspicion, and in 1982 Moon was given a prison sentence for income-tax evasion.[38]

A flood of immigration is not enough to bring about cultural transfers from the Pacific to the United States—particularly given the size of the immigrant groups relative to the total American population. A further necessary condition is for Pacific cultures to become better known in America. Prior to World War II very few departments in American universities specialized in this region of the globe, and even when they did the numbers of teachers and researchers were usually very small. The responsibilities falling on United States shoulders after its victory over Japan made it essential to cultivate a better understanding of other cultures to further the adoption of appropriate policies. The anticommunism of the 1950s made life difficult for some scholars. The Institute of Pacific Relations, seen as a hotbed of progressives, came under attack from those who held it responsible for the "loss of China." Owen Lattimore, one of the most respected specialists—though his arrogance was also famous—was investigated, and *Pacific Affairs*, an "international review of Asia and the Pacific," was forced to move from the United States to British Columbia. Scholars were obliged to avoid sensitive current issues. This most detrimental situation changed, starting at the end of the 1950s, mostly as a result of the Vietnam War. Headed by John K. Fairbank in Chinese studies, and Edwin O. Reischauer in Japanese studies, both at Harvard, a sizable brigade of scholars, increasingly well-versed in the far eastern languages, some of them of Asian origin but others of American extraction, brought about a more penetrating, and often critical, perception of relations with the western Pacific and Asia, while excellent scholarly journals proliferated. Chairs were created in many universities, generally on the initiative of administrators, and sometimes also with external financial support, an example of which was the funding provided by the Japan Foundation to further the cultural influence of Japan and encourage friendly relations among nations, based on improved mutual understanding. The American public has never had so much reliable information on the complex world of the Pacific available as it does today. This does not prevent old stereotypes from sometimes showing their faces, but it may eventually reduce their harmful effects. Even at the end of the twentieth century, when the primacy given to the teaching of Western civilization in the university curriculum is under attack from extreme multiculturalists, the cultures of the Pacific still play no more than a secondary role, if only for the reason that since the seventeenth century the part they have played in the creation of United States society has been such a minor one.

The civilizations dwelling on the far side of the ocean once seemed so archaic and even barbaric that very few Americans could ever have thought they had much to learn from the East. But the Japanese economic boom of the past two decades have made them ask if there are not certain things they could learn from a partner who is at the same time their competitor. Japanese management practices were studied in university departments which provide training for future senior executives. Ezra Vogel of Harvard concluded a study published in 1979 by declaring that Japan had now become "number one," and that the United States had much to learn from it if it wished to get its economy back on track, improve productivity, motivate workers, and improve relations between large companies and their suppliers. Vogel argued that certain aspects of the Japanese business culture could be transplanted to America to the great benefit of both management and workers. So it seemed that Japan, far from being a unique and inimitable case, could be imitated with success. This point of view, which reverses the direction of cultural flow across the Pacific, was rejected ten years later by commentator James Fallows.[39] The Americans, proclaimed Fallows, are unique, and instead of trying to imitate Japan they should try to be more like themselves! If they were to recover their past

economic leadership, they would have to return to their old individualist values, with the resultant social mobility and the possibility for
each to control his or her own destiny. To his mind, too much importance was being given to education and to obtaining paper qualifications, for this was leading to a kind of covert Confucianism, which
could only result in a hierarchical class system where there would
no longer be any upward or downward mobility. A cultural debate
couched in such terms demonstrates how attitudes have evolved. Cultural exchange no longer travels in one direction. Cultural symbiosis,
even if it still is far off, is no longer a pipe dream. The cultural frontier,
too, is in the process of disappearing.[40]

The United States and the Pacific in the Twenty-first Century

During the years we have surveyed, relations between the Americans and the other Pacific peoples have undergone a complete transformation. On the economic level, the frontier has disappeared, and the Pacific, once a marginal zone in United States trade, now holds pride of place, ahead of Western Europe. In this respect, the two oceans are on an equal footing at the start of the twenty-first century. On a political and strategic level, too, the outlines of the frontier have become totally blurred. The end of the Cold War as a confrontation between the United States and Soviet Russia has reduced the importance of the Atlantic as the heart of the free world. Economic disputes, which have now replaced military conflict as the major manifestations of the rivalry between major nations, have given the Pacific a new significance where America's vital interests are concerned. Finally, even if the symbiosis is not yet complete in the cultural domain (America still shares most of its values with the Atlantic world), the era of contempt and mutual incomprehension has been left behind. Is the Pacific the last remaining frontier of the United States? Today, we can answer this question in the negative, for the frontier has been as good as eliminated, except in the cultural domain. The Pacific is now as central as the Atlantic. The process we set out to describe is now complete.

What will happen in the twenty-first century? Will the Pacific itself not become the center of the world, with the Atlantic being shifted increasingly to the margins, bringing about, for the first time, a reversal of a hierarchy dating back to the great sixteenth-century voyages of discovery? After a twentieth century which the American journalist Henry Luce called the "American century," are we not about to see a "Pacific century" as the third millennium begins? Any such prediction can run aground on one of two shoals. The first is the temptation to project exponentially the trends of the past onto future decades, as if the fate of human societies were decided by some inexorable determinism. The second is the error of adhering too close to immediate reality, and seeing the prevailing conditions, which never fail to produce abrupt twists and turns, as delineating the way of the future. The events of the past few years should have taught us to be extremely circumspect. The collapse of Soviet communism was an event so unexpected that no rational prediction could have anticipated it. As for the prevailing conditions, an analysis of the media is enough to induce dizziness. By the end of the 1980s, everyone was convinced that the United States was in decline, in comparison with European dynamism and the extraordinary growth of the Japanese economy. America, everyone said, was being left behind in technology, and was even destined for underdevelopment. Suddenly, early in the 1990s—this was particularly obvious after Bill Clinton became president—the European Union, which had been struggling to bring itself into being, was surpassing every unemployment record for the years since World War II, and was struggling to reignite economic growth. Japan, too, whose wealth was estimated above that of the United States when its financial bubble was at its height, suffered a serious recession, and its leadership in high technology began to look doubtful. The economic crisis which devastated Southeast Asia starting in the summer of 1997, seemed to have undermined the foundations of the "Asian miracles" and put an abrupt end to an expansion which it once seemed nothing could stop. In America, on the other hand, euphoria seemed to reign. With the exception of the negative trade balance, all the signs looked positive as the century drew to a close. There was sustained growth, and very low unemployment. Inflation was under control, while productivity gains were being achieved once again. The dollar was strong. Dynamic high-technology industries were creating employment. There was a surplus in the federal budget. The stock markets rose so rapidly that they left the observer amazed, if not incredulous. In 1998 only Communist China in the Pacific region could still boast of a high growth rate, but even this was showing clear signs of a slowdown as serious problems arose in employment and in the financial viability of numerous state enterprises. If in 1988 we had attempted to predict the future by looking at past trends and the prevailing conditions, we might have prophesied the decline of the United States in the Pacific, because its dynamism was being stifled. But today, looking at the same factors, is it not tempting to envisage a radiant future for them?

The Pacific zone, as we have defined it in this study, is home to about a third of the human race. Its demographic growth is slightly slower than in the rest of the world, for in China, and particularly in Japan, the birthrate is lower than in the underdeveloped countries. If this trend continues, the relative importance of the Pacific region in the world economy can only grow if per capita productivity increases more than elsewhere. In spite of the recession which began in Thailand in July 1997 and has temporarily affected the entire Asian region, this is probably what will occur, in the longer term. A combination of quite strong population growth and strong economic expansion should propel the Pacific region as a whole, from the shores of far eastern Asia to the shores of Latin America, and the archipelagoes in between, to a position comparable to that of the great continental regions of North America and Europe. The latter is fated to decline, in relative terms, even if it does manage to recover a high rate of per capita productivity, for its population growth is negligible. As for the other geographic regions, they are either dropping behind, in relative terms, like Africa, or making modest progress, like the Latin American countries facing the Atlantic. This being the case, the United States can only take an increasing interest in the Pacific region, since the absolute value of its interests there will become more and more substantial.

It seems rather unlikely that we are moving toward the creation of self-sufficient economic blocs—a "fortress Europe" on the one hand, and a North American Free Trade Area on the other, with, in the East, an Asian zone under the aegis of Japan. Despite setbacks, multilateralism will endure, and will even be strengthened as the per capita production of countries converge. However, it is possible that in times of increased tension we will witness the creation of customs unions. Japan will extend its sphere of influence in the direction of Southeast Asia, in order to take advantage of lower labor costs, but its investments there will have to compete against the American capital which is already in place, or which will move there for the same reason. Japanese and American interests are now sufficiently intermeshed for there to be little likelihood that we shall see the development of two antagonistic currency zones, a yen zone in the West and a dollar zone in the East. Such a basically optimistic view of things is diametrically opposed to that of George Friedman and Meredith Lebard who, on the basis of inflexible geostrategic imperatives, prophesy a future war between the United States and Japan, whatever their conscious intentions may be, simply because they are both "reasonable nations living in a dangerous world."[1] However, this Hobbesian conception of international relations overlooks the commonality of interests, and the value of cooperation, in the name of an uncompromising defense of acquired or desired advantages. In any case, is it not very questionable to attribute to Japan the role of the United States' principal enemy, now, at the beginning of the twenty-first century? China is a much stronger military power in the Far East, and a much more likely aspirant to regional hegemony.

Strategically, the Americans will never withdraw to within their own frontiers. Only they are in any position to impose a relative peace on the Pacific, thanks to their status as a military superpower. They play an indispensable role in the security of the Pacific nations. No other country can challenge this supremacy without devoting enormous resources to the effort, thereby weakening itself economically. Washington has no desire to assume the role of policeman, but simply wishes to prevent the emergence of any dangerous power capable of terrorizing its neighbors in order to exploit them as it likes, and to avoid a situation where the fears arising from an imbalance of power in the region might lead to all kinds of compromises, including some dangerous to United States interests. Such a policy should not impose an excessively heavy drain on the federal budget.

There is every indication that the cultural influence of the United States will increase during this century. American values express a universal message which civilizations obsessed with their uniqueness are unable to offer the rest of the world. Once an initial stage has been passed, development encourages individualism and freedom from traditional restraints, while not necessarily implying a plunge into license or anarchy. By turning to the values of liberty tempered by collaboration, the Americans have an ability to react and adapt which enables them to progress, despite occasional setbacks.

Will the changes which can be expected to result from the new ethnic composition of the population modify these fundamental traits? It seems unlikely to do so to more than a superficial extent, for American culture is characterized by both dynamism and inertia.[2] The Pacific, in the twenty-first century, will probably resemble America more than it does today.

With its two seaboards, one facing the Atlantic and the other the Pacific, the United States enjoys the advantage of direct access to two of the three principal economic, political, and cultural regions of the world, while, in addition, its ever-closer links with Canada and Mexico and, in the future, other South American countries, make its situation unique. With the advances in transportation, the lowering of costs, and almost instantaneous communications, geographic location is bound to lose a good part of its influence in future configurations. Nevertheless, something will always remain to make things different. This is why the demise of the frontier makes future history such a fascinating enigma.

APPENDICES

1. Population of States and Territories on the Pacific Coast

	HI	AK	CA	OR	WA	Total	Pac./U.S.	CA/Pac.
			Population in Thousands					
1850			93	12	1	106	0.5	88
1860			380	52	12	444	1.4	86
1870		30	560	91	24	705	1.8	79
1880		33	865	175	75	1,148	2.3	75
1890		32	1,213	318	357	1,920	3.0	63
1900	154	64	1,485	414	518	2,635	3.5	56
1910	192	64	2,378	673	1,142	4,449	4.8	53
1920	256	55	3,427	783	1,357	5,878	5.5	58
1930	368	59	5,677	954	1,563	8,621	7.0	66
1940	423	73	6,907	1,090	1,736	10,229	7.7	68
1950	500	129	10,586	1,521	2,379	15,115	10.0	70
1960	633	226	15,717	1,769	2,853	21,198	11.8	74
1970	770	303	19,971	2,092	3,413	26,549	13.1	75
1980	965	402	23,668	2,633	4,132	31,800	14.0	74
1990	1,108	550	29,760	2,842	4,867	39,127	15.7	76

AK = Alaska; CA = California; HI = Hawaii; OR = Oregon; WA = Washington.
Pac./U.S. = Population of the five Pacific states as a percentage of the total United States population.
CA/Pac. = Population of California as a percentage of that of the five Pacific states.
Source: Historical Statistics of the United States; Statistical Abstract of the United States.

2. United States Trade with the Pacific

Year (1)	Imp (2)	Exp (3)	P/Ai (4)	P/Ae (5)	Ch/Pi (6)	Ch/Pe (7)	P/Acot (8)	Cot/P (9)
1821	3.6	2.1	6.6	3.8	86	43		
1822	6.3	2.2	7.8	3.6	84	39		
1823	7.8	3.1	10.7	4.6	84	34		
1824	6.5	2.9	9.0	4.2	87	29		
1825	8.3	4.2	9.2	4.6	91	25		
1826	9.6	3.5	12.3	4.8	78	26	16	5
1827	4.7	3.9	6.6	5.3	77	34	43	13
1828	7.4	4.8	9.1	7.5	72	21	65	14
1829	6.0	3.1	8.9	4.5	78	25	47	19
1830	5.3	3.0	8.4	4.2	73	22	31	14
1831	5.0	3.2	5.2	4.4	62	29	36	13
1832	7.7	2.7	8.1	3.4	69	29	36	16
1833	10.0	3.9	9.9	4.5	75	29	31	20
1834	9.8	3.0	9.0	2.9	80	21	29	21
1835	8.6	2.3	6.3	2.0	70	21	23	29
1836	10.8	2.7	6.1	2.1	68	29	32	27
1837	12.9	2.9	9.9	2.6	69	17	48	48
1838	7.3	3.2	7.6	3.0	65	25	44	52
1839	7.1	3.1	4.5	2.7	52	18	53	51
1840	10.0	3.4	10.2	2.7	66	16	49	52
1841	6.8	3.3	5.5	2.9	59	24	40	38
1842	8.7	3.5	9.1	3.5	57	24	58	49
1843	6.5	3.7	15.3	4.5	67	50	61	53
1844	7.7	3.4	7.5	3.2	64	35	47	40
1845	10.2	4.8	9.0	4.6	71	44	58	52
1846	10.2	4.0	8.7	3.7	64	30	65	57
1847	9.5	4.6	7.8	2.9	59	39	71	63
1848	11.5	5.1	7.7	3.7	70	41	58	64
1849	9.4	5.3	6.7	3.8	58	30	63	59
1850	11.0	5.1	6.4	3.5	60	31	51	47
1851	12.0	8.8	5.7	4.7	59	27	50	41
1852	16.5	8.4	8.0	5.0	64	31	56	51
1853	16.2	12.8	6.1	6.3	65	25	55	38
1854	19.7	9.7	6.7	4.1	53	13	42	24
1855	20.2	11.2	7.8	5.1	51	9	45	23
1856	20.2	14.2	6.5	5.0	52	14	41	20
1857	20.0	12.2	5.7	4.2	42	17	38	19

2. United States Trade with the Pacific (*cont.*)

Year (1)	Imp (2)	Exp (3)	P/Ai (4)	P/Ae (5)	Ch/Pi (6)	Ch/Pe (7)	P/Acot (8)	Cot/P (9)
1858	21.3	11.7	8.1	4.3	50	26	47	23
1859	22.3	12.3	6.7	4.2	48	32	47	32
1860	24.1	17.3	6.8	5.2	56	33	57	36
1861	24.7	13.0	8.5	3.2	46	32	46	28
1862	14.0	10.9	7.4	5.7	54	20	28	8
1863	18.2	10.1	7.5	4.9	61	29	5	1
1864	21.8	12.0	6.9	7.6	48	15	16	1
1865	15.8	9.6	6.6	5.8	32	15	10	2
1866	22.0	12.4	5.1	3.6	46	19	50	5
1867	28.1	14.7	7.1	5.0	43	18	54	12

(1) Fiscal year ending September 30 up to 1842, and then on June 30 (e.g., 1821: from October 1, 1820, to September 30, 1821; 1860: from July 1, 1859, to June 30, 1860. In 1843, therefore, the year is reduced to nine months, from October 1, 1842, to June 30, 1843.
(2) Imp= imports in millions of gold dollars.
(3) Exp = exports and reexports in millions of gold dollars.
(4) P/Ai = percentage of American imports from the Pacific.
(5) P/Ae = percentage of American exports and reexports to the Pacific.
(6) Ch/Pi = percentage of American imports from the Pacific originating in China.
(7) Ch/Pe = percentage of American exports and reexports to the Pacific going to China.
(8) P/Acot = percentage of exports of American cotton textiles destined for the Pacific.
(9) Cot/P = percentage of cotton textiles in American exports to the Pacific.
Source: Foreign Commerce and Navigation of the United States, annual.

3.   Ships Engaged in the Fur Trade on the Northwest Coast

Year	Total	American	British	Spanish	French	Portuguese	Swedish
1788	7 (2)	2★	5	(2)			
1789	9 (2)	4	5	(2)			
1790	4 (3)	2	2	(3)			
1791	12★ (6)	6★	4	(6)	I		I★
1792	21★ (10)	6	9(3)	(7)	I	4	I★
1793	14³ (8)	7³	5³ (3)	(5)	I	I	
1794	11 (5)	5	6 (2)	(3)			
1795	7	3	4				
1796	7	3	4				
1797	5	3	2				
1798	7 ?	5?	2				
1799	9	6	3				
1800	9 ???	8??	I?				
1801	21 ?	18	3?				
1802	16	14	2				
1803	11	11					
1804	7	7					
1805	8	8					
1806	11	11					
1807	11	10	I				
1808	9	7	2				
1809	8	7	I				
1810	10	10					
1811	14	14					
1812	12	12					
1813	10	10					
1814	6	3	3				
1815	8	6	2				
1816	12	10	2				
1817	16	14	I		I		
1818	12	10	I		I		
1819	7	7					
1820	9	9					
1821	9	9					

3. Ships Engaged in the Fur Trade on the Northwest Coast (*cont.*)

Year	Total	American	British	Spanish	French	Portuguese	Swedish
1822	12	12					
1823	10	10					
1824	11	11					
1825	11	10	1				

* Possibly one questionable ship should be added.
³ Add a tender.
? Questionable voyage.
( ) = number of ships present not trading in furs—to be added to the genuine fur traders.

Sources: F. Howay, "A List of Trading Vessels in Maritime Fur Trade," *Transactions of the Royal Society of Canada*, Section II, vol. XXIV, May 1930, pp. 111–134, (1785–1794); vol. XXV, May 1931, pp. 117–149 (1795–1804); vol. XXVI, May 1932, pp. 43–86 (1805–1814), vol. XXVII, May 1932, pp. 119–147 (1815–1819); vol. XXVIII, May 1934, pp. 11–49(1820–1825); Derek Pethick, *The Nootka Connection, Europe and the Northwest Coast, 1790–1795,* (Vancouver, British Columbia: Douglas & McIntyre, 1980); Adele Odgen, *The California Sea-Otter Trade, 1784–1848,* (Berkeley, Calif.: University of California Press, 1941), pp. 155–182.

4. Trade between the United States and Pacific Countries in 1859–1860, Showing Percentage of Commodities

	Imports	Exports
Central America	hides (38), coffee(36), wood (14)	cotton textiles (33), flour (8), manufactures of iron (7)
Colombia*	cigars and tobacco (22), hides (21), textiles (11), coffee (7), tree-bark (7), indigo (6), rubber (4), wood (3), cocoa (3), cochineal (2)	manufactures of iron (12), grains (10), meat (9), cotton textiles (8), medicines (6), wood (4), shoes (2)
Ecuador		wood (21), manufactures of iron (20), glass (13), ice (13), flour (12)
Peru	guano (30), wool (16), copper (11), wood (9), saltpeter (8)	cotton textiles (30), wood and furniture (13), manufactures of iron (11), sugar (6), meat (6), cereals (4), quicksilver (3)
Chile	copper (58), wool (17), hides (13), saltpeter (3)	cotton textiles (41), sugar (12), manufactures of iron (7), meat (7), wood and furniture (5), tallow (4), quicksilver (2)
Japan	rape seed oil (30), lacquered articles (11), tea (8)	alcoholic beverages (18), manufactures of iron (11)
China	tea (65), raw silk (8), silks (6), sugar (5), cassia (2), matting (2), shawls (2)	cotton textiles (68), cereals (6), ginseng (5), meat (3), manufactures of iron (2), tobacco (2), coal (2)
Dutch East Indies	coffee (61), sugar (12), spices (9), rubber (6)	cotton textiles (61), flour (13), manufactures of iron (3)
Philippines	hemp (58), sugar (27), indigo (6), cigars (3)	cotton textiles (46), flour (26), copper (8), manufactures of iron (2)

4. Trade between the United States and Pacific Countries in 1859–1860, Showing Percentage of Commodities (*cont.*)

	Imports	Exports
Australia and New Zealand	coal (31), alcoholic beverages (10), gums (5), wool (2)	cereals and flour (26), tobacco (25), wood and furniture (18), manufactures of iron (10), vehicles (4), shoes (3), cotton textiles (1)
Hawaii	sugar and molasses (31), hides (18), whale oil (8), copra oil (8)	cotton textiles (11), manufactures of iron (9), clothing (8), meat (5), cereals (4)
Other Islands in Oceania	whale oil (46), oranges (19), guano (18)	wood (17), cereals (14), manufactures of iron (10), cotton textiles (8)

*A large proportion of the trade with Colombia was actually in transit across the Isthmus of Panama.

Source: *Foreign Commerce and Navigation of the United States, 1860.*

5. Whaling Ships Sailing to the Pacific

	NB	NA	SM	NNE	RI	CT	NY	OTH	Total
1791	1	4							5
1792		1		1			1		3
1793		2							2
1794		2							2
1795	1	2					1		4
1796									0
1797	7	8					3		18
1798	2	3							5
1799	2	4							6
1800	4	3				1			8
1801	1	1							2
1802	1	7							8
1803		4					2		6
1804	2	7							9
1805									0
1806	3	7							10
1807	2	4							6
1808	6	6							12
1809	2	18							20
1810	5	12							17
1811	4	17							21
1812		4							4
1813									0
1814									0
1815	5	20					2		27
1816	1	9	1						11
1817	4	14					2		20
1818	10	23	2						35
1819	6	18							24
1820	15	31	2	1	2		6		57
1821	19	25	2	3	2	2	4		57
1822	15	26	1	2	1	4	1		50
1823	4	12	2				2		20
1824	14	14	2	1	1	1	3		36
1825	18	25	3	1	2	2			51
1826	12	19	1				2		34
1827	19	13	3	1	1	7	1		45

5.   Whaling Ships Sailing to the Pacific (*cont.*)

	NB	NA	SM	NNE	RI	CT	NY	OTH	Total
1828	27	21	3	1	4		1		57
1829	12	17		1	2		1		33
1830	27	13	3	1	4	3	1		52
1831	44	15	7		4	3	5		78
1832	24	27	7	6	8	3	4		79
1833	39	16	6	11	11	3	11		97
1834	29	15	4	4	7	2	5	1	67
1835	40	17	4	2	6	1	4		74
1836	27	25	7	4	7		4	1	75
1837	34	18	4	5	11	1	2	1	76
1838	43	17	4	3	7	3	1		78
1839	58	19	8	2	7	5	11	3	113
1840	43	25	7	2	9	3	8		97
1841	61	29	8	2	10	8	17	1	136
1842	41	13	5	2	8	9			78
1843	65	15	6	2	13	20	19		140
1844	71	14	10	2	10	41	27		175
1845	82	29	8	2	11	42	32		206
1846	61	14	5		7	15	14		116
1847	68	12	10		5	20	15		130
1848	68	17	8		6	24	8		131
1849	51	6	3		6	15	6		87
1850	82	13	6		2	22	5		130
1851	122	15	11		4	37	16	2	207
1852	60	9	5		6	3	2	8	93
1853	98	11	15	1	4	14	5	7	155
1854	88	8	8	1	2	12	10	5	134
1855	77	11	8		3	2	1	8	110
1856	80	7	9		9	12	3	9	129
1857	86	1	9		1	13	5	4	119
1858	59	3	7			6		3	78
1859	48	3	4			2			57
1860	45	5	4	1	1				56
1861	9		2		1			1	13
1862	27	1	3			4			35
1863	30					1			31
1864	28		1			1	1		31
1865	39		2			1	1	1	44

5. Whaling Ships Sailing to the Pacific (*cont.*)

	NB	NA	SM	NNE	RI	CT	NY	OTH	Total
1866	28	3	3	1		1			36
1867	22		1	1	2				26
1868	19	2				1	3	1	26
1869	24	1	3					4	32
1870	20				1			5	26
1871	23				1			1	25
1872	17		1			1			19
1873	3		1					1	5
1874	10						1		11
1875	16							1	17
1876	11		1					2	14

NB = New Bedford and Fairhaven; NA = Nantucket; SM = other ports in southern Massachusetts; NNE = Northern New England (northern Massachusetts, New Hampshire, and Maine); RI = Rhode Island; CT = Connecticut; NY = New York and New Jersey; OTH = other Ports (including San Francisco).

Source: Senate Miscellaneous Documents, 44th Congress, 1st Session, no. 107, U.S. Commission of Fish and Fisheries, Part IV, Report of the Commissioner for 1875–1876, Appendix A: *The Sea Fisheries. I. History of the American Whale-Fishery from Its Earliest Inception to the Year 1877,* by Alexander Starbuck, pp. 180–659 (Washington, D.C.: Government Printing Office, 1878).

6. The Voyages of the *Lagoda*

No	Yr	Dur	Dest	SO	WO	Wb	C	NR	NRP	RRP
1	1841–43	719	Ind/Pac	575	2,100	17,000	28,919	37,498	29.7	56.6
2	1843–46	931	Pac	113	2,990	12,185	13,653	30,115	120.6	104.6
3	1846–49	1,024	Pac	68	2,734	28,400	17,425	29,092	67.0	62.8
4	1850–53	1,028	Pac/Arc	400	2,900	36,744	19,042	52,781	177.2	145.1
5	1853–56	935	Pac/Arc	—	2,500	35,010	31,635	62,570	97.8	97.8
6	1856–60	1,440	Pac	228	2,144	26,500	24,134	47,518	96.9	106.9
7	1860–64	1,331	Pac/Arc	270	2,175	26,026	20,959	97,159	363.6	153.8
8	1864–68	1,402	Pac/Arc	272	3,769	54,767	37,168	118,632	219.2	328.8
9	1868–73	1,777	Pac/Arc	720	2,860	34,000	36,328	78,204	115.3	156.0
10	1873–77	1,543	Pac	1,398	238	1,509	30,530	33,492	9.7	−2.0
11	1877–81	1,398	Atl	2,144	70	326	25,367	40,929	61.3	49.4
12	1882–86	1,511	Pac	1,395	450	478	19,429	13,363	−31.2	−5.0
Total		15,039		7,583	24,930	272,945	304,589	641,353	110.6	100.8
13	1887–90		Pac/Arc	75	130	3,650				

No = number of voyage; Yr = year; Dur = duration of voyage, in days; Dest = destination (Arc = Arctic; Ind = Indian Ocean; Pac = Pacific; Atl = Atlantic); SO = barrels of sperm oil; WO = barrels of whale oil; Wb = whalebone, in lbs; C = cost of initial fitting out, in current dollars; NR = net revenue to owners in current dollars; NRP = {[NR/C] −1} x 100 = nominal rate of profit, as a percentage; RRP = real rate of profit = {[NR/wholesale price index for month of arrival]/[C/wholesale price index for month of departure] −1} x 100. For the monthly wholesale price index, see George F. Warren and Frank A. Pearson, *Prices* (New York: Wiley, 1933), pp. 12–13.

Sources: Old Dartmouth Historical Society, New Bedford, *Jonathan Bourne Papers, Manuscripts 18, Box 7–8*. The figures sometimes differ from those given by Benjamin Baker, *History of the Bark Lagoda of New Bedford, Mass.: One of New Bedford's Most Successful Whaling Vessels*, in *Old Dartmouth Historical Sketches*, No. 45, 1916. pp. 33–47, and by Joseph Dias, *New Bedford Whaling Ships, 1783–1906*, Baker Library, Harvard Business School. The difference arises because for the fourth, fifth, and sixth voyages Baker neglected to include the quantities sent home in advance. Also, in converting the gallons of oil into barrels at a rate of 31.5 gallons per barrel, some errors, probably minor in nature, crept in. Figures for the thirteenth voyage are taken from Reginald B. Hegarty, *Returns of Whaling Vessels Sailing from American Ports. A Continuation of Alexander Starbuck's "History of the American Whale-Fishery"* (New Bedford, Mass.: ODHS, 1959).

7.   Chronology of Treaties with Pacific Countries (1818–1867)

	Latin America	Islands	Far East	North America
1818				Great Britain
1819				Spain
1824	Colombia			Russia
1825	Central America			
1826		Sandwich Is.		
1827				Great Britain
1831	Mexico			
1832	Chile			
1833			Siam	
1836	Peru-Bolivia Confederation.			
1839	Ecuador	Samoa		
1840		Fiji		
1844			China	
1846	Colombia			Great Britain
1848	Mexico			
1849	Guatemala	Sandwich Is.		
1850	El Salvador, Gt. Britain	Brunei		
1851	Costa Rica, Peru			
1853	Mexico (Gadsden)			
1854			Japan, Ryukyu	
1856			Siam	
1857			Japan	
1858	Bolivia		China, Japan	
1864	Honduras		Japan	
1866			Japan	
1867	Nicaragua			Russia (Alaska)

Source: Charles I. Bevans. *Treaties and Other International Agreements of the USA, 1776–1949,* 13 vols., Washington, D.C., 1968–1976.

8. American Trade with the Pacific (Percentages of Total Accounted for by the Pacific regions)

Year	Imp	Exp	P/Ai	P/Ae	Lat-Am		Chi		Jap		SEA		Oce	
					I	E	I	E	I	E	I	E	I	E
(1)	(2)	(3)	(4)	(5)	(6)	(7)	(8)	(9)	(10)	(11)	(12)	(13)	(14)	(15)
1868	29	13	8.2	4.7	22	40	39	22	8	4	26	2	4	32
1869	34	16	8.3	5.5	23	40	38	24	9	6	25	1	4	28
1870	40	14	9.1	3.6	22	49	37	19	8	3	30	1	4	28
1871	52	14	9.9	3.0	24	58	39	14	10	3	23	2	3	24
1872	67	18	10.7	4.1	15	58	40	14	10	4	27	1	3	21
1873	68	18	10.6	3.5	15	56	40	12	12	6	26	1	8	24
1874	54	18	9.5	3.0	23	56	34	11	12	5	25	3	7	24
1875	62	19	11.6	3.7	27	47	24	17	12	8	28	6	5	22
1876	55	18	11.9	3.3	16	41	23	23	28	5	27	4	8	26
1877	54	24	12.1	3.9	18	34	23	20	25	5	26	11	5	30
1878	56	28	12.9	4.0	20	31	32	24	13	8	28	6	8	31
1879	57	30	12.9	4.2	19	31	32	20	17	9	25	6	7	33
1880	78	26	11.7	3.1	17	35	31	15	19	10	23	12	10	27
1881	80	31	12.4	3.4	15	29	31	27	18	5	26	6	10	32
1882	83	39	11.4	5.2	19	27	27	23	17	9	23	10	14	33
1883	82	41	11.4	4.9	19	30	27	19	18	8	21	7	15	34
1884	76	41	11.4	5.6	18	35	23	19	15	6	27	7	17	32
1885	69	42	11.9	5.7	17	28	25	19	17	7	23	7	18	33
1886	78	44	12.3	6.5	15	26	26	25	19	7	22	7	18	33
1887	85	42	12.2	5.9	19	30	24	22	20	8	19	8	17	32

8. American Trade with the Pacific (Percentages of Total Accounted for by the Pacific regions) (cont.)

Year	Imp	Exp	P/Ai	P/Ae	Lat-Am		Chi		Jap		SEA		Oce	
					I	E	I	E	I	E	I	E	I	E
(1)	(2)	(3)	(4)	(5)	(6)	(7)	(8)	(9)	(10)	(11)	(12)	(13)	(14)	(15)
1888	88	44	12.1	6.3	19	31	21	18	21	10	21	8	18	33
1889	91	46	12.3	6.3	18	27	20	20	18	10	23	8	21	35
1890	93	45	11.8	5.3	17	29	18	16	23	12	24	6	18	36
1891	97	56	11.6	6.3	20	28	21	24	20	9	18	6	21	33
1892	101	46	12.7	4.5	19	32	21	23	24	7	20	5	17	33
1893	108	38	12.5	4.4	16	35	20	22	25	8	23	5	16	29
1894	91	40	13.9	4.5	18	29	20	25	21	10	25	7	16	29
1895	97	41	13.3	5.1	22	33	22	19	24	11	19	5	13	32
1896	114	56	14.6	6.4	18	30	21	21	22	14	22	4	17	30
1897	109	73	14.3	7.0	17	22	20	24	22	18	23	4	18	31
1898	112	75	18.2	6.1	16	18	19	22	22	27	23	3	21	29
1899	124	84	17.8	6.9	16	15	17	26	22	21	27	3	18	35
1900	158	117	18.6	8.4	15	13	18	20	21	25	29	5	18	35
1901	118	98	14.4	6.6	24	21	17	18	25	20	29	7	6	32
1902	127	110	14.0	8.0	20	16	18	30	30	20	26	7	6	26
1903	151	108	14.8	7.6	19	17	19	26	29	20	27	6	6	31
1904	158	108	15.9	7.4	22	22	20	22	29	23	23	7	5	26
1905	178	183	15.9	12.0	20	16	17	35	29	29	27	5	7	15
1906	189	173	15.4	9.9	23	24	16	29	28	23	27	5	7	17
1907	216	170	15.0	9.0	22	29	17	20	32	24	21	7	8	19

8. American Trade with the Pacific (Percentages of Total Accounted for by the Pacific regions) (*cont.*)

Year	Imp	Exp	P/Ai	P/Ae	Lat-Am		Chi		Jap		SEA		Oce	
					I	E	I	E	I	E	I	E	I	E
(1)	(2)	(3)	(4)	(5)	(6)	(7)	(8)	(9)	(10)	(11)	(12)	(13)	(14)	(15)
1908	210	178	17.6	9.6	21	28	13	18	32	24	18	9	16	20
1909	210	143	16.0	8.6	20	29	15	19	34	19	23	11	8	21
1910	219	151	14.0	8.7	24	33	15	15	30	15	22	14	9	23
1911	236	197	15.5	9.6	25	30	16	14	33	19	21	13	5	24
1912	250	236	15.1	10.7	25	28	13	15	32	23	24	13	5	20
1913	290	253	16.0	10.3	25	29	15	13	32	23	23	13	6	21
1914	302	256	15.9	10.8	25	29	14	14	36	20	17	14	8	22
1914	298	230	16.7	10.9	24	28	13	13	36	19	19	13	8	24
1915	386	297	21.7	8.4	26	25	15	10	28	16	22	12	10	22
1916	672	559	28.1	10.2	26	21	13	8	27	20	24	8	9	15
1917	985	663	33.3	10.6	27	23	14	9	27	29	28	11	4	12
1918	1,171	710	38.6	11.5	24	21	12	11	28	40	27	12	9	15
1919	1,204	993	30.8	12.5	18	17	15	13	35	39	24	13	7	13
1920	1,560	1,191	29.6	14.5	21	22	15	14	27	33	28	15	5	14
1921	686	701	27.3	15.6	21	18	16	18	37	35	20	13	5	16
1922	895	617	28.7	16.1	17	17	17	20	40	37	21	10	5	17
1923	1,111	760	29.3	18.2	18	18	19	17	32	37	26	9	5	19
1924	1,049	785	29.1	17.1	21	20	13	16	33	33	29	11	5	20
1925	1,393	815	33.0	16.6	16	23	13	13	28	29	38	12	6	23
1926	1,482	928	33.5	19.3	17	23	10	13	27	29	41	12	5	23

8. American Trade with the Pacific (Percentages of Total Accounted for by the Pacific regions) (cont.)

Year	Imp	Exp	P/Ai	P/Ae	Lat-Am		Chi		Jap		SEA		Oce	
					I	E	I	E	I	E	I	E	I	E
(1)	(2)	(3)	(4)	(5)	(6)	(7)	(8)	(9)	(10)	(11)	(12)	(13)	(14)	(15)
1927	1,324	876	31.6	18.0	16	22	13	12	31	30	37	14	4	22
1928	1,239	980	30.3	19.1	19	22	12	16	31	30	33	13	4	18
1929	1,403	994	31.9	19.0	20	23	13	14	31	27	32	15	4	19
1930	954	659	31.2	17.2	23	25	12	16	30	26	33	16	3	16
1931	662	479	31.7	19.7	24	20	11	23	31	33	31	15	3	9
1932	422	348	31.9	21.6	24	15	7	19	32	39	35	17	2	11
1933	461	358	31.8	21.4	19	17	9	17	28	41	41	16	3	10
1934	524	503	31.7	23.6	20	18	9	15	23	43	45	13	3	11
1935	653	492	31.9	21.6	17	19	11	10	24	42	43	14	4	15
1936	749	531	30.9	21.6	15	20	11	10	23	39	46	16	5	15
1937	1,037	741	33.6	22.1	15	19	11	9	20	41	47	17	7	13
1938	603	676	30.8	21.8	21	21	8	8	21	38	47	19	3	14
1939	743	732	32.0	23.0	19	24	9	10	22	34	46	22	4	11
1940	1,013	840	38.6	20.9	18	28	9	11	16	28	53	21	3	11
1941	1,297	931	38.8	18.1	21	35	7	14	6	7	54	31	12	13

Imp = imports, in millions of current dollars; Exp = exports and reexports, in millions of current dollars; P/Ai = proportion of total U.S. imports coming from the Pacific, as a percentage; P/Ae = proportion of U.S. total exports going to the Pacific, as a percentage. The figures for 1891 and 1892 have been corrected (cf. J. Heffer, "A New York: les importations en provenance du Brésil en 1890 and 1891–92," in F. Mauro, ed. *Transport et commerce en Amérique Latine*, Paris: L'Harmattan, 1990, pp. 168). Lat Am = Latin America; Chi = China and Hong Kong; Jap = Japan; SEA = Southeast Asia; Oce = Oceania. The totals of the five regions' shares in imports from the Pacific (I) and exports to the Pacific (E) do not always total 100, for the percentages are rounded off, and the far eastern part of Russia is not included. Before 1914, the year was the fiscal year, from July 1 to June 30, while from 1914 on, it is the calendar year.

Source: Foreign Commerce and Navigation of the United States.

9.   American Trade with the Pacific, by Product 1905–1909, in Percentages

	Imports			*Exports*	
1.	Silk	24	Machinery		19
2.	Sugar	8	Cotton textiles		13
3.	Nitrates	7	Petroleum		12
4.	Tea	6	Cereals and meat		11
5.	Coffee	6	Cotton		8
6.	Wool	5	Wood		5
7.	Tin	5	Tobacco		4
8.	Hemp	5	Leather		3
9.	Hides	3			
10.	Copper	3			
11.	Silks	3			
12.	Bananas	3			
13.	Matting	2			
14.	Gums	2			
15.	Rubber	2			
16.	Spices	1			
17.	Cocoa	1			
	Other products	14	Other products		25
	Total	100	Total		100

Source: Foreign Commerce and Navigation of the United States, 1909.

10. Structure of American Trade with the Pacific, by Country,
1905–1909, in Percentages

	Imports	Exports
Central America (Guatemala to Costa Rica)	coffee (48.5), bananas (36.1), rubber (4.9)	machinery (18), cereals (16.7), cotton textiles (16.5), meat (6.6), wood (5.3), leather (3.7), petroleum (1.9).
Panama	bananas (56.1), rubber (8.7), hides (7.3)	machinery (30.2), wood (11.4), meat (8.3), vehicles (7.8) cotton textiles (4.9), coal (4.3), cereals (3.9), petroleum (2)
Colombia	coffee (61.2), hides (12.3), bananas and coconuts (10.3), rubber (4.8)	cotton textiles (22.6), machinery (22.1), cereals (11.6), meat (6.9), leather (4.1), petroleum (3.5), wood (2.6)
Ecuador	cocoa (48.5), rubber (19.4), plant ivory (9.6), hides (9)	machinery (26.2), meat (19.3), cereals (14.9), cotton textiles (6.6), petroleum (4.9), leather (3.3), wood (2.6)
Peru	copper (43.6), sugar (13.7), raw cotton (11.9), nitrates (9), wool (6), hides (5.1), rubber (3.3)	machinery (34.2), wood (15.5), cereals (10.3), meat (5.2), vehicles (3.4), petroleum (2.9), cotton textiles (2.5)
Bolivia		machinery (47.9), cotton textiles (15.7), vehicles (10.3)
Chile	nitrates (84.6), copper (12.2)	machinery (25.5), petroleum (13.4), wood (11.2), cotton textiles (9.7), vehicles (4.8), agricultural machinery (3.9), cereals (3), meat (2.8)
Russian Asia	hides and furs (26.2), wool (21.2), licorice (20.7)	cereals (38.7), agricultural machinery (28), machinery (7.2)
Japan	silk (58.5), tea (11.7), silks (8), matting (4), camphor and sulfur (2.3), porcelain (2.2), copper (1.8)	raw cotton (31.1), machinery (23.7), cereals (11.2), petroleum (10.9), leather (5.1), tobacco (1.5), cotton textiles (1.1)

10. Structure of American Trade with the Pacific, by Country, 1905–1909, in Percentages (*cont.*)

	Imports	*Exports*
Korea		machinery (7.3), petroleum (17.5)
China	silk (33.8), tea (15.6), hides and furs (12), wool (11.3), hats (4.6), opium (4.5), matting (3.1), vegetable oil (2.3), pig bristles (2), rice (1.9), fireworks (1.3), silks (1)	cotton textiles (45.9), petroleum (20.9), copper (10.3), cereals (6.7), tobacco (5.1), machinery (3.4)
Hong Kong	matting (26.4), vegetable oil (10.1), cement (7.6), rice (7.3), spices (6.2)	cereals (47.9), petroleum (13.4), ginseng (12.8), machinery (6.7), cotton textiles (2.2)
French Indochina		petroleum (88.4)
Siam	rice (63.3), spices (25.1)	petroleum (32.4), machinery (23.4), lamps (13.2)
Malaysia	tin (59.2), gums (10.3), spices (9.9), rubber (8.4), wood (3.9), cereals (3.7)	petroleum (22.6), tobacco (21), machinery (18.3), meat (11.7), cereals (9.4), cotton textiles (1)
Dutch East Indies	sugar (79.3), coffee (7.7), spices (2.1), gums (1.5)	petroleum (63.4), machinery (21.4)
Philippines	hemp (87.1), sugar (10)	machinery (23.6), cotton textiles (11.2), petroleum (8.7), explosives (6.2), cereals (5.8), leather (5.8), meat (4.8), wood (4), coal (3.1), instruments (2.4), vehicles (1.8)
Australia	wool (52.1), copper (24.7), hides (9.9), coal (6.4), tin (1.7)	machinery (25.9), wood (11.1), petroleum (9.4), tobacco (9), cotton textiles (4.2), leather (3.9), agricultural machinery (3.3), vehicles (2.9), chemicals (2.6), instruments (2.1)

10. Structure of American Trade with the Pacific, by Country,
    1905–1909, in Percentages (*cont.*)

	Imports	*Exports*
New Zealand	gums (43), wool (24), hides (17.8), vegetable fibers (10.6)	machinery (30.3), tobacco (9), petroleum (8.2), wood (5.3), cotton textiles (4.4), fruits (4.3), leather (3.8), agricultural machinery (3.5), vehicles (3.3), chemicals (3)
British Oceania	guano and phosphates (67.5)	wood (25)
French Oceania	copra (43.7), chromate of iron (37.2)	cereals (26.2), cotton textiles (14.4), wood (13.9), machinery (7.3)

Source: Foreign Commerce and Navigation of the United States, 1909.

11. Trade with the Pacific Region from American Ports in 1880 and 1900 (in Tons)

	Boston	New York	NE	South	San Francisco	OPP	Total
Incoming	*1880*						
SA	6,347	10,058	4,103	–	34,623	7,454	62,585
CA	–	2,347	–	–	90,015	1,262	93,624
FE	–	74,889	–	–	118,672	14,190	207,751
SEA	17,268	68,774		–	2,635	–	88,677
Islands	2,986	3,916	758	–	173,902	80,836	262,398
BC	–	–	–	–	185,906	155,836	341,742
Total	26,601	159,984	4,861	–	605,753	259,578	1,056,777
Outgoing	*1880*						
SA	3,934	9,325	461	–	9,315	6,134	29,169
CA	–	877	5,763	–	99,827	340	106,807
FE	–	52,420	6,000	–	70,971	4,598	133,989
SEA	1,177	70,837	–	3,926	–	–	75,490
Islands	14,752	29,807	528	–	67,386	23,381	135,854
BC	–	–	–	–	171,739	171,308	343,047
Total	19,863	162,816	12,752	3,926	419,238	205,761	824,356
Incoming	*1900*						
SA	2,182	71,142	34,129	5,471	21,639	26,502	161,065
CA	–	–	–	–	100,901	67,478	168,379
FE	–	96,420	5,119	–	192,056	382,986	676,581
SEA	66,137	146,830	80,205	2,269	123,563	17,733	436,737
Islands	2,867	19,991	11,171	–	314,448	173,796	522,273
BC	–	–	1,430	–	415,790	1,031,797	1,449,017
Total	71,186	334,383	132,054	7,740	1,168,397	1,700,292	3,414,052
Outgoing	*1900*						
SA	–	30,451	14,286	6,974	5,694	25,644	83,049
CA	–	–	–	240	95,491	29,878	125,609
FE	–	216,500	91,384	43,310	178,561	215,861	745,616
SEA	–	30,374	4,959	31,083	110,863	49,556	226,835
Islands	–	200,265	9,468	8,009	219,667	143,516	580,925
BC	–	–	–	–	420,545	1,025,101	1,445,646
Total	–	477,590	120,097	89,616	1,030,821	1,489,556	3,207,680

NE = Other ports in the Northeast (Baltimore, Philadelphia); OPP = other Pacific coast ports.
SA = South America (from Ecuador to Chile); CA = Central America (from Mexico to Colombia); FE = Far East (from China to Russian Asia); SEA = Southeast Asia (Dutch East Indies, Philippines); Islands = Australasia, Hawaii, the archipelagoes; BC = British Columbia.
Source: *Foreign Commerce and Navigation of the United States.*

12   Direct American Investments in the Pacific Region (Book Value in Millions of Dollars)

	1897	1908	1914	1919	1929	1940
Central America	11	38	90	110	250	190
South America	12	61	240	410	700	550
Asia★	20	70	110	140	360	350
Oceania	2	10	17	53	162	120
Total	45	179	457	713	1,472	1,210
Percent of total investments	7	11	17	18	20	17

Source: Based on Mira Wilkins, *The Emergence of Multinational Enterprise. American Business Abroad from the Colonial Era to 1914* (Cambridge, Mass.: Harvard University Press, 1970), p. 110; Idem., *The Maturing of Multinational Enterprise: American Business Abroad from 1914 to 1970* (Cambridge, Mass.: Harvard University Press, 1974), pp. 31, 55, 57–59, 182–183; Cleona Lewis, *America's Stake in International Investments* (New York: Arno Press, 1976), pp. 578, 579, 583, 590, 595.
★The Dutch East Indies are included in Asia.

13. Structure of Direct American Investment in the Pacific in 1929
(in Percentages)

	Manufacturing	Trade	Petroleum	Mining	Agriculture	Public Utilities	Total
Central America	3	★	2	3	53	39	100
South America	2	3	11	66	3	15	100
Asia	18	9	32	3	18	20	100
Oceania	31	14	51	4	–	–	100
Total	9	5	19	34	14	19	100
Central America	5	1	1	2	61	36	17
South America	11	27	28	95	9	38	48
Asia	46	43	42	2	30	26	24
Oceania	38	29	29	1		–	11
Total	100	100	100	100	100	100	100

Source: As for Table 12.
★ Less than 0.5.

14. Net U.S. Foreign Aid

	1945–1960	*1961–1975*	*1976–1990*
Total (millions of dollars)	78,626	120,091	161,040
Pacific	12,134	56,483	17,936
Pacific's share (percent)	15.4	47.0	11.1
Japan	3.6	0.8	-1.0
South Korea	3.6	8.5	0.6
Taiwan	2.2	3.5	-0.4
Indochina	2.2	23.2	0.0
Indonesia	0.6	1.6	1.1
Philippines	1.3	1.1	2.2
Thailand	0.3	1.4	0.5
Trust territories	0.1	0.5	1.2
Pacific Lat. Am. America	1.4	4.6	6.5
Others	0.2	1.9	0.4

Source: Statistical Abstract of the United States.

15. American Trade with the Pacific

Year	Imp	Exp	P/Ai	P/Ae	Lat-Am		Jap		SEA		4SD		Oce	
					I	E	I	E	I	E	I	E	I	E
(1)	(2)	(3)	(4)	(5)	(6)	(7)	(8)	(9)	(10)	(11)	(12)	(13)	(14)	(15)
1942	711	714	25.9	8.8	45	30	0	0	20	8	2	1	32	51
1943	628	853	18.6	6.6	58	27	0	0	1	0	0	0	39	67
1944	543	710	13.9	5.0	74	35	0	0	0	0	0	0	24	58
1945	577	850	13.9	8.7	68	38	0	0	2	8	0	0	30	42
1946	1,002	1,627	20.4	16.7	39	30	8	6	25	25	3	4	18	7
1947	1,347	2,348	23.5	16.3	39	37	3	3	37	28	6	6	12	7
1948	1,650	2,260	23.2	17.9	40	32	4	14	39	30	5	10	10	14
1949	1,532	2,389	23.1	19.8	42	32	5	20	37	27	4	11	8	7
1950	2,107	1,804	23.8	18.7	37	39	9	23	37	20	4	9	10	8
1951	2,777	2,437	25.3	18.1	33	34	7	25	41	27	5	6	16	7
1952	2,592	2,410	24.2	19.1	40	35	9	26	39	23	5	9	9	10
1953	2,443	2,490	22.5	21.4	47	34	11	27	32	23	4	9	8	9
1954	2,292	2,585	22.5	21.2	50	35	12	27	29	20	3	10	7	7
1955	2,547	2,743	22.4	19.8	42	35	17	24	33	20	3	11	7	9
1956	2,685	3,000	21.3	17.8	41	36	21	31	30	22	4	5	8	10
1957	2,649	3,748	20.4	19.9	39	29	23	33	29	18	4	13	8	8
1958	2,478	2,935	19.4	18.5	36	32	27	29	26	17	4	14	8	7
1959	3,230	2,966	21.2	18.8	29	29	32	32	25	17	5	12	10	8
1960	3,301	3,777	22.5	20.0	29	26	35	36	23	15	6	11	8	11
1961	3,181	4,333	21.6	22.6	30	24	33	41	22	16	6	11	10	13
1962	3,727	4,075	22.8	20.9	26	25	37	35	19	16	7	13	12	12

15. American Trade with the Pacific (cont.)

Year	Imp	Exp	P/Ai	P/Ae	Lat-Am		Jap		SEA		4SD		Oce	
					I	E	I	E	I	E	I	E	I	E
(1)	(2)	(3)	(4)	(5)	(6)	(7)	(8)	(9)	(10)	(11)	(12)	(13)	(14)	(15)
1963	3,983	4,596	23.2	21.9	25	24	38	38	18	16	7	12	13	11
1964	4,419	5,130	23.7	20.9	25	23	40	37	17	14	8	11	10	15
1965	5,354	5,540	25.1	21.1	23	23	45	37	15	14	9	11	8	15
1966	6,284	6,106	24.6	21.0	20	24	47	38	13	15	10	11	9	13
1967	6,465	6,706	24.1	22.0	19	19	46	40	13	15	12	13	9	14
1968	8,152	7,884	24.5	22.8	17	19	50	37	12	16	14	17	9	13
1969	9,538	8,622	26.5	22.7	13	17	51	40	11	14	16	19	9	12
1970	11,109	10,590	27.8	24.5	13	16	53	44	10	14	17	17	8	11
1971	12,864	9,959	28.2	22.6	10	17	56	41	9	15	19	19	7	12
1972	16,338	11,297	29.4	22.7	9	14	55	44	9	16	22	20	7	9
1973	19,404	18,629	28.1	26.1	9	12	50	45	11	13	24	21	8	9
1974	26,498	25,509	26.2	25.9	11	14	47	42	16	15	22	19	6	11
1975	25,251	24,011	26.1	22.3	11	16	45	40	18	15	22	22	6	10
1976	34,244	25,281	28.4	22.0	10	15	45	40	17	15	25	23	5	11
1977	41,286	27,329	28.0	22.5	10	16	45	39	17	14	25	24	4	11
1978	52,042	34,007	30.3	23.7	9	15	47	38	15	14	26	25	5	10
1979	59,140	45,983	28.7	25.3	10	13	44	38	16	15	26	26	5	9
1980	69,546	58,278	28.4	27.0	9	14	44	36	18	16	25	25	5	8
1981	80,485	61,123	30.8	26.2	7	14	47	36	16	14	26	25	4	11
1982	79,757	58,823	32.7	27.7	7	12	47	36	14	17	28	26	4	10
1983	91,072	57,937	35.3	28.9	7	11	45	38	15	17	30	29	3	8

Appendices

15. American Trade with the Pacific (cont.)

Year	Imp	Exp	P/Ai	P/Ae	Lat-Am		Jap		SEA		4SD		Oce	
					I	E	I	E	I	E	I	E	I	E
(1)	(2)	(3)	(4)	(5)	(6)	(7)	(8)	(9)	(10)	(11)	(12)	(13)	(14)	(15)
1984	119,301	62,207	36.6	28.6	6	10	48	38	13	15	30	28	3	9
1985	133,202	60,050	38.6	28.2	5	9	52	38	11	13	29	28	3	11
1986	153,502	65,967	40.1	34.6	5	9	53	41	9	20	30	35	2	6
1987	171,124	74,052	42.2	29.3	4	9	49	38	10	13	34	32	2	9
1988	186,707	98,557	42.3	30.8	4	7	48	38	11	11	34	34	3	8
1989	189,208	108,033	40.0	29.7	5	8	49	41	8	8	33	36	3	9
1990	188,245	115,974	38.0	29.4	5	8	48	42	9	9	32	35	3	9
1991	204,225	131,387	41.8	31.1	5	8	45	37	14	16	29	35	3	7
1992	226,335	142,154	42.5	31.7	5	9	43	34	16	17	28	34	2	7
1993	249,634	151,979	43.0	32.7	5	11	43	32	17	18	26	34	2	7
1994	284,604	169,358	42.9	33.0	5	10	42	32	19	19	25	35	2	7
1995	315,466	206,852	42.4	35.4	5	9	39	31	20	19	26	36	2	6
1996	319,026	209,373	40.3	35.5	5	9	36	32	21	18	26	36	2	7
1997	348,872	223,953	40.0	32.5	6	10	35	29	21	22	25	35	2	6
1998	362,650	196,346	39.7	28.7	6	12	34	29	20	20	24	32	2	7

IMP = imports, in millions of current dollars; EXP = exports and reexports, in millions of current dollars; P/Ai = percentage of total U.S. imports originating in the Pacific; P/Ae = percentage of US total exports destined for the Pacific, as a percentage; Lat-Am = Latin America; Jap = Japan; SEA = Southeast Asia; 4SD = South Korea, Taiwan, Hong Kong, Singapore; Oce = Oceania. The totals of the five regions' shares in imports from the Pacific (I) and exports to the Pacific (E) do not always total 100, for Singapore is included both in SEA and 4SD, and China is not included.
Source: Statistical Abstract of the United States.

16. Structure of Trade between the United States and the Pacific in 1961

General Categories	Australia EX	Australia IM	Japan EX	Japan IM	Taiwan EX	Taiwan IM	Korea EX	Korea IM	Philippines EX	Philippines IM	Indonesia EX	Indonesia IM	Malaysia EX	Malaysia IM	Chile EX	Chile IM	Peru EX	Peru IM
Foodstuffs	1	66	12	7	19	46	27	6	13	53	25	18	9	1	9	4	10	51
Beverages and tobacco	5	–	2	–	2	–	–	–	z	1	1	z	19	–	1	–	1	–
Raw materials	9	22	40	3	34	6	25	32	9	22	6	46	2	49	4	22	4	20
Fuels	3	–	9	z	2	–	1	–	2	–	4	33	5	–	3	–	2	–
Animal and vegetable oils	z	z	2	z	3	–	1	–	z	8	–	1	–	–	2	z	2	z
Chemicals	11	3	9	2	6	6	15	3	9	z	10	z	8	–	11	2	9	z
Manufactured articles 1	16	7	4	43	8	28	12	39	25	7	14	1	11	49	11	71	14	26
Manufactured articles 2	8	z	2	26	2	13	2	15	4	8	3	–	8	z	5	–	5	z
Machinery and Transport	46	z	20	17	21	z	15	z	35	–	37	–	36	–	51	–	51	–
Others	z	2	z	2	3	1	2	4	1	z	z	z	2	z	3	z	2	z

z = Less than 1 percent.

EX = Exports from the U.S. to the Pacific; IM = general imports from Pacific countries.

Manufactured goods are divided into two categories:

1: By raw material (leather, rubber, wood, paper, textiles, nonmetallic minerals, iron and steel, and nonferrous metals).

2: Various (plumbing supplies, heating and lighting, furniture; leather wares; clothing; shoes; scientific and professional instruments, photography, optics, clock and watch making, etc.).

Source: U.S. Bureau of the Census, Foreign Commerce and Navigation of the United States, 1946–1963, Tables 19 and 20, House Documents, 89th C., 1st Session, 1965, vol. 32, no. 219 (Washington, D.C.: GPO, 1965).

17. Principal Commodities Involved in Trade between the United States and the Pacific in 1961–1963 (as percentages)

Country (Japan = 100)	Exports	Imports
Colombia (14.6; 20.4)	nonelectrical machinery (25), electrical machinery (7), transport (18), cereals (5), others (45)	coffee (81), petroleum (11), others (8)
Ecuador (3.1; 5.2)	nonelectrical machinery (15), electrical machinery (5), transport (15), textiles (6), others (59)	bananas (50), coffee (17), cocoa (11), sugar (9), fish meal (8), others (5)
Peru (11.3; 15.4)	nonelectrical machinery (24), electrical machinery (8), transport (20), cereals (7), others (41)	sugar (30), copper and lead (26), ores (11), animal foodstuffs (9), coffee (9), raw cotton and wool (7), others (8)
Chile (11.6; 14.4)	nonelectrical machinery (27), electrical machinery (7), transport (17), cereals (7); others (42)	copper (71), ores (14), nitrates (7), others (3)
Thailand (4.7; 3.0)	nonelectrical machinery (23), electrical machinery (10), transport (10), tobacco (9), others (48)	rubber (57), fruits (18), plant fibers (10), others (15)
Vietnam (5.9; 0.3)	nonelectrical machinery (14), electrical machinery (7), dairy products (12), cereals (11), raw cotton (6), transport (5), pharmaceuticals (5), iron and steel (5), others (35)	rubber (65), others (35)
Malaysia (1.7; 13.2)	nonelectrical machinery (31), tobacco (19), others (50)	tin (49), rubber (48), others (3)
Singapore (3.1; 1.0)	nonelectrical machinery (36), tobacco (11), others (53)	rubber(61), tin (11), petroleum (6); others (22)
Indonesia (7.5; 10.5)	cereals (23), nonelectrical machinery (18), electrical machinery (5), transport (15), raw cotton (5), others (34)	rubber (36), petroleum (33), coffee and tea (9), spices (8), others (14)

17. Principal Commodities Involved in Trade between the United States and the Pacific in 1961–1963 (as percentages) *(cont.)*

Country (Japan = 100)	Exports	Imports
Philippines (18.9; 25.5)	nonelectrical machinery (20), electrical machinery (6), textiles (11), transport (9), cereals (7), raw cotton (7), others (40)	sugar (47), oil-producing grains (13), vegetable oils (8), clothing (8), fruits (6), wood (6); others (12)
South Korea (12.6; 1.1)	cereals (25), raw cotton (17), fertilizer (10), nonelectrical machinery (9), others (39)	manufactured wood products (21), textiles (15), silk (14), clothing (12), fish (6), ores (6), others (26)
Hong Kong (8.2; 12.4)	raw cotton (12), tobacco (9), nonelectrical machinery (8), plastics (6), pharmaceuticals (6), fruits (6), textiles (5); others (48)	clothing (35), textiles (12), others (53)
Taiwan (8.4; 4.0)	raw cotton (21), cereals (18), nonelectrical machinery (11), electrical machinery (7), oil-producing grains (9), others (34)	sugar (25), manufactured wood products (17), fruits and vegetables (16), clothing (11), textiles (8), others (23)
Japan (100; 100)	nonelectrical machinery (15), scrap metal (12), raw cotton (10), cereals (9), oil-producing grains (8), petroleum (5), others (41)	electrical machinery (12), textiles (12), iron and steel (10), clothing (7), fish (6), manufactured wood (5), others (48)
French Oceania (0.3; 0.05)	petroleum (22), nonelectrical machinery (21), wood (8), others (49)	spices (40), fertilizers (16), others (44)
New Zealand (4.1; 11.4)	nonelectrical machinery (30), textiles (9), tobacco (6), others (55)	meat (47), wool (32), hides (10); others (11)
Australia (24.2; 20.2)	nonelectrical machinery (27), electrical machinery (6), textiles (6), others (47)	meat (53), wool (16), sugar (7), others (24)

17. Principal Commodities Involved in Trade between the United States and the Pacific in 1961–1963 (as percentages) (*cont.*)

Country (*Japan = 100*)	Exports	Imports
Trust territories (0.2; 0.1)	cereals (12), nonelectrical machinery (12), tobacco (7), others (69)	copra (63), others (37)

This table is based on more specific categories (same source as for Table 16). Only commodities amounting to more than 5 percent of trade with a given country are listed. The figures given in parentheses under the names of the countries are the indexes of exports and imports by comparison with Japan, so as to give a better impression of the scale. In 1961–1963, all the countries listed, apart from Japan, equaled 140 percent of the amount of U.S. exports destined for Japan, and 158 percent of the amount of U.S. imports from Japan.

18. Structure of Trade between the United States and the Newly Industrialized Countries of the Asia-Pacific, in 1984–1986 (in Percentages)

	Foodstuffs		Raw Materials		Fuels		Manufactured Goods		Chemicals		Partly Finished		Machinery		Transport Equipment		Consumer Goods		Others	
	E	I	E	I	E	I	E	I	E	I	E	I	E	I	E	I	E	I	E	I
Hong Kong	17	1	4	★	2	★	75	97	12	★	9	5	36	20	6	★	12	72	2	2
Taiwan	22	2	14	★	6	★	58	96	15	1	4	15	29	24	7	4	3	52	2	★
South Korea	14	2	23	★	5	★	57	97	11	1	6	18	31	18	7	3	2	57	1	★
Singapore	5	2	1	1	4	7	87	87	8	1	4	2	54	59	17	3	4	22	2	3
Malaysia	5	9	1	10	★	2	92	77	8	★	4	3	74	58	5	★	1	15	1	2
Philippines	18	28	3	2	1	★	76	68	10	2	7	4	51	30	3	★	5	33	2	2
Indonesia	18	7	15	11	2	69	64	13	20	★	3	7	29	1	10	★	2	5	1	★
Thailand	9	26	9	11	1	3	79	59	16	1	4	9	45	17	10	★	4	32	2	1
China	8	5	14	4	★	19	77	71	16	4	7	15	35	1	18	★	1	51	★	1

E = Exports from the U.S.; I = imports to the U.S.
★ = Less than 1%.
Source: C.I.A., OECE Trade with Asia. A Reference Aid, Washington, D.C., March 1988, pp. 41–144.

19. American Direct Investments in the Pacific (in Millions of Dollars)

	1950	1960	1970	1980	1990	1992
Central America	597	636	1,875	4,209	10,004	12,148
Colombia	194	424	698	1,012	1, 647	2,077
Ecuador	14	nd.	nd	322	278	310
Peru	140	446	688	1,665	594	466
Chile	530	738	748	536	1,876	2,446
Japan	19	254	1,483	6,243	22,511	26,213
Hong Kong				2,078	5,994	8,544
Taiwan				498	2,214	2,870
South Korea				575	2,677	2,779
Singapore	nd	147	1,450	1,204	3,183	6,631
Malaysia				632	1, 513	1,714
Thailand				361	1, 789	2,459
Indonesia	58	178		1,314	3,175	4,278
Philippines	149	414	701	1,259	1,355	1,565
China	0	0	0	0	356	469
Australia	198	856	3,304	7,662	14,997	16,697
New Zealand	25	53	184	578	3,099	3,008
Oceania (others)	nd	85	nd	nd	nd	nd
Total Pacific	1,924	4,231	11,131	30,148	77,262	94,674
Percent of Pacific	16.3	12.9	14.2	14.0	18.1	19.5

nd = No data available.
Investments valued on the basis of the historical cost.
Source: U.S. Department of Commerce, Survey of Current Business, December 1952, August 1962, November 1972, August 1982, July 1993.

20. Direct Investments from Pacific Countries in the United States
(in Millions of Dollars)

	1980	1990	1992
Japan	4,225	83,091	96,743
Australia		6,542	7,140
New Zealand	253	157	108
Panama	707	4,188	4,732
Hong Kong		1511	1,714
South Korea		-1,009	-496
Malaysia		56	54
Philippines	325	77	59
Singapore		1,289	847
Taiwan		836	1,154
Others Asia/Pacific		398	402
Total for Pacific	5,510	97,136	112,457
Percent from Pacific	8.1	24.6	26.8

Source: Survey of Current Business, August 1982, July 1993.

21. Sino-American Trade

	I	E	R	%I	%E
1972	33	64	194	0.2	0.6
1973	66	690	1045	0.3	3.7
1974	116	807	696	0.4	3.2
1975	160	305	191	0.6	1.3
1976	203	136	67	0.6	0.5
1977	204	176	86	0.5	0.6
1978	329	826	251	0.6	2.4
1979	597	1,725	289	1.0	3.8
1980	1,059	3,756	355	1.5	6.4
1981	1,897	3,613	190	2.4	5.9
1982	2,287	2,945	129	2.9	5.0
1983	2,245	2,194	98	2.5	3.8
1984	3,067	3,027	99	2.6	4.9
1985	3,865	3,876	100	2.9	6.5
1986	4,772	3,136	66	3.1	4.8
1987	6,295	3,522	56	3.7	4.8
1988	8,514	5,054	59	4.6	5.1
1989	11,990	5,755	48	6.3	5.3
1990	15,237	4,806	32	8.1	4.1
1991	18,976	6,287	33	9.3	4.8
1992	25,676	7,470	29	11.3	5.3
1993	31,535	8,767	28	12.6	5.8
1994	38,781	9,287	24	13.6	5.5
1995	45,543	11,754	26	14.4	5.7
1996	51,495	11,978	23	16.1	5.5
1997	62,558	12,862	21	18.0	5.7
1998	71,156	14,258	20	19.6	7.3

I = Imports in millions of dollars; E = exports in millions of dollars; R = ((E/I) x 100): ratio of exports to imports expressed as a percentage; %I = China's share of U.S. imports from the Pacific; %E = China's share of U.S. exports to the Pacific.
Source: Statistical Abstract of the United States and Table 15.

22. Immigration from the Pacific

	Total (Thousands)	Origin by Percent			Pacific's Share in Total
		Asia	Latin America	Oceania	
1961–1970	559	55	42	3	17
1971–1980	1,513	78	20	2	34
1981–1990	2,787	73	26	1	38
1961–1990	4,859	73	26	1	32

Source: Statistical Abstract of the United States, 1992, p. 11.

23. Proportion of American Population Originating in the Asia-Pacific (in Thousands)

	1940	1950	1960	1970	1980	1990	% 5 States, 1990
Chinese	77	118	237	435	806	1,645	50
Filipinos	46	62	176	343	775	1,407	58
Japanese	127	142	464	591	701	848	72
Koreans					355	799	41
Vietnamese					262	615	51
Laotians					48	149	
Cambodians					16	147	
Hmong					5	90	
Thai					45	91	
Hawaiians					167	211	86
Samoans					42	63	83
Natives of Guam					32	49	65
Others (1)	5	49	213	721	70	344	
Total	255	371	1,095	1,369	3,324	6,458	
% of American Population	0.2	0.2	0.6	0.7	1.5	2.6	

(1)The figures for 1940–1970 include Aleutians, Asian Indians, Inuit, Hawaiians, Indonesians, Koreans, and Polynesians. Asian Indians are excluded from the figures for 1980 and 1990, when they numbered 362,000 and 815,000.
% 5 States = percentage residing in the five states bordering the Pacific (California, Oregon, Washington, Alaska, and Hawaii).
Source: Statistical Abstract of the United States, 1979, p. 36; 1992, pp. 17, 24–25; CQ Researcher, vol. 1, no. 30, December 13, 1991, p. 948.

1. American Shipping in Canton, 1784–1833

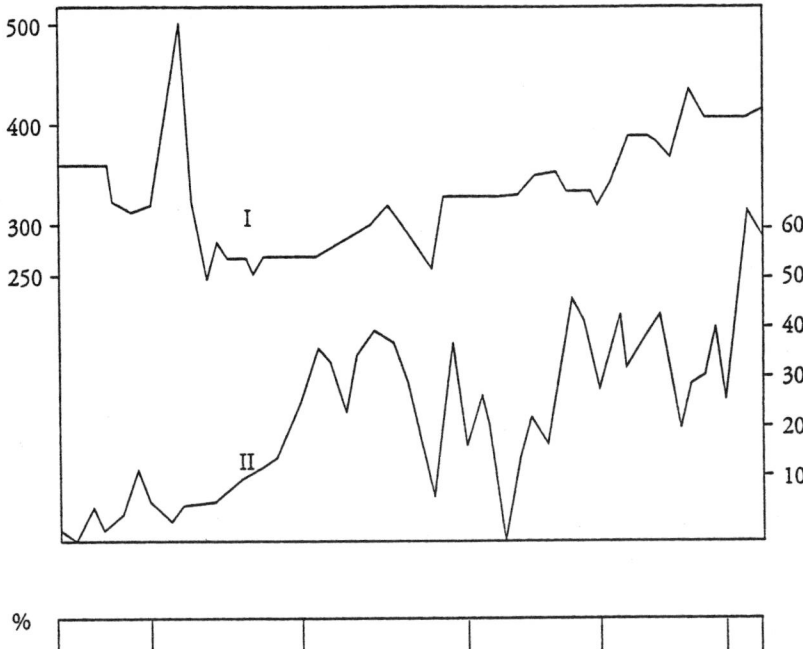

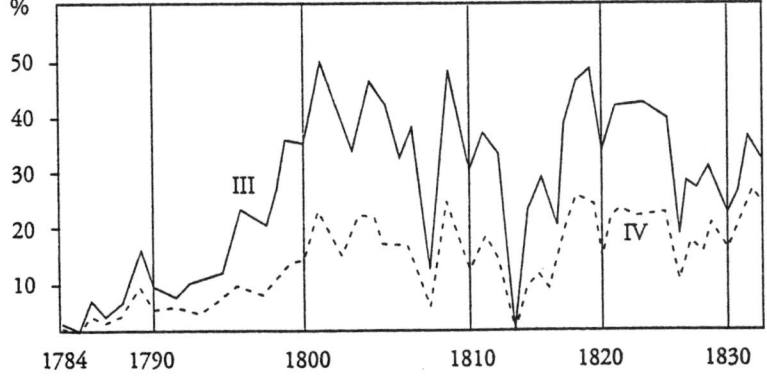

I: Average tonnage.
II: Number of vessels.
III: American vessels as percentage of all shipping.
IV: American tonnage as percentage of total.

Source: Louis Dermigny, *La Chine et l'Occident. Le commerce à Canton au XVIIIe siècle, 1719–1833*, vol. II, pp. 523–525.

2. Imports of Whaling Products to the United States, 1804–1892

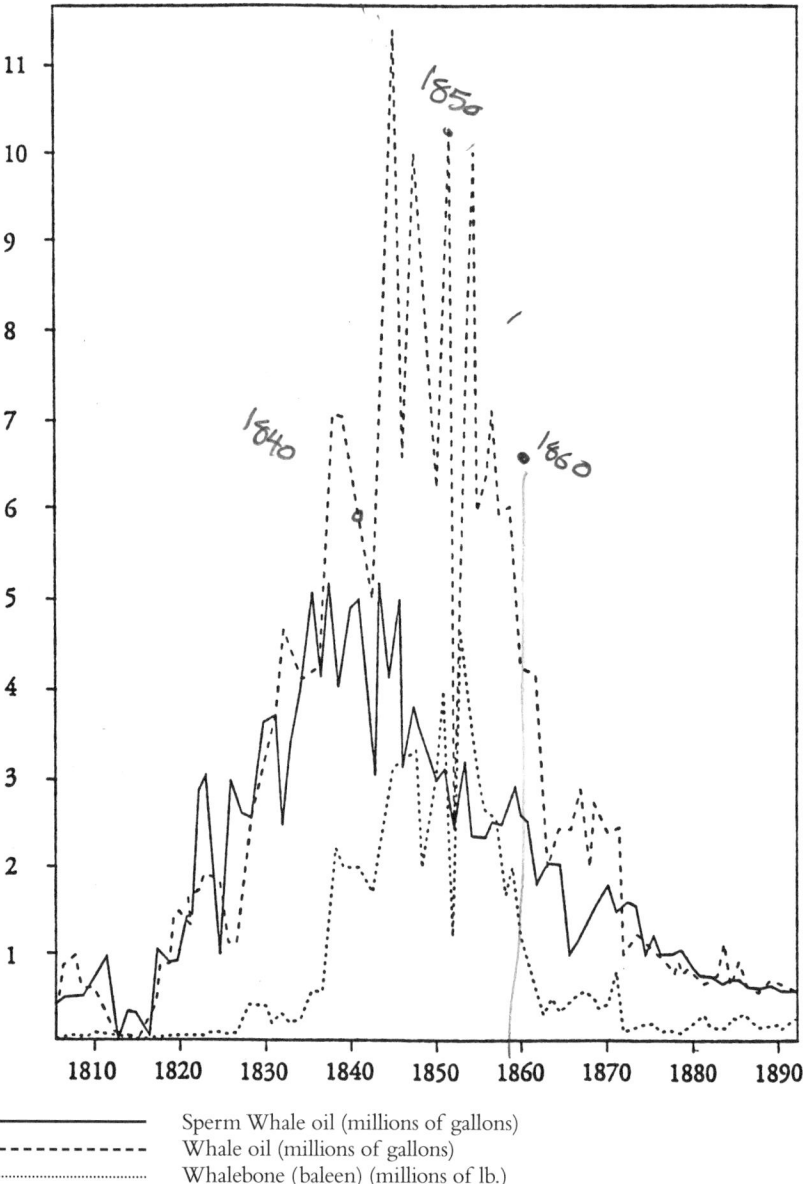

Sperm Whale oil (millions of gallons)
Whale oil (millions of gallons)
Whalebone (baleen) (millions of lb.)

Source: Alexander Starbuck, *History of the American Whale-Fishery,* Report of Commissioner of Fish and Fisheries, Appendix J (for the years 1804–1875); *Whalemen's Shipping List* (for the years 1876–1892).

3. Nominal Prices of Whaling Products, 1804–1892

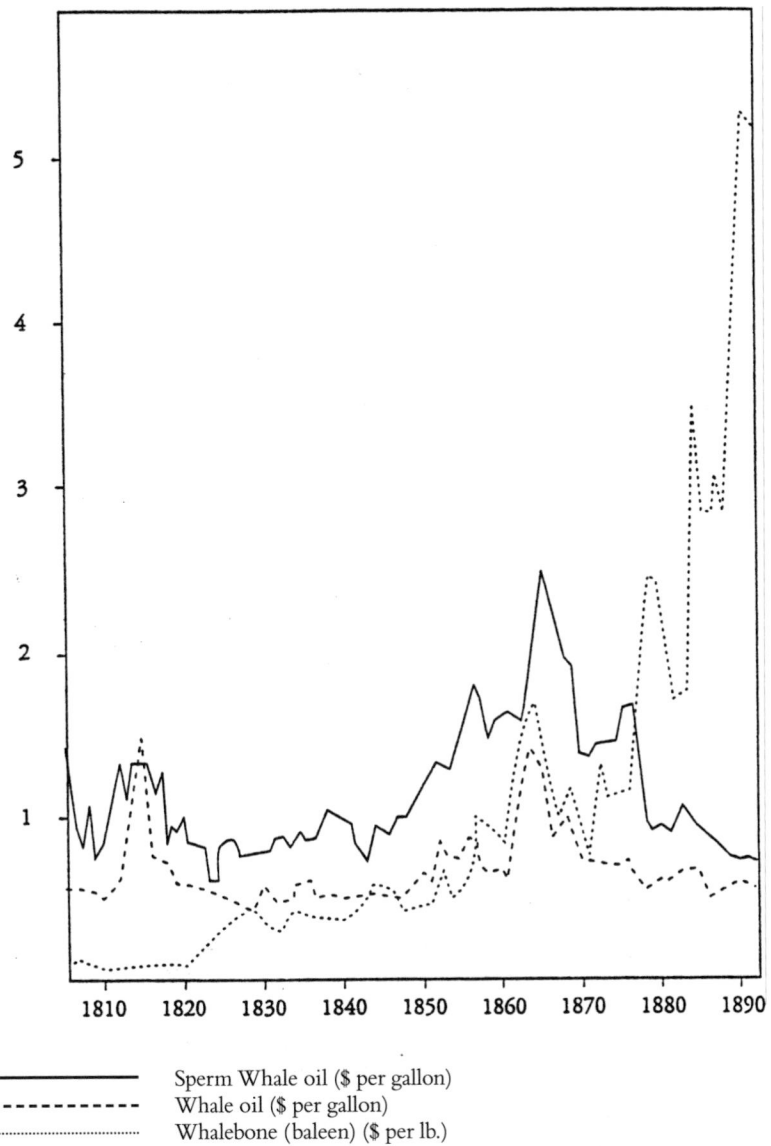

Sperm Whale oil ($ per gallon)
Whale oil ($ per gallon)
Whalebone (baleen) ($ per lb.)

N.B.: These figures apply to all the oceans combined.

Source: Alexander Starbuck, History of the American Whale-Fishery, Report of Commissioner of Fish and Fisheries, Appendix J (for the years 1804–1875); Whalemen's Shipping List (for the years 1876–1892).

4. American Foreign Trade with the Pacific by Volume (in 1982 dollars)

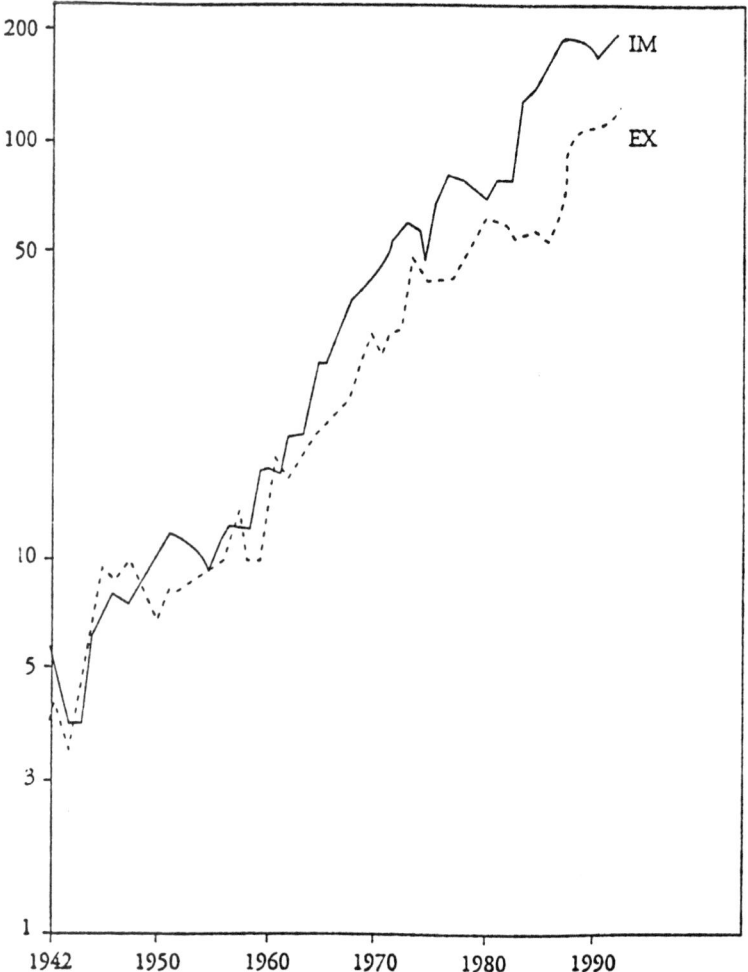

Source: Table 15 (columns 2 and 3, divided by import and export price indices ([1982 = 100]), in *Economic Report of the President,* 1991, p. 291.

5. The United States' Share of Japanese Foreign Trade, 1950–1990 (as a percentage)

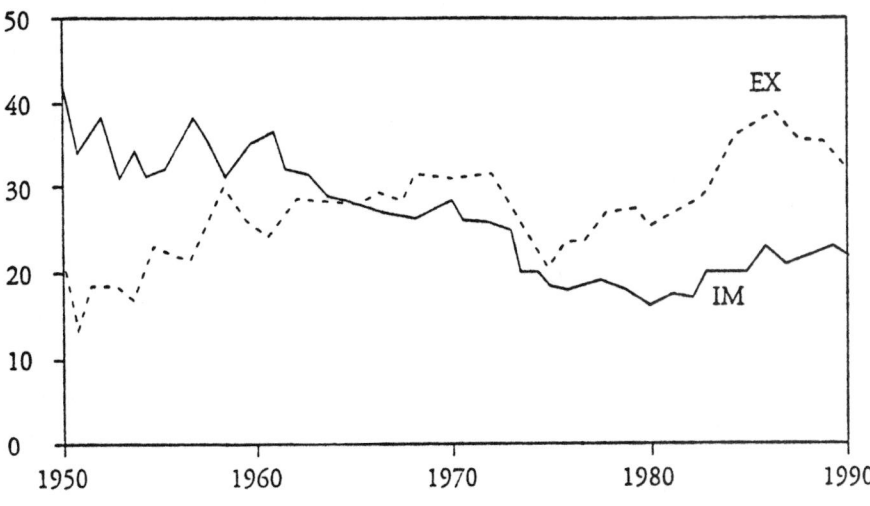

Source: Japan Statistical Yearbook

6. Ratio of Exports to Imports in U.S.–Japanese Trade, 1962–1993 (expressed as a percentage)

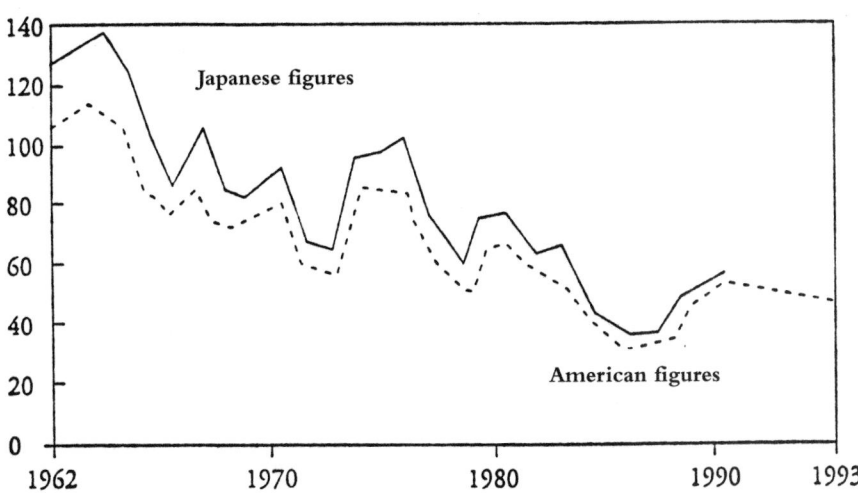

Sources: Statistical Abstract of the United States, Japan Statistical Yearbook.

7. Yen-Dollar Rate of Exchange

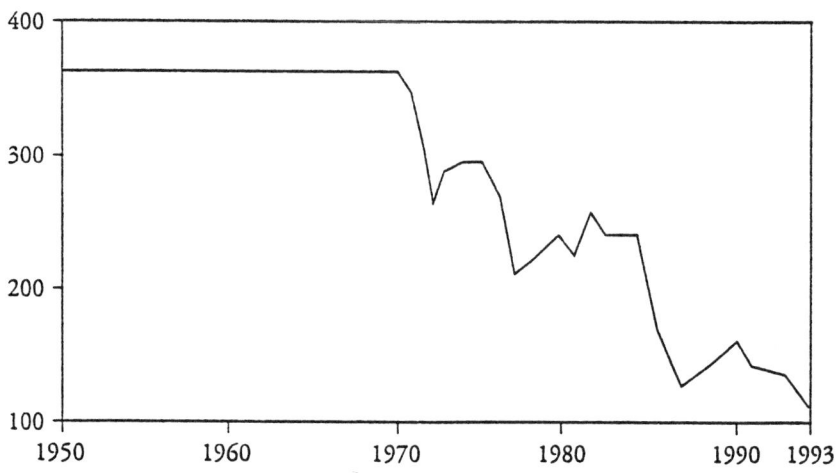

Source: Economic Report of the President, January 1993, p. 470.

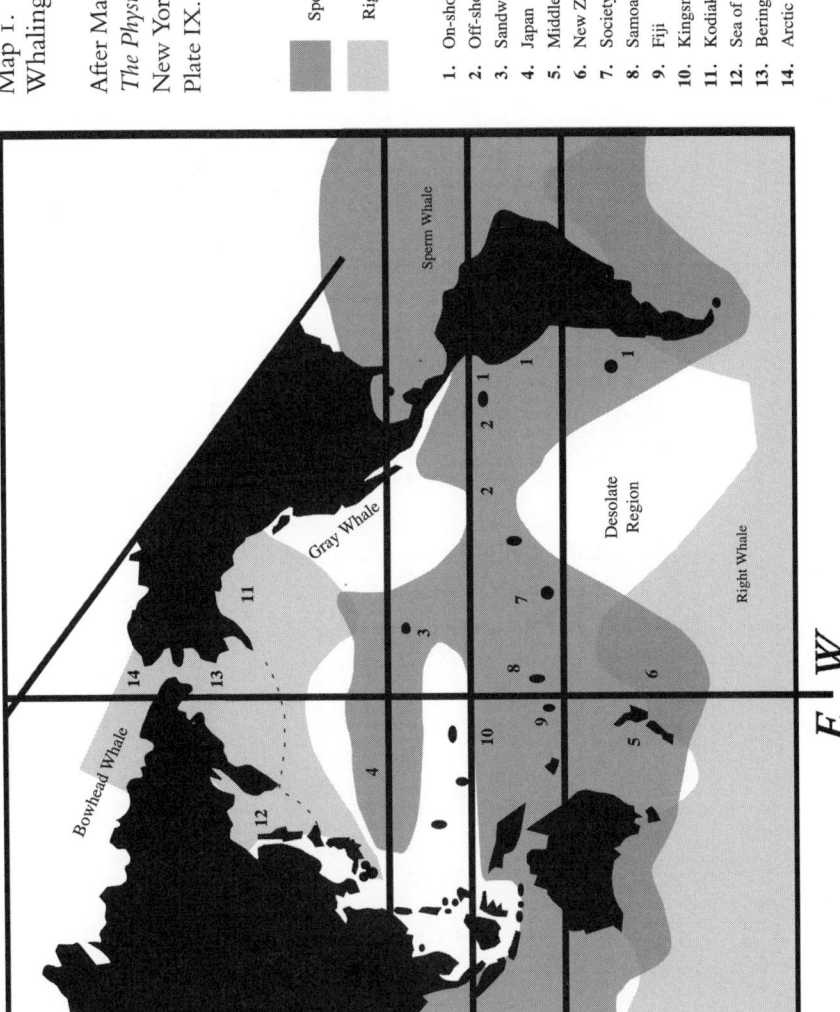

Map 1.
Whaling Grounds

After Matthew Fontaine Maury,
The Physical Geography of the Sea,
New York, Harper & Brothers, 1857,
Plate IX.

Sperm Whale

Right Whale

1. On-shore
2. Off-shore
3. Sandwich
4. Japan
5. Middle ground
6. New Zealand
7. Society Island
8. Samoa
9. Fiji
10. Kingsmill
11. Kodiak
12. Sea of Okhotsk
13. Bering Sea
14. Arctic

Map 2.
The Cruise of the
Lagoda, 1860–1864

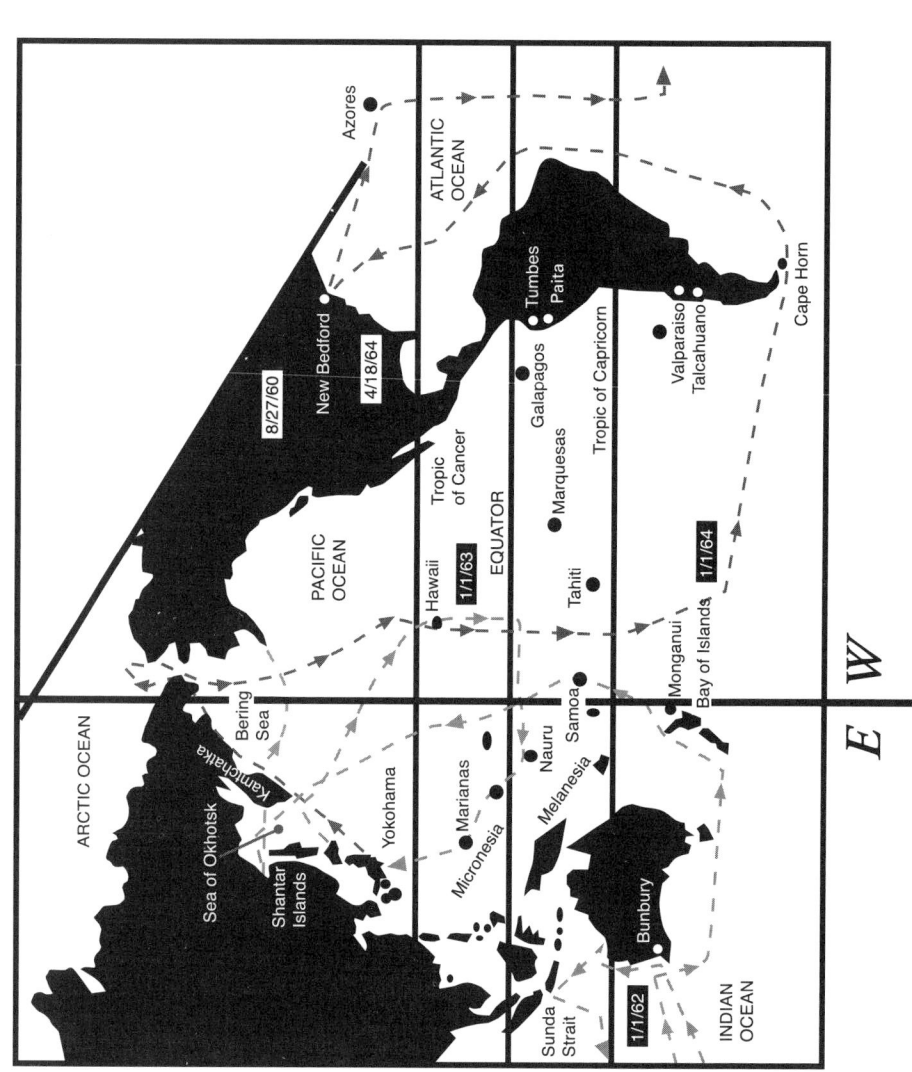

Map 3.
The American Expedition to the
Pacific, 1838–1842 (Charles Wilkes)

Courses followed by:

——— the *Vincennes* (Wilkes)

- - - the *Porpoise*

•••••• the *Peacock*

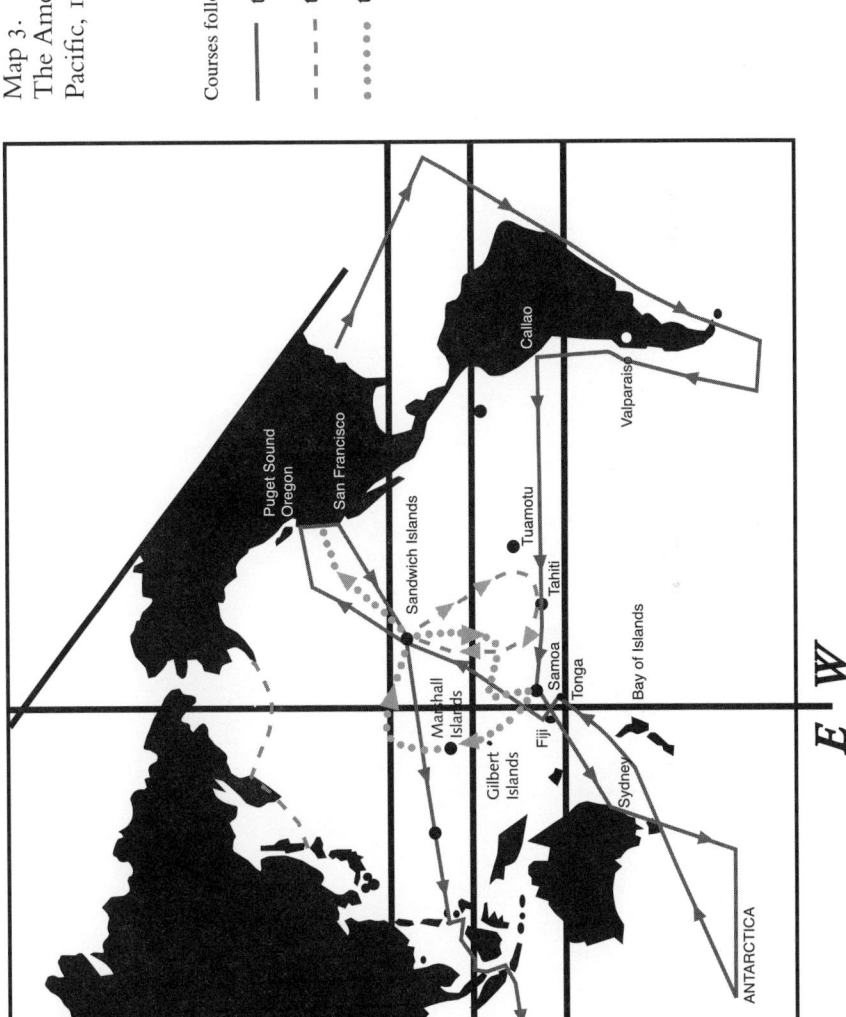

NOTES

INTRODUCTION

1. Frederick J. Turner (1963). Titles, place of publication, and publishers are listed in the bibliography.

2. Jean Heffer, "Pourquoi n'y a-t-il pas de socialisme américain?" ["Why is there no American socialism?"] *Histoire*, no. 4 (March 1980), pp. 39–55.

3. Cape Horn is 67° W, the Bering Strait 169° W.

4. Herman Melville, *Moby Dick*, 1851, chap. CXI.

ONE. THE CHINESE MAGNET

1. *Encyclopaedia Britannica* (Edinburgh, 1771), vol. III: 449: "Pacific Ocean, that vast ocean which separates Asia from America: it is called Pacific, from the moderate weather the first mariners who sailed in it met with between the tropics; and it was called the South Sea, because the Spaniards crossed the Isthmus of Darien from north to south when they first discovered it: though it is properly the western ocean, with regard to America."

2. Beaglehole (1966); Friis (1967), pp. 237–249; O. E. Allen (1980); Spate (1983, 1988).

3. Chaunu (1960); Spate (1983), pp. 284–289.

4. Dermigny (1964), pp. 522–523.

5. Dermigny (1964); Fairbank (1953); Jane Kate Leonard, "Geopolitical Reality and the Disappearance of the Maritime Frontier in Qing Times," *American Neptune* 48, no. 4 (Fall 1988): 230–236.

6. Dermigny (1964), p. 754; Chaunu (1960), pp. 16–17.

7. May and Fairbank (1986). For the Canton trade, see Morse (1926), vols. II–IV; Dermigny (1960), II: 523–525.

8. The figures showing these trends have been arrived at using the series of indices of import and export prices for the period 1821–1860 (1830 = 100), taken from Douglass C. North, *The Economic Growth of the United States, 1790–1860* (New York: Norton, 1966), p. 280. For imports during the period 1861–1867, the list was reconciled with the indices published by Jean Heffer in "Les termes de l'échange américain, 1860–1900," in Fischer, McInnis, Schneider (1986), p. 567. For exports, the figures apply only to 1821–1861, because of the serious distortions for the Civil War period, depending on whether a Laspeyres or Paasche index is used.

9. The average annual levels of imports and exports were $10.9 and $5.3 million, respectively. The average ratio of exports to imports was 50.8% between 1821 and 1867, increasing on average by 0.39% per annum. It was particularly low (around 25%) in 1835–1937, and rose to around 60% just before the Civil War.

10. Rodris Roth (1961).

11. Philip Freneau, "On the First American Ship That Explored the Route to China and the East Indies, after the Revolution 1784," *The Poems of Philip Freneau, Poet of the American Revolution* (Princeton, N.J.: Princeton University Library, 1903), II: 261–262.

12. Philip C. F. Smith (1984); *The Journals of Major Samuel Shaw, the First American Consul at Canton, with a Life of the Author by Josiah Quincy* (Boston: Wm. Crosby and H. P. Nichols, 1847), pp. 133–213.

13. National Archives [of France], 8 AQ 351 (not 350): "Notes explicatives sur 14 cargaisons expédiées de l'Amérique septentrionale aux Indes et à la Chine," New York, February 3, 1786.

14. Dermigny (1960) II: 523–525.

15. Butel (1989), pp. 107–110, 116, 182–183. See also the articles by Yen Ping Hao and Robert P. Gardella, in May and Fairbank (1986). A complete set of statistics for tea imports between 1790 and 1895 (excepting 1792–1794 and 1820) can be found in *Finance, Commerce, and Navigation of the United States*, 54th Congress, 1st session, 1895–96, U.S. House Documents, vol. 83, doc. 91, November 1895 (Washington, D.C.: GPO, 1896): 624–626. The figure for 1850 should be corrected to 18,743,376 lb. instead of 28,743,376 lb. as wrongly given in *Historical Statistics*, Series U-298.

16. The percentage of tea imports which was re-exported was 32.8% in 1800–1822, 19.6% in 1821–1825, 15.2% in 1840–1845, and 18.4% in 1855–1860. Figures for tea exports from China to the United States can be found in Morse (1926) and in Rev. Charles Gützlaff, *A Sketch of Chinese History, Ancient and Modern: Comprising a Retrospect of the Foreign Intercourse and Trade with China* (London: Smith, Elder, 1834), vol. II, table V.

17. Figures based on *Finance, Commerce and Immigration (1895)*, pp. 624, 627–628, and, for population figures, *Historical Statistics*, vol. I, Series A-7. These

figures differ slightly from those given by Timothy Pitkin, *A Statistical View of the Commerce of the United States; Its Connection with Agriculture and Manufactures; and an Account of the Public Debt, Revenues and Expenditures of the U.S.* (New York: James Eastburn, 1817) 1st ed. and (New Haven: Durrie and Peck, 1835) 2nd ed., pp. 75, 247–248.

18. Thomas Weiss, "Economic Growth before 1840; Revised Conjecture," paper presented at the Second World Cliometrics Congress, Santander, 24–27, June 1989, p. 17.

19. The price elasticity of demand for tea was around 1. In 1821–32, the average consumption per head was 85 oz., with an average import price of $.30 per lb., and in 1833–39, 131 oz. and $.32 per lb. But we have to take into account the abolition of customs duties as of March 1, 1833, duties that doubled the import price.

20. House of Representatives, 61st Cong., 2nd. sess, no. 671, *Tariff Acts Passed by the Congress of the United States from 1789 to 1909* (Washington, D.C.: GPO, 1909), pp. 14, 16, 42, 43, 60, 101.

21. Victor Clark (1949), I: 575–577.

22. Dermigny (1960); Hugh Rockoff, "Money, Prices and Banks in the Jacksonian Era," in Robert W. Fogel and Stanley L. Engerman, eds., *The Reinterpretation of American Economic History* (New York, Harper & Row, 1971), pp. 452–454; Greenberg (1951); Charles Gützlaff, *A Sketch of Chinese History*, vol. II, table IV.

23. Gützlaff, *A Sketch of Chinese History*, table IV; Seaburg and Paterson (1971); Stelle (1940, 1941); Downs (1968). For statistics of imports to China, see Morse (1910), I: 209–210.

24. His research and publications of logbooks make Judge Frederick W. Howay the principal authority on matters concerning the Northwest. Apart from the recent synthesis by Gibson (1992), other sources for table 3 should also be mentioned, such as Dermigny (1960), III: 1145–1160, and Foucrier (1990), chap. I.

25. In *John Ledyard's Journal of Captain Cook's Last Voyage*, ed. James Kenneth Mumford (Corvallis: Oregon State University Press, 1963), pp. xvii–xxxix. Sinclair H. Hitchings mentions two other Americans who took part in the expedition, one being John Gore, a native of Virginia, who became commodore after the deaths of Cook and Clerke.

26. F. W. Howay, *Voyages of the Columbia to the North West Coast, 1787–1790 and 1790–1793* (Boston: Massachusetts Historical Society, 1941), pp. 95–96.

27. Ibid., pp. 396–397.

28. Those unable to consult the original logs may refer to a number which have been published, or to accounts of voyages often written much later: Fanning (1838), pp. 137–151 (on the massacre of the crew of the *Tonquin* in 1811); Cleveland (1842), pp. 69–94 (March–June 1799); Ingraham (1971); S. W. Reynolds (1970), for the viewpoint of a sailor before the mast.

29. Ingraham (1971), p. 58; Ogden (1941), chaps. III–IV; M. E. Wheeler (1971); Howay, Sage and Angus (1970), chaps. 1–3; Gough (1980); Ronda (1990).

30. Delano (1970), p. 306. For his advice to sealers, see p. 255; Dermigny (1960), p. 1193; Dodge (1986), pp. 169–172; Fanning (1838); Pereira Salas (1971), chaps. VIII–X, XVIII, XX, XXIII, XXVI, XXVII.

31. Bradley (1942), chaps. I–II, Gützlaff (1834), II: table IV; Fanning (1838), pp. 53–63, 124–126; Shineberg (1974); I. C. Campbell (1989), pp. 59, 61–63; W. Patrick Strauss (1964).

32. Abby Jane Morrell (1970), pp. 55–70; Wilkes (1845), III: 103–105, 137, 218–222; I. C. Campbell (1989), pp. 64–66.

33. Heffer (1990).

34. Neumann (1963), pp. 7–10.

35. Owen (1984). See also his article: "Americans in the Abaca Trade: Peele, Hubbell & Co., 1856–1875," in Stanley (1984), pp. 201–230.

36. Thomas Dunbabin, "New Light on the Earliest American Voyages to Australia," *The American Neptune*, January 1950, pp. 52–64; Grattan (1961), pp. 70–81.

37. Grattan (1961), pp. 95–97.

38. Cleveland (1842).

39. Clayton (1986).

40. Dana (1965); Robinson (1970); William H. Davis (1889); Ogden (1929); Fritzsche (1968).

41. Bradley (1942); Stevens (1968). See also the article by Sumner J. La Croix and James Roumasset, "The Evolution of Private Property in Nineteenth-Century Hawaii," *Journal of Economic History* 50, no. 4, (December 1990), pp. 829–852.

42. Robert B. Forbes, *Personal Reminiscences, 2nd ed. rev., to which is added rambling recollections connected with China* (Boston: Little Brown, 1882), pp. 168–169.

43. Ibid. p. 46.

44. J. Jemison, Jr., experienced the same misfortune when, hoping to make his fortune on the Mexican coast, he sailed from Canton in July 1831. The supercargo lacked the courage to trade. Jemison wrote, at sea off San Blas: "Unless the vessel enters here & I see not hope of it, our destiny is an endless crusade along the coast of South America. To me this is desperation, madness. I should prefer a pilgrimage at the tomb of Mahomet with peas in my shoes. . . . It is Hell." Letter from J. Jemison, Jr., to Robert B. Forbes, January 7, 1832. (Forbes Collection Papers, Baker Library, Harvard Business School, Box 2, Folder 15.)

45. Philo White (1965), pp. 17, 25, 49.

46. Larkin (1951–1955); Hague and Langum (1990).

47. Lockwood (1971).

48. Forbes Collection Papers, Baker Library, Harvard Business School, Paul S. Forbes Papers, Box 2, Folder 19, January 12, 1861. The sharing of the profits in 1858–60 is found in the same collection: letter from Edward Cunningham to P. S. Forbes, June 28, 1863; Edward Waldo Emerson, "John Murray Forbes," *Atlantic Monthly* 84 (1899): 383–385.

49. The figures are listed in the annual reports: *Foreign Commerce and Navigation of the United States*. For 1821–1833, only data concerning tonnage is available. From 1834 on, information is given on tonnage and the number of ships, and the crews, making a distinction between American and foreign vessels.

50. However, *direct* traffic between the coastal ports and San Francisco is not included in these statistics, which refer only to shipping between the United States and foreign ports.

51. Morison (1921), chaps. VI, VII, IX, XIV, XV, XVI, XVII.

52. Kemble (1943,1950). However, the route via Nicaragua must not be forgotten, although it was taken by only a fifth the number of passengers and conveyed only a sixteenth the quantity of gold. See David I. Folkman, Jr., *The Nicaragua Route* (Salt Lake City: University of Utah Press, 1972), p. 173. In addition, some emigrants and equipment were conveyed by clipper ship around Cape Horn (Delgado, 1990).

53. Of the arrivals, 25.4% and 25.7% of the departures. For China, the percentages come to 25.3% of the arrivals and 14.8% of the departures. British vessels accounted for only 6.1% of the arrivals and 8.2% of the departures.

54. Cutler (1984), chap. XVII. According to this author, 129 American clippers conveyed tea from China to London between 1851 and 1860, and 95 between 1851 and 1855. In his Appendix, pp. 561–568, he refers to 139 voyages between 1850 and 1860, with varying figures in certain years. Whipple (1981).

TWO. THE WHALERS' PACIFIC

1. Starbuck (1878), pp. 186–187. After nineteen months at sea, the five vessels brought back 4,150 barrels of sperm oil and 1,520 barrels of right whale oil.

2. Davis, Gallman and Hutchins (1988).

3. Webb (1988), p. 36. A single cotton mill in Lowell consumed 6,772 gallons in 1851, i.e., 220 barrels of sperm oil. Three sperm whale would have been needed to provide this amount.

4. Wilkes (1845), vol. V: 488–489.

5. Webb (1988), pp. 36–40.

6. Starbuck (1878), pp. 155–156.

7. Dr. Collineau, the article "Corset" in *La Grande Encyclopédie*, p. 1107.

8. Bockstoce (1986), p. 346. Article in the *Whalemen's Shipping List*, May 14, 1849.

9. Starbuck (1878), p. 113, suggests four causes: (1) the scarcity and timid nature of the whales led to longer and more costly voyages; (2) the exorbitant expenditures for equipment at the start and in the course of the voyage; (3) the type of men taken on as crew; (4) the introduction of mineral oils.

10. According to Scammon (1874), 10% of the sperm whales harpooned and 20% of the right whales were lost (cited by Starbuck [1878], p. 661). Davis, Gallman,

and Hutchins (1988), pp. 569–575, consider these estimates too high (see their table 1, p. 577).

11. Edward Mitchell, "Potential of Whaling Logbook Data for Studying Aspects of Social Structure in the Sperm Whale, *Physeter macrocephalus*, with an example—the ship *Mariner* to the Pacific, 1836–1840," in: International Whaling Commission, *Historical Whaling Records Including the Proceedings of the International Workshop* in *Historical Whaling Records* (Sharon, Mass., September 12–16, 1977), ed. Michael F. Tillman and Gregory P. Donovan, Reports of the International Whaling Commission, Special Issue 5, Cambridge, 1983, p. 70.

12. Davis, Gallman, and Hutchins (1987).

13. Davis, Gallman, and Hutchins (1991). A connection is sometimes made between the decrease in the whale stock and a deterioration in the quality of the crews, the suggestion being that the former resulted directly in increased whaling costs, and that owing to competition from other products, these higher production costs could not be passed on to the consumer. As a result, crews received less pay, and consequently were less competent.

14. Davis, Gallman, and Hutchins (1989); (1987, pp. 390–392).

15. There is an immense bibliography on American whaling in the nineteenth century: see Honore Forster, *The South Sea Whaler: An Annotated Bibliography of published historical, literary and art material relating to Whaling in the Pacific Ocean in the nineteenth century* (Sharon: The Kendall Whaling Museum, 1985). Also, Elmo P. Hohman, *The American Whaleman: A Study of Life and Labor in the Whaling Industry* (New York: Longmans, Green, 1928); A. B. C. Whipple (1980), chaps. 3 and 4.

16. Dana, (1965), pp. 69, 75, 80, 89–94, 104, 115, 117. The *Lagoda* was of 340 $^{62}/_{95}$ tons. Her measurements were: length, 107 ft., 9 in.; breadth, 26 ft. 612 in.; depth, 13 ft. 314 in.

17. Barbara Clayton and Kathleen Whitley, *Guide to New Bedford* (Montpelier: Capital City Press, 1979), 186.

18. Benjamin Barker, "The Jonathan Bourne Office and Some of Those Connected with It," vol. 2, pp. 150, 168, Old Dartmouth Historical Society. Overall, Bourne earned more than the Wing brothers, who carried on business between 1852 and 1916 (cf. Martin Joseph Butler, *J. & W. R. Wing of New Bedford: A Study of the Impact of a Declining Industry upon an American Whaling Agency*, Ph.D. diss, Pennsylvania State University, Ann Arbor, University Microfilms, 1973.

19. If $7,500, representing the *Lagoda*'s initial cost, is subtracted, the net rate of profit from the first voyage comes to 92%. The low price paid for the ship is explained by the fact that she was secondhand, already sixteen years old in 1841. A new vessel of 351 tons would have cost $31,225, according to Joseph Grinnell, speaking before the House of Representatives on May 1, 1844 (*Congressional Globe*, 28th Congress, 1st Session, vol. XIII, p. 570).

20. The logbooks of the *Lagoda*'s voyages nos. 1, 4, 5, 7, 9 and 12 have been preserved. The seventh is available on microfilm at the Free Public Library in New Bedford.

21. According to *Ship Registers of New Bedford, Massachusetts* (Boston: The National Archives Project, 1940), J. Bourne's share was only $^7/_{16}$, the other two shares going to Almy B. Covell, for himself, and as guardian of Clement A. Covell.

22. Old Dartmouth Historical Society, Jon. Bourne's Papers, Manuscript 18, Series B, s. 115, Box 8, Folder 14. The ship was valued at $6,000.

23. The captain was entitled to $^1/_{16}$ of net sales, the first officer to $^1/_{25}$, the second to $^1/_{38}$, the third to $^1/_{55}$, and the fourth to $^1/_{75}$. These shares, or "lays" were close to the average worked out by B. Barker, p. 242, for forty expeditions carried out between 1837 and 1877, i.e., $^1/_{14.5}$ for the captain, $^1/_{24}$ for the first officer, $^1/_{37}$ for the second, $^1/_{56}$ for the third, $^1/_{68}$ for the fourth and for the cooper, $^1/_{87}$ for the harpooners, $^1/_{116}$ for the steward, $^1/_{135}$ for the cook, $^1/_{170}$ for the ordinary sailors, $^1/_{200}$ for the "green hands."

24. *Whaling Manuscript Crew Lists. Roll 172.* New Bedford Free Public Library. The library possesses only the *Whalemen's Shipping Papers* on microfilm for the years 1841–1858 and 1866 (rolls 179–186). When the information is complete, which is far from always the case, it gives the origin, age, height, color, and hair color. On the preceding voyage (1856–1860), New Yorkers were already more numerous (9), with residents of Connecticut (6) and Philadelphia (3) in second place. On the tenth voyage, on the other hand, the crew was essentially made up of foreigners, with natives of the Azores (8), Cape Verde (5), Germany (5), Sweden (2), Scotland (1), and New Brunswick (1). There were only eight Americans—four from New York State, three from Massachusetts, and one from Pennsylvania. No information is given for two crew members.

25. On each of its voyages, the *Lagoda* suffered a large number of desertions. The largest number occurred on the third voyage, in June 1848, when fifteen men made off in three rowboats, ending up on the Japanese coast, where they were held as prisoners, ill-treated, and finally picked up by the master of the *Preble*, James Glynn, in April 1849.

26. The day was counted from midday to noon of the following day. The captain on the fifth voyage, B. B. Lamphier, also died, but in the Sea of Okhotsk.

27. On September 25, 1862, Vanderipe sketched a woman, an American flag, and the slogan: "The Union forever. It must and shall be preserved."

28. The *Lagoda* played her part in the rescue of whalers in the Arctic in 1871, bringing home 170 men. The government paid a reward of $23,611, but not before 1891.

29. The log shows E. Eastwood as captain. It is not this name, but that of David Burns, which appears in the final accounts.

30. The *Lagoda* took twenty-four right whales and six sperm whales.

31. In addition to these eight, three crew were recruited after sailing, and eleven taken on at Hawaii for the return voyage.

32. *Historical Statistics of the United States, 1975*, Series D 735, p. 165.

33. After all of these deductions, the owners actually paid out a total of only $7,924 to the officers and crew.

420 Notes to Pages 60–66

34. For the sake of comparison, at the end of the first voyage, in fall 1843, sperm oil was worth 78¢ a gallon, whale oil was worth 38¢, and whalebone 43¢ a pound.

35. The *Lagoda* did not carry out its final voyage—the thirteenth—under Bourne's flag. The vessel was sold in 1886 for $2,475 to John McCullough, who passed her on to a new group of owners. On May 10, 1887, she sailed once again for the Pacific, and made two unsuccessful voyages out of San Francisco in 1889 and 1890. On August 7, 1890, she was declared unserviceable in Yokohama, auctioned off, and acquired by the Canadian Pacific Co., who converted her for storing bunker coal. In 1899 the *Lagoda* was sold to the Japanese, and met her end in a fire. See C. F. Keith, *The Story of the Lagoda*, manuscript, Old Dartmouth Historical Society, New Bedford, Mass., 1950, pp. 10–12.

36. The Pacific accounted for at least 50% of the sailings from American ports in only 20 years out of 86 between 1791 and 1876: 1797–1799; 1810–1811; 1825; 1843–1851; 1853–1857. If we look at the tonnage, rather than the number of vessels, the proportion is somewhat higher, since whaling vessels sailing to the Pacific were heavier than those heading for the North Atlantic. In 1845, the Pacific accounted for 69% of sailings, but 79% of the tonnage, and in 1865, 32% and 45%, respectively. If a threshold of 40% of sailings is taken as representative of a majority of sailings, the Pacific was the destination of more than 50% of the vessels in 1791, 1797–1799, 1806, 1809–1812, 1818, 1820–1822, 1824–1828, 1831, 1833, 1839, 1841, 1843–1858, and 1860. The busiest decades were the 1820s, 1840s, and 1850s.

37. R. Gerard Ward (1966).

38. Frances L. Williams (1963).

39. *Whalemen's Shipping List*, April 26, 1853.

40. Stephen Curtis, Jr., *Brief Extracts from the Journal of a Voyage Performed by the Whaleship M. of New Bedford, Mass., commencing May 25, 1841, and terminating August 1, 1844* (Boston: S. N. Dickinson, 1844), p. 45.

41. Herman Melville, *Typee*, chap. I.

42. The *Acushnet*, under Captain Pease, with Melville on board, sailed from Fairhaven on January 3, 1841, and returned to the same port on May 13, 1845, carrying 850 barrels of sperm oil, 1,350 barrels of whale oil, and 13,500 lb. of whalebone. By deserting, Melville missed the adventure in the North Pacific.

THREE. AMBITION AND MODESTY

1. James D. Richardson, ed., *Messages and Papers of the Presidents* (Washington, D.C.: GPO, 1897), vol. 1, pp. 353–354; Guiness (1933).

2. Ray Allen Billington, *Westward Expansion: A History of the American Frontier*, 3rd ed. (New York: Macmillan, 1970), p. 449; Meriwether Lewis and George Clark, *The Journals of Lewis and Clark* (numerous editions).

3. Pletcher (1973), pp. 101–110, 214–226, 229–253, 296–351, 380–381, 402–417; Graebner (1955), chap. VII; Merk (1967); Heffer (1987), pp. 131–134.

4. Dangerfield (1965), chap. 2. Because of King Ferdinand VII's reluctance, the treaty was not ratified until two years later, on February 22, 1821.

5. Charles Griffin (1937).

6. G. Dangerfield (1965), chaps. 5 and 6; Ernest R. May (1975); Elise Marienstras, *Naissance de la République fédérale, 1783–1828* (Nancy: Presses universitaires de Nancy, 1987), pp. 142–144.

7. Texts of treaties can be found in Charles I. Bevans, *Treaties and Other International Agreements of the USA, 1776–1949* (Washington, D.C.: GPO, 1968–1976), 13 vols.

8. Mary Williams (1965); Georges Fischer (1979); LaFeber (1989).

9. Rea (1977); Swisher (1922); Tong Te-kong (1964); Jiang (1988), chap. 1.

10. Treat (1917, 1963); Neumann (1963), chaps. I–III; Francis L. Hawks, *Narrative of the Expedition of an American Squadron to the China Seas and Japan, Performed in the Years 1852, 1853 and 1854, under the Command of Commodore M. C. Perry, United States Navy, by Order of the Government of the U.S. Compiled from the Original Notes and Journals of Commodore Perry and his Officers at His Request and under His Supervision* (New York: D. Appleton, 1856), vol. I; Morison (1967), chaps. XX–XXVIII; Wiley (1990); Trautmann (1990).

11. Albion (1954).

12. In 1800, under the command of Captain Edward Preble, the *Essex,* a frigate of 860 tons, armed with 32 cannons, had escorted merchant ships to Sunda Strait to protect them from French piracy. See Charles Oscar Paullin, "Early Voyages of American Naval Vessels to the Orient," *USNIP,* vol. 36, no. 2 (June 1910): 433–442.

13. Porter (1986), p. 82.

14. Ibid., p. 274.

15. Ibid., p. 403.

16. Robert B. Johnson (1980), p. 6.

17. Ibid., pp. 211–219.

18. Billingsley (1967).

19. Philo White (1965); Brooke (1962).

20. Bockstoce (1986), chap. VI. The *Alabama,* commanded by Captain Semmes, destroyed seven Union ships off Singapore and in the Indian Ocean between December 1863 and January 1864; Logan (1962).

21. Before the *Congress,* apart from the *Essex,* the *Peacock* reached Sunda Strait in 1815, where she captured four British ships, although peace had been signed six months earlier. The *Congress* was therefore the first American warship to enter the Pacific via the Cape.

22. The corvette *Vincennes* made six voyages to the Pacific between 1826 and 1856, two of which were voyages of exploration.

23. On the Asia Squadron, see Robert E. Johnson (1979); Henson (1982); Paullin (1971); Schroeder (1985), chaps. 3, 8, 9; E. Mowbray Tate, "American Merchant and Naval Contacts with China, 1784–1850," *American Neptune*, July 1971, pp. 177–191; James M. Merrill, "The Asiatic Squadron: 1835–1907," *American Neptune*, April 1969, pp. 106–117. For a comparison with England, see Gerald S. Graham, *The China Station: War and Diplomacy, 1830–1860* (Oxford: Clarendon Press, 1978).

24. Henry S. Commager, *Documents of American History*, 7th ed. (New York: Appleton-Century-Crofts, 1963), pp. 242–244.

25. The combined tonnage of the six vessels was 2,358. There were two corvettes (the *Vincennes* and the *Peacock*), a brig (the *Porpoise*), a supply ship (the *Relief*), and two tenders, formerly New York pilot vessels (the *Sea-Gull* and the *Flying Fish*). 665 people took part in the expedition, for varying lengths of time. On the 1838–1842 expedition: Wilkes (1844); Stanton (1975); Tyler (1968); Viola and Margolis (1985).

26. Allan B. Cole, "The Ringgold-Rodgers-Brooke Expedition to Japan and the North Pacific, 1853–1859," *Pacific Historical Review* 26, no. 2 (May 1947), pp. 152–162; Cole (1947).

27. Graebner (1955).

28. Foucrier (1992); Harlow (1982).

29. According to Cox (1973), the proposal to annex Taiwan had nothing to do with the "manifest destiny," but grew out of the imaginations of a little group of merchant adventurers who had recently arrived in China (William Robinet and Giden Ney). Peter Parker is said to have viewed this as an opportunity to evangelize the island. See also Gulick (1973). On the guano islands, see O'Donnell (1993); R. G. Ward (1966).

30. Charles S. Stewart, *A Visit to the South Seas in the U.S. Ship* Vincennes, *during the Years 1829 and 1830; with Scenes in Brazil, Peru, Manila, the Cape of Good Hope, and St. Helena* (New York: Praeger, 1970), vol. II, pp. 252–253.

31. Bradley (1942); Stevens (1968); Kuykendall (1938, 1953); Tare (1967).

32. Kushner (1975); S. Frederick Starr, *Russia's American Colony. A Special Study of the Kennan Institute for Advanced Russian Studies of the Woodrow Wilson International Center for Scholars* (Durham, N.C.: Duke University Press, 1987), (USA-USSR Conference at Sitka, 1978. See particularly the papers by N. N. Bolkhovitinov, James R. Gibson, and H. I. Kushner); Mary E. Wheeler (1971); S. B. Okun, *The Russian-American Company* (Cambridge: Harvard University Press, 1951); Petr Aleksandrovich Tikhemev, *A History of the Russian-American Company* (Seattle: University of Washington Press, 1978), vol. I; Michel Poniatowski, *Histoire de la Russie d'Amérique et de l'Alaska* (Paris: Perrin, 1977).

33. Anatole G. Mazour, "The Russian-American and Anglo-American Conventions 1824–25: An Interpretation," *Pacific Historical Review* 14, no. 3 (September 1945), pp. 303–310; Irby C. Nichols, Jr., "The Russian Ukase and the

Monroe Doctrine: A Re-evaluation," *Pacific Historical Review* 36, no. 1 (February 1967), pp. 13–26; Irby C. Nichols, Jr. and Richard A. Ward, "Anglo-American Relations and the Russian Ukase: A Reassessment," *Pacific Historical Review* 41, no.4 (November 1972), pp. 444–459. The 1824 treaty did not define the frontier between Russian America and the Oregon territory. In the Anglo-Russian treaty of 1825 the parallel at 54°40' was fixed as the borderline.

34. Charles Vevier, "The Collins Overland Line and American Continentalism," *Pacific Historical Review* 28, no. 3 (August 1959), pp. 237–253; Ahvenainen (1981), pp. 25–32.

35. Bolkhovitinov (1990).

36. Paolino (1973), p. 114; Walter G. Sharrow, "William Henry Seward and the Basis for American Empire, 1850–1860," *Pacific Historical Review* 36, no. 3 (August 1967), pp. 325–342.

37. Ronald J. Jensen, *The Alaska Purchase and Russian-American Relations* (Seattle: University of Washington Press, 1975).

FOUR. CULTURES IN COLLISION

1. Charles Pickering, *The Races of Man; and Their Geographical Distribution* (London: H. G. Bohn, 1851). (The modern term is "Negritos," rather than "Negrillos.")

2. Delano (1970), pp. 421–422.

3. Fenimore Cooper, *Cooper's Novels*, vol. 14: *The Sea Lions* (New York: D. Appleton, n.d.), p. 334 (chap. XXI).

4. Dana (1965), p. 86.

5. Henry T. Cheever, *The Whaleman's Adventures in the Southern Ocean, as Gathered by the Rev. H. T. Cheever, on the Homeward Cruise of the "Commodore Preble," edited by the Rev. W Scoresby*, 3rd ed. (London: W. Kent, 1859), pp. 302–303.

6. There were also virtuous sailors. Stephen Curtis Jr. (1844), p. 33, reports that at Nuku-Hiva in July 1842 : "Every night, at sunset, a boat was sent ashore, for the laudable purpose of bringing girls on board of the ship; consequently, scenes which ought to make man blush with shame were constantly before the eyes of the more virtuous part of the crew; for there were some who could claim a share of virtue, although they were sailors." This debauchery, what is more, he adds, had the captain's approval.

7. Porter (1986), pp. 292–293.

8. Ibid., p. 295.

9. Ibid., p. 308.

10. *Letters and Papers of the American Board of Commissioners for Foreign Missions*, vol. 236, Houghton Library, Harvard University. See also Hezel (1983), pp. 132–142, 147.

11. *Letters and Papers of the ABCFM*, vol. 237 (February 16, 1858).

12. Melville, *Typee*, chaps. XVIII, XXV. The Pacific as a place to look for sex was not just of interest to male Americans. Some ships' captains brought their wives along, but in such cases there was every likelihood of the ship being a "moral" one. In *Mourning Becomes Electra* (*The Haunted*, act I, scene 2), by Eugene O'Neill (1931), supposedly set in 1865–1866, Orin Mannon reproaches his sister Vinnie with having been too susceptible to the charms of native men during their voyage to the South Seas (Hawaii?). "But they turned out to be Vinnie's islands, not mine. They only made me sick—and the naked women disgusted me. I guess I'm too much of a Mannon, after all, to turn into a pagan. But you should have seen Vinnie with the men—! . . . If we'd stayed another month, I know I'd have found her some moonlight night dancing under the palm trees—as naked as the rest!" A playwright cannot be blamed for taking liberties with chronology.

13. Cooper, *The Crater*, in *Cooper's Novels* (New York & London: D. Appleton, n.d.). In other respects the novel is quite mediocre, with characters lacking any psychological depth.

14. Goldstein (1978).

15. Stuart Creighton Miller (1969), chaps. 1–6.

16. Opinions about the Malays of Malaya and Indonesia were considerably more negative: they were seen as scoundrels and pirates who had to be watched constantly. Heffer (1990), p. 56. There is one exception: Walter M. Gibson, the author of *The Prison of Weltevreden and a Glance at the East Indian Archipelago* (1855) (see Denys Lombard, "Dans les geôles de Weltevreden ou les tribulations d'un Américain aux Indes néerlandaises de 1852–1853," *Archipel* 40 [1990]: 65–77).

17. Dana (1965), p. 144 (chap. 21). For the image of the Latino and the contrasting descriptions of the males and females, see Hunt (1987), pp. 58–62.

18. Hezel (1983), pp. 141–142.

19. Melville, *Typee*, chap. XVII.

20. Horatio Hale, *Ethnography and Philology* (Philadelphia: Lea & Blanchard, 1846), pp. 13–17, 37–39, 45–50, 72–73.

21. Sidney B. Ahlstrom, *A Religious History of the American People* (New Haven: Yale University Press, 1972), chap. 24–27.

22. Hutchinson (1987); Andrew (1976); Phillips (1969), chap. 1. Valentin H. Rabe, "Evangelical Logistics. Mission Support and Resources to 1920," in Fairbank, ed. (1974), p. 58, does not see the missionary effort as the doing of a church militant, but as the initiative of a handful of determined pioneers whom the churches subsequently decided to support.

23. According to C. J. Phillips (1969), p. 29, before 1840, out of the 170 evangelists recruited, 133 joined up with the intention of going overseas: 66.2% came from New England (Yale, Amherst, Williams, Dartmouth, Middlebury), 17.3% from New York State and Ohio—regions inhabited by Yankees—and 9% from New Jersey (Princeton, Rutgers). The South provided very few.

24. Hiram Bingham, *A Residence of Twenty-One Years in the Sandwich Islands, or The Civil, Religious, and Political History of Those Islands* (1847) (Rutland: Charles C. Turtle, 1981), pp. 57–59.

25. Ibid., pp. 60–61. Emphasizing the inadequate geographical knowledge of the time, James A. Field, Jr., "Near East Notes and Far East Queries," in Fairbank, ed. (1974), p. 30, points out the part played by chance in the choice of Hawaii, in the form of a young Hawaiian orphan who attended Yale College.

26. Bingham, p. 70.

27. Ibid., p. 153. In the case of foreign words, particularly those occurring in the Scriptures, additional consonants—*b, d, f, g, r, s, t, v, z*—were introduced. Since double consonants were not used, "Boston" became "Bosetona."

28. Ibid., p. 327.

29. See the graph showing the annual expenditures of the ABCFM by region, in James A. Field, Jr., in Fairbank, ed. (1974), p. 32.

30. In *Typee*, chap. XXVI, Melville violently attacks the missionary's work in Hawaii. He blames them for destroying the heathens along with heathenism, of behaving like abominable colonialists (the "small remnant of the natives" having been "civilized into draught horses, and evangelized into beasts of burden"), and of living in the lap of luxury.

31. Quoted in Phillips (1969), p. 253.

32. James A. Field, Jr., in Fairbank, ed. (1974), p. 33, tallies 54 pastors, 27 doctors and associate missionaries, 89 women, making a total of 170 sent in 1812–1860 to Hawaii and Micronesia, i.e., fewer than to the Levant (268) and India-Ceylon (218).

33. Fred W. Drake, "Protestant Geography in China: B. C. Bridgman's Portrayal of the West," in Barnett and Fairbank, eds. (1985), p. 9.

34. There were missions in Siam (1831–1850), Singapore (1834–1843), and Borneo (1838–1852).

35. Stuart Creighton Miller, "Ends and Means: Missionary Justification of Force in Nineteenth Century China," in Fairbank, ed. (1974), pp. 251–257.

36. Latourette (1929), pp. 247–257, 365–372.

37. Peter Duus, "Science and Salvation in China. The Life and Work of W. A. P. Martin (1827–1916)," in Liu, ed. (1970), pp. 11–26. In Shanghai also, results were disappointing. The Southern Methodist mission, after six years activity, in 1866, had only twenty Chinese members (see Adrian A. Bennett and Kwang-Ching Liu, "Christianity in the Chinese Idiom: Young J. Allen and the Early Chia-hui hsin-pao, 1868–1870," in Fairbank, ed. (1974), p. 165. The results were the same for the Baptist Tarleton P. Crawford in Shanghai and Shandong. See Hyatt Jr. (1976), chap. 1.

38. Paul A. Cohen, chap. 11, in John K. Fairbank, ed., *The Cambridge History of China*, vol. 10, *Late Ching, 1800–1911* (Cambridge, Cambridge University Press, 1978): 543–590.

39. Carlson (1974), pp. 68–75.

40. Communist Chinese historians see the work of the missions as "cultural imperialism" in the service of economic interests. But Harris (1991) shows that its evolution was much more complex. The ABCFM followed two different policies in China between 1830 and 1860: first, a program of secular culture, supported by businessmen, aiming to show the Chinese the superiority of Western civilization, and then an education program (in Chinese) to reserve the services of converts for themselves—according to Harris, the only real policy of cultural imperialism. Harris concluded: "Understanding cultural imperialism in structural terms may therefore make it necessary to reverse the old dictum that 'Christianizing' mission strategies were less imperialistic than 'civilizing' strategies. One purpose of abandoning a 'civilizing' strategy was to assert even greater control over dependent and unequal Chinese converts" (p. 338).

41. Hezel (1983), pp. 142–170; Hanlon (1988); idem., "God versus Gods: The First Years of the Micronesian Mission on Ponape, 1852–1859," *Journal of Pacific History* 19, no. 1 (January 1984): 41–59.

42. Hezel (1983), pp. 200–210.

FIVE. MORE INTENSIVE ECONOMIC RELATIONS

1. Heffer (1988), pp. 140, 142–144.

2. The growth rates in constant dollars of imports and exports during this period were 4.1% and 4.4%, respectively.

3. Between 1868 and 1941, the population grew at an average rate of 1.8% per annum, compared to 3% between 1821 and 1867. The *per capita* growth of trade with the Pacific in 1913 constant dollars was 3.6% for imports (1.1% in 1821–1867) and 4.5% for exports (1.4%).

4. Growth of 0.72% per annum raised it, theoretically, from 38% to 90% between 1868 and 1941.

5. For a statistical study of trade relations between Pacific Rim countries (excluding Latin America), between 1920 and 1934, see Wright (1935).

6. Heffer (1986), chap. 4. Between 1901–1906 and 1925–1928, consumption fell by 1.6%, and the real price by 0.9% per annum on average. With an elasticity of approximately + 2, tea was apparently an inferior good.

7. *Statistical Abstract of the United States, 1906,* p. 482; *1928,* p. 559; *1946,* p. 692.

8. Clark (1949); Heffer (1986), pp. 134–137; William L. Li, "The Silk Export Trade and Economic Modernization in China and Japan," in May and Fairbank (1986), pp. 77–99; Haru Matsukata Reischauer (1986). The amount of silk by weight was relatively low, but the sums of money involved were considerable because of the high unit cost:

	Tons	Dollar Value (Millions of Dollars)	Price per Pound
1910–1914	10,946	$77	$3.25
1921–1925	23,971	$348	$6.70
1926–1930	34,653	$368	$4.90
1931–1935	32,093	$115	$1.65

9. Heffer (1986), chap. V.

10. In 1890–94, 4% of the exports from the United States to Japan were cotton and 60% petroleum.

11. *Bulletin* no. 74, p. 35. See Melvin Thomas Copeland, *The Cotton Manufacturing Industry of the United States* (Cambridge: Harvard University Press, 1912), chap. XII, and Bruce L. Reynolds, "The East Asian 'Textile Cluster' Trade, 1868–1973," in May and Fairbank (1986). There are numerous references to the trade in cottons in Northern China at the end of the nineteenth and at the beginning of the twentieth centuries in *Monthly Summary of Commerce and Finance*, for example in the issues dated December 1899, pp. 1608–09, 1677, and June 1901, pp. 2829, 2854–55. For a good example of American activity in Latin America, see the report by W. A. Tucker, "Textiles Markets of Bolivia, Ecuador, and Peru," published by the U.S. Department of Commerce, Bureau of Foreign and Domestic Commerce, Special Agents Series, no. 158 (Washington, D.C.: GPO, 1918).

12. Williamson and Daum (1959), pp. 236–237, 492–494, 661–676, and Williamson, Andreano, Daum, and Klose (1963), *passim*. Chu-yuan Cheng, "The U.S. Petroleum Trade with China, 1876–1949," in May and Fairbank (1986), pp. 205–233.

13. In 1890–94, these commodities accounted for only 2% of exports to China and 5% to Japan.

14. George Sweet Gibb, *The Saco-Lowell Shops. Textile Machinery Building in New England, 1813–1949* (Cambridge: Harvard University Press, 1950), pp. 478–481, 668, 671.

15. Wright (1935), pp. 76, 219, 259–261. He points out (p. 181) the "absurdity" of the duty on Chinese tungsten—208% *ad valorem* in 1922 and 130% in 1930, at a time when the United States was producing very little of this ore, essential to the manufacture of rapid-cutting steel tools.

16. Ibid., pp. 193–200.

17. John G. B. Hutchins, *The American Maritime Industries and Public Policy, 1789–1914: An Economic History* (Cambridge: Harvard University Press, 1941), 627 pp.; Jeffrey J. Safford, "The United States Merchant Marine in Foreign Trade, 1800–1939," in Yui and Nakagawa (1985), pp. 93–101, and "The Decline of the

American Merchant Marine, 1850–1914: An Historiographical Appraisal," in Fischer and Tague (1983).

18. Heffer (1986), pp. 112–113.

19. Kemble (1943); E. M. Tate (1986); Liu (1962); Cochran and Ginger (1954).

20. McCullough (1977) provides a remarkable treatment of all aspects of the subject. Siegfried (1940), pp. 145–178, is still a reliable source. See also Georges Fischer (1960); LaFeber (1989); Folkman (1972); Lael (1987); Randall (1992).

21. On the Japanese fleet, see the articles by K. Nakagawa, "Japanese Shipping in the Nineteenth and Twentieth Centuries: Strategy and Organization," pp. 1–33; Ryoichi Miwa, "Maritime Policy in Japan; 1868–1937," pp. 123–152; Kazuo Sugiyama, "Shipbuilding Finance of the Shasen Shipping Firms: 1920's–1930's," pp. 255–272; William D. Whay, "NYK and the Commercial Diplomacy of the Far Eastern Freight Conference, 1896–1956," pp. 279–305, all in Yui and Nakagawa (1985). Also of great interest is the article by Mariko Tatsuki, "Competition and Steamlining on the Pacific Routes in the 1920s and 1930s," in Fischer and Nordvik (1990).

22. Radius (1944), p. 89.

23. Ibid., p. 112.

24. Nakawaga, in Yui and Nakagawa (1985), pp. 13–14. Data on the transportation of silk may be found in the article by Tatsuki, in Fischer and Nordvik (1990).

25. Ahvenainen (1981), pp. 158–174, 178–185, 204–205; idem., "Telegraphs, Trade and Policy. The Role of the International Telegraphs in the Years 1870–1914," in Fischer, McInnis, and Schneider (1986), pp. 505–518; Daniel R. Headrick, *The Invisible Weapon. Telecommunications and International Politics, 1851–1945* (New York: Oxford University Press, 1991), chap. 6, pp. 99–102 and chaps. 10 and 11, for the years between the wars.

26. Clayton (1985), chap. 12; Richard Daley, *Juan Trippe and His Pan American Empire* (New York, 1980); Wesley Phillip Newton, *The Perilous Sky: U.S. Aviation Diplomacy and Latin America, 1919–1931* (Coral Gables, Fla.: University of Miami Press, 1978).

27. *Pacific Islands Monthly*, April 1991, pp. 24–27. Earhart's plane never reached Howland Island, where it was expected. Various theories have been suggested: an error by the navigator, Noonan, said to have drunk too much; capture by the Japanese after an espionage mission over the Marshall Islands (the extremely improbable theory developed by Randall Brink, *Lost Star* [New York: Norton, 1994]); a crash into the sea after running out of fuel; or death on Nikumaroro.

28. Matthew Simon, "The United States Balance of Payments, 1861–1900, in *Trends in the American Economy in the Nineteenth Century, Studies in Income and Wealth*, vol. 24 (Princeton, N.J.: Princeton University Press, 1960), pp. 699–705; *Historical Statistics of the United States*, Part 2, 1975, pp. 866–867.

29. Gavin Wright, "The Origins of American Industrial Success, 1879–1940," *American Economic Review*, September 1990, pp. 651–668.

30. Investments overseas have been studied in depth by Cleona Lewis (1938) and Mira Wilkins (1970, 1974).

31. Stewart (1946).

32. Clayton (1986), chap. 7.

33. Frederick B. Pike (1977), pp. 159–168.

34. McCormick (1967), pp. 74–76, 87–89, 133–134.

35. Hunt (1973).

36. Hunt (1983), pp. 209–216.

37. Thomas F. O'Brien, "Rich beyond the Dreams of Avarice: The Guggenheims in Chile," *Business History Review* 63 (Spring 1989): 122–159; Sater (1990), chap. 5. Until 1929, nitrates were a more valuable export than copper.

38. Irvine H. Anderson Jr. (1975), Appendix A, pp. 202–214; Gibb and Knowlton (1956); Peter Reed (1958); Allen and Donnithorne (1957), p. 289.

39. Sherman Cochran, "Commercial Penetration and Economic Imperialism in China: An American Cigarette Company's Entrance into the Market," in May and Fairbank, eds. (1986), pp. 151–203; idem (1980), chap. 1–4. On unsuccessful American ventures in China, see Young (1968), chaps. 2–3.

40. Cleona Lewis (1938), p. 347.

41. Ibid., pp. 624–625.

42. Marichal (1989), chaps. 7–8; Paul W. Drake (1989); Stallings (1987).

SIX. CLOSER CULTURAL CONTACTS

1. Akira Iriye (1992), pp. 45–47; Liu (1963), p. 30, indicates that before 1901 only 57 students of Chinese nationality attended American universities. They numbered 2,406 between 1901 and 1920, and 5,474 in 1921–1940.

2. Lutz (1971).

3. Beauchamp and Iriye (1990); Burks (1985); Jones (1980); Dulles (1965); Huffman (1987); Roden (1986); Schwantes (1955), chaps. 2 and 4; *Kodansha Encyclopedia of Japan* (Tokyo: Kodansha, 1983): article "Foreign Employees of the Meiji Period," (Beauchamp), vol. 2, pp. 310–311, and articles on the *yatoi*. According to Umetani Noboru, cited in the last-mentioned publication, in 1874, Americans represented 9.3% of the 503 foreigners employed, many fewer than the English, (53.5%), or French (21.5%), but more than the Germans (7.4%). Between 1881 and 1898, in the Public Service, they accounted for 21.7% of 1,242 teachers (fewer than there were English or German) and 3.9% of the 541 engineers. In the private sector, they accounted for 55.7% of the 3,886 teachers and 8.2% of the 3,809 engineers.

4. Cited in Fujita Fumiko, "Encounters with an Alien Culture: Americans Employed by the *Kariakushi*," in Beauchamp and Iriye (1990), p. 94.

5. John M. Maid, "William Smith Clark, *yatoi*, 1826–1886," in Beauchamp and Iriye (1990), chap. 5.

6. In China also there were very few American military advisers. However, we should mention Philo McGiffin , who served in the Imperial Navy from 1885 to 1894, taking part in the naval battle of Yalu against the Japanese, and Homer Lea, chief of staff to Sun Yat-sen in 1912. See Richard H. Bradford, "That Prodigal Son: Philo McGiffin and the Chinese Navy," *American Neptune*, 1978, pp. 157–169; Anschel (1984).

7. Rosenstone (1988); Chisholm (1963).

8. Thomson, Stanley, and Perry (1981).

9. Ninkovich (1984), p. 803; Rosenberg (1982); Keenan (1977).

10. Thomson (1969), chap. 6.

11. Drake (1989), chap. 3; Seidel (1972).

12. Stanley (1974, 1984); Owen (1971). Peter W. Stanley (1984) speaks of "the hubristic illusion" of Americans who "imagine that, in the colonial era, they liberalized, modernized, or, for that matter, exploited the Philippines in any large, systemic, or lasting way," for in fact "they accomplished little. Underestimating the resilience of Hispano-Philippine institutions and the social and economic rhythms of Philippine life, overestimating the power of their own example, Americans achieved much less than they said they did. Sometimes progressive and sometimes not, the American empire was above all naively pretentious in its self-evaluation" (p. 2).

13. Wiest (1988); Breslin (1980). Very early on, the Mormons tried to convert the natives in Hawaii. The adventures of Walter Murray Gibson, expelled from the church in 1864, who occupied the position of prime minister from 1882 to1887, are well known.

14. Alfred D. Chandler Jr., *The Visible Hand: the Managerial Revolution in American Business* (Cambridge, Mass.: Belknap Press, 1977).

15. Rabe (1978); idem, "Evangelical Logistics: Mission Support and Resources to 1920," in Fairbank (1974), pp. 56–90; Hutchinson (1987).

16. Harlan P. Beach, *A Geography and Atlas of Protestant Missions: Their Environment, Forces, Distribution, Methods, Problems, Results and Prospects at the Opening of the Twentieth Century,* 2 volumes: 1. *Geography;* 2. *Statistics and Atlas* (New York: Student Volunteer Movement for Foreign Missions, 1901–1903); H. P. Beach and Burton St. John, eds., *World Statistics of Christian Missions; containing a directory of missionary societies, a classified summary of statistics, and an index of mission stations throughout the world* (New York: Committee of Reference and Counsel of the Foreign Missions Conference of North America, 1916); H. P. Beach and Charles H. Fahs, eds., *World Missionary Atlas* (New York: Institute of Social and Religious Research, 1925); James S. Dennis, *Centennial Survey of Foreign Missions. A Statistical Supplement to "Christian Missions and Social Progress"* (Edinburgh: Oliphant, Anderson & Ferrier, 1902).

17. Jean Meyer, *Les Religions chrétiennes en Amérique latine* (Paris: Desclée de Brouwer, 1991).

18. The principal source dealing with this area is found in the ABCFM archives, in the Houghton Library at Harvard. Regarding the transfer of the Ocean Island mission to the LMS, see documents 81, 83, and 86 in vol. 784. The German, R. Deeken, *Die Karolinen*, 1912, p. 35, recognized the spiritual hold of the Americans over the natives: "It will of course be a long time before our missionaries can spread the German way of thinking and eliminate the natives' preference for the Americans. We can hardly expect such a result before the next generation." In 1884, the ABCFM added a mission on Ruk to those on Kosrae and Ponape.

19. James A. Field, Jr., "Near East Notes and Far East Queries," in Fairbank (1974), p. 34, shows that from 1885 on the Far East surpassed India as the primary missionary field for the four major American Protestant denominations.

20. Winburn T. Thomas (1959); Paik (1970).

21. Field, in Fairbank (1974), p. 36. In 1890, out of 175 ordained Protestant missionaries, Americans accounted for 146 (83%). In China, on the other hand, with 198 (37%), they were outnumbered by the British.

22. Each year the ABCFM received a report from its mission in Japan, some traces of which can be found in the archives of the Houghton Library. In 1926, there were 665 students attending Kobe College, 41 of them in the Music Department. It employed 12 American instructors, 9 of them women. Doshisha enjoyed strong growth in the 1920s. In 1928, over 4,000 students and pupils were enrolled in the university, in the colleges (mostly commercial in the girls' case), in the secondary school, and in the "Academy."

23. Fairbank (1974); Varg (1958); Rabe (1978); Latourette (1929); Liu (1966); Lutz (1971); the chapters of the *Cambridge History of China*, written by Paul Cohen (vol. 10, pp. 573–589) and Albert Feuerwerker (vol. 12, pp. 165–177). The richest source for the history of American Protestantism in Asia is the *Chinese Recorder and Missionary Journal*, a periodical founded in May 1868 at Fuzhou by the American Methodist S. E. Baldwin. After an interruption in publication from June 1872 to December 1873, it appeared monthly from 1886 until December 1941, with a circulation of up to 3,000 copies. It is easy to use, thanks to the index prepared by Kathleen Lodwick: *The Chinese Recorder Index. A Guide to Christian Missions in Asia, 1867–1941* (Wilmington, Del.: Scholarly Resources, 1986), 2 vols.

24. Pearl, Buck, *The Good Earth* (New York: John Day, 1931), p. 128.

25. Hyatt (1976), chaps. 1–3. Crawford (1821–1902) remained in China, with brief periods of leave in the United States, from 1852 to 1900.

26. Hyatt (1976), chaps. 7–9.

27. Latourette (1929), pp. 513–517, puts the losses among foreign Protestants at 187 (134 adults and 53 children), more than half belonging to the CIM, and

among the Chinese converts at 1,912. Among the Catholics, the missionaries (47 dead) were, relatively speaking, more fortunate than the converts (30,000 dead).

28. Albert Feuerwerker, who provides these figures in the *Cambridge History of China*, vol. 12, p. 175, puts the figure at 12 percent out of a total of 21,000 in 1925, whereas 16–17% might be expected.

29. Hunter (1984).

30. William Ernest Hocking, *Re-Thinking Mission: A Laymen's Inquiry after One Hundred Years*, Laymen's Foreign Missions Inquiry. Commission of Appraisal, William Ernest Hocking, chairman (New York: London, Harper & Brothers, 1932).

31. Grant Wacker, "A Plural World: The Protestant Awakening to World Religions," in William R. Hutchinson, ed., *Between the Times: The Travail of the Protestant Establishment in America, 1900–1960* (Cambridge: Cambridge University Press, 1989), pp. 253–277; Paul Varg (1958), chap. 10. On the missions and fundamentalist Protestantism, see Carpenter and Shenk (1990).

32. Janet B. Heiminger, "Private Positions Versus Public Policy: Chinese Devolution and the American Experience in East Asia," *Diplomatic History* 6, no. 1 (Summer 1982), pp. 287–302.

33. Edmund John Michael Rhoads, "Lingnam's Response to the Rise of Chinese Nationalism: The Shakee Incident (1925)," in Liu (1966), pp. 183–214; Shirley Stone Garrett, "Why They Stayed: American Church Politics and Chinese Nationalism in the Twenties," in Fairbank (1974), pp. 283–310.

34. See the articles by William A. Brown, "The Protestant Rural Movement in China (1920–1937)," pp. 217–248, and Fox Butterfield, "A Missionary View of the Chinese Communists (1936–1939)," pp. 249–301, in Liu (1966); Paul Varg (1958), chaps. 11–15. On Walter Judd, a medical missionary and hospital director in China from 1925 to 1931 and 1934 to 1938, and later Republican Congressman from Minnesota from 1943 to 1963, an ardent defender of the Nationalist cause, see Edwards (1990).

35. Masland (1941), p. 290. Investments are put at $50 million (p. 279). To this should be added independent organizations such as the YMCA, the Rockefeller Foundation, the Association for Christian Colleges in China (numbering 13), to which were affiliated 4 foundations and 27 prestigious American colleges. The 2,500 missionaries in 1938 should be compared to the figures of 200 in 1870 and 1,000 in 1900.

36. James Reed (1983), chap. 1.

37. Franz Boas, Preface to Margaret Mead, *Coming of Age in Samoa*, in Mead (1963), p. 295.

38. Derek Freeman (1983). For a critique of Freeman, see Holmes (1987); Richard Handler, "Review Essay, Ruth Benedict, Margaret Mead, and the Growth of American Anthropology," *Journal of American History* 71, no. 2 (September 1984): pp. 364–368. M. Mead, entrusted by the Social Sciences Research Council, under the New Deal, to prepare a study on the possible contribution of

ethnology to a better understanding of cooperation and competition in human societies, including the United States of the time, collected thirteen analyses, five of them dealing with peoples of the Pacific, under the title *Cooperation and Competition among Primitive Peoples* (1937).

39. Takaki (1990), p. 314.

40. *Historical Statistics* (1975), Series C104 and 106. Between 1854 and 1883, 288,000 Chinese immigrants were admitted. The ten years are: 1854, 1869–1870, 1873–1877, 1881–1883 (the record year was 1882, with 40,000 admitted), with an average of 18,500 per annum. In the other twenty years the average was 5,000. During this period, the proportion of Europeans was 26 times higher. Between 1884 and 1914, only an average of 1,500 Chinese per annum were allowed in. The Japanese represented a significant number starting only in 1889, and averaging 17,000 between 1902–1908 (1907 was the record year, with 30,000).

41. Jean Heffer, "Du pull et du push," in Robert Rougé, ed., *Les Immigrations européennes aux États-Unis (1880–1910)* (Paris: Presses de l'université de Paris-Sorbonne, 1987), pp. 21–48.

42. *Historical Statistics*, Series A 97; Hunt (1983), chap. 1.

43. Yasuo Wakatsuki, "Japanese Emigration to the United States, 1866–1924: A Monograph," *Perspectives in American History* 12 (1979): 387–516. Robert C. Schmitt, *Historical Statistics of Hawaii* (Honolulu: University Press of Hawaii, 1977), table 1.12.

44. Stuart C. Miller (1969), chap. 1, pp. 6–9.

45. Among the most important legal decisions are *Fong Yue Ting* vs. *U.S.* (1893), *Li Sung* vs. *U.S.* (1901), *Chin Bak Kan* vs. *U.S.* (1902), *Chin Yow* vs. *U.S.* (1907); see Kim (1986).

46. *Gung Lum vs. Rice.*

47. R. Takaki (1990), pp. 197–210; Mitziko Sawada, "Culprits and Gentlemen: Meiji Japan's Restrictions of Emigrants to the U.S., 1891–1909," *Pacific Historical Review* 60, no. 3 (August 1991): 339–359.

48. Takaki (1990), pp. 205–206.

49. The basic study is the Ph.D. thesis (University of Wisconsin, 1957) by Thompson (1978). See also Wu (1982).

50. Madison Grant, *The Passing of the Great Race* (New York: Arno Press, 1970 [c1918]); Lothrop Stoddard, *The Rising Tide of Color against White World-supremacy* (Westport, Conn.: Negro Universities Press, 1971 [c1920]).

51. Peter B. Kyne, *The Pride of Palomar* (New York: Cosmopolitan Book Co., 1922), p. 316.

52. Pierton W. Donner, *Last Days of the Republic* (San Francisco: Alta California Publishing House, 1880), p. 257.

53. Lea (1909), p. 158. On Homer Lea, in addition to Anschel (1984), see Valerie M. Hudson and Eric Hyer, "Homer Lea's Geopolitical Theory: Valor of Ignorance?" *Journal of Strategic Studies* 12, no. 3 (September 1989): 324–348.

54. Bywater (1925).

55. Lancaster (1963); Frank Lloyd Wright, *Writings and Buildings* (Cleveland: Meridian Books, 1960), pp. 298–300; Donald Hoffmann, *Frank Lloyd Wright. Architecture and Nature* (New York: Dover Publ., 1986), pp. 72–73. On "Japonism" in America, see Meech and Weisberg (1990).

56. Meetings were held at Honolulu (1927), Kyoto (1929), Shanghai (1931), Banff (1933), Yosemite Park (1936), and Virginia Beach (1939). They resulted in voluminous publications dealing with the diverse problems of the Pacific: 27 volumes in 1929, 20 in 1931, 12 in 1933, and 15 in 1936.

57. Valerie Noble, *Hawaiian Prophet: Alexander Hume Ford* (Smithtown: Exposition Press, 1980).

SEVEN. IMPERIALISM AND THE OPEN DOOR

1. LaFeber (1963); Field (1978). For a bibliographic overview, see Edward P. Crapol, "Coming to Terms with Empire: The Historiography of Late-Nineteenth-Century American Foreign Relations," *Diplomatic History* 16, no. 4 (Fall 1992): 573–597.

2. On Mahan, see Seager (1977).

3. The interpretation of imperialism and anti-imperialism has given rise to an enormous amount of literature. We will mention only May (1961, 1968); Pratt (1936); Healy (1970); Thomas G. Paterson (1990); Beisner (1968); Rydell (1983); and the chapter by Laurent Cesari in Heffer and Weil (1994).

4. The United States had claimed a number of atolls—the guano islands (1856) and Midway (which was claimed by Captain Reynolds on instructions from Washington on April 28, 1867)—before taking an interest in Samoa, but these islands were uninhabited.

5. Kennedy (1974); Ryden (1933).

6. E. Mowbray Tate (1986), pp. 49–61.

7. On Hawaii, Kuykendall (1953, 1967); Stevens (1968); Merze Tate (1967, 1968); Tabrah (1980).

8. Adler (1966).

9. On the 1893 revolution, see Russ (1943, 1959); Pratt (1932); Weigle (1947), pp. 41–58; Campbell (1976), chap. 10; Merze Tate (1965), pp. 155–193.

10. Russ (1961); William A. Morgan, "The Anti-Japanese Origins of the Hawaiian Annexation Treaty of 1897," *Diplomatic History* 6, no. 1 (Winter 1982): 23–44.

11. Cited in Charles S. Campbell (1976), p. 279.

12. Ephraim K. Smith, "A Question from Which We Could Not Escape: William McKinley and the Decision to Acquire the Philippine Islands," *Diplomatic History* 9, no. 4 (Fall 1985): 363–375.

13. Earl S. Pomeroy, "American Policy Respecting the Marshalls, Carolines, and Marianas, 1898–1941," *Pacific Historical Review* 17 (February 1948): 43–53.

14. Ernest Dupuy and Trevor N. Dupuy, *Military Heritage of America* (Fairfax: Hero Books, 1984), p. 330.

15. Stuart Creighton Miller (1982); Welch (1974, 1979); Agoncillo (1974); Smythe (1962); Salamanca (1968); Russell Roth (1981); Linn (1989); Gates (1973, 1984).

16. Livermore (1944). This station remained in use until 1924.

17. Charles S. Campbell (1976), pp. 168–176; Pike (1963), chap. 3; Goldberg (1986); Sater (1990).

18. Livermore (1943); E. Taylor Parks and J. Fred Rippy, "The Galapagos Islands: A Neglected Phase of American Strategy Diplomacy," *Pacific Historical Review* 9 (March 1940): 37–46; Braisted (1971), pp. 49–57, 231–245, 522–525; Wilson (1979).

19. Busch (1985); Thomas A. Bailey, "The North Pacific Sealing Convention of 1911," *Pacific Historical Review* 4, no. 1 (March 1935): 1–14; John Bassett Moore, *History and Digest of the International Arbitrations to which the United States has been a Party*, vol. I (Washington, D.C.: GPO, 1898), pp. 755–961; Charles S. Campbell, "The Anglo-American Crisis in the Bering Sea, 1890–1891," *Mississippi Valley Historical Review*, 48 (1961): 393–414; Gray (1987); D. C. Paterson (1977); D. G. Paterson and Wilen (1977); Gerald O. Williams (1984).

20. Georges Fisher (1960); Brands (1992); Stanley (1974, 1984); Owen (1971); Gleeck (1976); Glenn A. May (1980).

21. Cited in Owen (1971), p. 5.

22. After 1906, the Filipinos held the majority, when one of the members previously nominated became elected.

23. Owen (1971), pp. 16–17.

24. Cameron Forbes, *The Philippine Islands* (Boston, 1928), I: 141.

25. The text appears in Henry S. Commager, *Documents of American History*, vol. II (New York: Appleton-Century-Crofts, 1968): 116–119. An amendment proposed by Senator James P. Clark of Arkansas, granting independence five years later, was rejected.

26. The bill was brought forward by Senators Hawes (Democrat, Missouri) and Cutting (Republican, New Mexico) and Congressman Hare (Democrat, South Carolina). It is noticeable that a key role was played by southerners, eager to be rid of colored races.

27. Commager, ed., *Documents of American History*, 1968, vol. II: 9–10.

28. Ibid., p. 11.

29. Marilyn Blatt Young (1968), p. 115. See also Israel (1971).

30. Cited in M. B. Young (1968), p. 131.

31. Kennan (1951), an eminent representative of the "realist" position, defended this theory.

32. McCormick (1967).

33. Hunt (1983).

34. The inequality was with respect to the lack of freedom to set tariffs, and to the rights to extraterritoriality granted foreigners.

35. Hunt (1983), chap. 4.

36. Esthus (1959); Vevier (1955).

37. His attachment to the open door did not prevent Hay from supporting the Navy's demand for a naval base on the Chinese coast.

38. B. Young (1968), p. 218; Trani (1971).

39. John Chay, "The First Three Decades of American-Korean Relations, 1882–1910, Reassessments and Reflections," pp. 15–33, and Andrew C. Nahm, "U.S. Policy and the Japanese Annexation of Korea," pp. 34–53, in Kwak (1982); Yur-Bok Lee, "Korean-American Diplomatic Relations, 1882–1905," pp. 12–45, and Fred Harvey Harrington, "An American View of Korean-American Relations," 1882–1905," pp. 46–67, in Yur-Bok Lee and Wayne Patterson (1986); Frederick Drake (1984).

40. William Appleman Williams (1962), p. 38. On the Wisconsin School, see the article by Laurent Cesari in Heffer and Weil (1994).

41. On *corporatism*, or *associationalism*, see Michael J. Hogan, "Corporatism," *Journal of American History* 77, no. 1 (June 1990): 153–160; Thomas J. McCormick, "Drift or Mastery: A Corporatist Synthesis for American Diplomatic History," *Reviews in American History* 10 (December 1982): 318–330.

EIGHT. THE PACIFIC PROBLEM

1. Académie diplomatique internationale, *Dictionnaire diplomatique*, vol. 2: 277.

2. Esthus (1988) considers that Roosevelt was attempting above all to maintain a balance of power on the continent of Asia. It is also possible that he was trying to divert Japanese expansionism toward China and Korea instead of Hawaii and California.

3. The studies by Akira Iriye are indispensable (1972, 1990, 1992).

4. Halford J. Mackinder, *Democratic Ideals and Reality* (1919).

5. Iriye (1990), chaps. I–III.

6. Ienaga (1978), Part I.

7. See chap. 7.

8. Zhang Yongjin (1991) draws a picture of a more activist China than is generally thought. On the role of private citizens, see Chong (1984).

9. The text of the memorandum appears in the article by Andrew C. Nahm, "U.S. Policy and the Japanese Annexation of Korea," in Kwak (1982), pp. 51–53.

10. *Foreign Relations of the United States, 1908*, pp. 510–511.

11. Griswold (1938), p. 129; Neu (1967); Neumann (1963).

12. Reckner (1988); Hart (1965). After passing through the Straits of Magellan, the fleet put in to South American ports before reaching California; on the return voyage it called at Hawaii, Samoa, and Australia, sailing around the last before heading on to the Philippines, visiting Japan and China, and returning home through the Suez Canal.

13. Cited in Iriye (1972), p. 226.

14. On Knox's diplomacy, see the article by Walter Scholes in Graebner (1961), pp. 59–78. The United States was nevertheless able to exclude Japan from the eastern waters of the Pacific. When, in 1911, Japanese traders purchased a private property in the Bay of Maddalena in Mexico, Senator Lodge put through an amendment to the Monroe Doctrine which urged the American government not to tolerate the occupation of any port or region of the Western hemisphere by a private association with links to a government on another continent.

15. The text of the Twenty-One Demands is given as an appendix in Reed (1983), pp. 203–205. On American opposition to the Chinese republican revolution, see Crane and Breslin (1986).

16. On the options available to Wilson, see Reed (1983), pp. 159–162.

17. *Lansing Papers, 1914–1920*, in *Foreign Relations of the United States, vols. 116 & 117* (Washington, D.C.: G.P.O., 1870–1946); Beers (1962).

18. Braisted (1971); Morley (1957); Unterberger (1956).

19. Fifield (1952); Pugach (1969).

20. Bywater (1921, 1925).

21. Iriye (1990).

22. Japan objected to the communiqué of December 15, 1921, which specified that the status quo would not apply to the islands which are part of Japan proper. On January 10, 1922, Kato asked that the islands in question should not be named in the text of the treaty, but in a separate declaration—a kind of return to secret diplomacy. Hughes rejected this source of future ambiguity. On the Five-Power Treaty, see *Papers Relating to the Foreign Relations of the United States, 1922*, vol. 1 (Washington, D.C.: GPO, 1938): 53–266. For the report of the American delegation (February 9, 1922), idem, pp. 298–378. On the Washington Conference, see Thomas H. Buckley (1970) and Dingman (1976).

23. *Foreign Relations, 1922*, vol. 1, p. 23. On the Four-Power Treaty, see pp. 1–53.

24. Ibid., pp. 271–298.

25. Ibid., p. 276.

26. Kwantung is the territory leased by China to Russia, taken over by Japan in 1905.

27. On the attitude of the United States toward China during this period, see Borg (1947); Bernard Cole (1983).

28. O'Connor (1962).

29. See chapter 3.

30. Lance C. Buhl, "Maintaining 'An American Navy,' 1865–1889," in Hagan (1984), pp. 145–173.

31. Thanks to these modern ships, the United States was able to bring Chile to heel in 1892 and Commodore Dewey to destroy the Spanish squadron in Manila Bay in 1898. Such successes would scarcely have been possible before 1883.

32. From 1919 until January 1941, the U.S. fleet was in fact a Pacific fleet. The Asiatic fleet and the Atlantic fleet did not include any battleships.

33. On the role of the Navy in the Pacific between 1898 and 1922, the two volumes by Braisted (1958, 1971) are fundamental.

34. The most complete study of Plan Orange is by Edward S. Miller (1991). However, Miller tends to exaggerate the continuity between the Orange plans and the strategy adopted during World War II.

35. It was the "thrusters" who inspired the Orange plans of March 1914 and January 1925 (Admiral Robert C. Coontz, the "through ticket"), while the "cautionaries" had an impact on the plans of September 1922 (Clarence S. Williams) and July 1934 (the "Royal Road").

36. Pelz (1974).

37. Pelz (1974), p. 221. The naval forces in the Pacific in the fall of 1941 were as follows:

	United States	Other Western Nations	Japan
Battleships	9	2	10
Carriers	3	0	10
Cruisers	24	12	36
Destroyers	80	20	113
Submarines	56	13	63

38. Rappaport (1963); Doenecke (1984). For 1933–1937, see Borg (1964); Thorne (1972).

39. The principal American diplomatic sources are found in the two volumes published by the U.S. Department of State, *Foreign Relations of the United States: Japan, 1931–1941* (Washington, D.C.: Government Printing Office, 1943). On the defeat of the Japanophiles, such as Sidney Gulick, see Taylor (1984).

40. Pelz (1974), pp. 7–63, gives a good analysis of Japanese responsibility; Iriye (1987); Barnhart (1987); Ienaga (1978), Part 1; Nobriya Bamba, "Japan's Search for Its National Identity: Towards Pearl Harbor," in Ion and Hunt (1988), p. 140.

41. Crowley (1966); Thorne (1972).

42. Heinrichs (1966).

43. However, in July 1940 it was the Army which wanted to expand the war into Indochina, while the Navy was much more reticent, fearing it might have to face the Americans before it was completely prepared.

44. Coox (1985).

45. See the six articles on the Japanese-Soviet rapprochement and Nazi Germany's reaction in David Wingeate Pike (1991), Part 5.

46. On Cordell Hull, see his *Memoirs*, and also Utley (1985), who gives a very precise analysis of Hull's policies and his struggles with the Washington bureaucracy. Spykman (1942) is an excellent illustration of the importance the

Americans gave the Pacific:"The Pacific Ocean, in contrast to the Atlantic Ocean, was for the United States not merely a passageway, a route of communication to a profitable market; it was the zone in which lay her colonial domain and her insular possessions. It has a territorial and strategic significance which the other sea lacked and this accounted, at least in part, for some of the differences in attitude toward the power struggles in Europe and Asia" (p. 141).

47. Hess (1987). Spykman (1942): A "Japanese Greater East Asia Co-Prosperity Sphere" would mean the final destruction of the balance of power in the transpacific zone which would have ultimate repercussions on our power position in the Western Hemisphere" (p. 155). Spykman (pp. 196–197) attaches considerable significance to a Japanese plan to restrict the United States sphere of influence to North America. This would have meant the end of the Monroe Doctrine and the encirclement of the country by totalitarian powers. The United States would have been deprived of certain strategic raw materials: tin, tungsten, abaca (Manila hemp), quinine, rubber, silk, and kapok (p. 297).

48. David Reynolds (1981). The Americans did not consider Singapore to be of fundamental strategic importance, believing that the British were only developing a base there to allow them better control over the empire.

49. Ruth P. Harris (1981); Lewin (1982); U.S. Department of Defense, *The "Magic" Background of Pearl Harbor* (Washington, D.C.: GPO, 1978), 8 vols.

50. Utley (1985), pp. 155–156, accuses Acheson of having deliberately misinterpreted Hull's wishes. Anderson Jr. (May 1975), pp. 201–231, sees it as a matter of "bureaucratic reflex."

51. Butow (1961).

52. Iriye (1987); Heinrichs (1988), dealing with the year 1941, from March to December; Utley (1985); Borg and Okamoto (1973); Feis (1950).

53. For a different interpretation, see Ion and Hunt (1988), p. 14.

54. Stephan (1984).

55. Thailand lay outside this perimeter.

56. John W. Chapman, "The Pacific in the Perceptions and Policies of the German Navy, 1919–1945," in Ion and Hunt (1988), pp. 121–124, points out that Hitler would have preferred that the Japanese offensive not be launched against the USSR after the German offensive, but against Southeast Asia, so as to divert American attention to the Pacific.

57. The book by Kinoaki Matsuo, which typifies this mentality, is the translation of a work published in Japan in October 1940, *How Japan Plans to Win: The Three-Power Alliance and a United States-Japanese War* (Boston: Little, Brown, 1942). See also the articles by Shoichi Saeki and Kimitada Miwa in Iriye (1975), pp. 100–137. It seems to me that historians tend to make too much of the so-called Japanese inferiority complex and of their desire to regain an honor which they had never lost.

NINE. WAR IN THE PACIFIC

1. James W. Morley has published several volumes of translations taken from *Taiheiyo senso e no michi, kaisen gaiko shi* (1976, 1980, 1983, 1984). On the progress of the war, see Ike (1967). The book by Kinoaki Matsuo (1942) is of particular interest. Matsuo was an ultranationalist information officer, who in October 1940 developed a war plan in his work (see above no. 57).

2. There is a vast bibliography devoted to the Pacific war. In December 1987, John S. Sbrega (1989) had already identified over 5,200 books and articles. Each arm of the services has its own official history in several volumes. For the Navy, Morison (1947–1960). Of the fifteen volumes, nine deal with the Pacific: 3, 4, 5, 6, 7, 8, 12, 13, and 14. A condensed version exists (1963). For the Army, *The United States Army in World War II: the War in the Pacific* (11 vols.), *The China-Burma-India Theater* (3 vols.), *The War Against Japan. Pictorial Record* (1 vol.). For the Air Force, Craven and Cate (1948–1958) are extremely useful, as are the volumes published by the Office of the Chief of Military History, Department of the Army, dealing with strategy and logistics: Matzloff and Snell (1953, 1959); Leighton and Coakley (1954, 1968); Morton (1962). For the Marines, *United States Marine Corps Operations in World War II*, 5 vols., 1958–1971. For shorter studies, see Spector (1987); Costello (1981); Willmott (1982, 1983); and in the *West Point Military History Series*, edited by Thomas E. Griess, the analysis by John J. Bradley and Jack W. Dice, *The Second World War: Asia and the Pacific* (Wayne, N.J.: Avery Publishing Group Inc.,1984), accompanied by an atlas containing 52 pages of maps.

3. Iriye (1981), p. 119.

4. Dower (1986), Part 3.

5. For a leftist interpretation of Japan during the war, see Ienaga (1978).

6. The geopolitical lines of force of the "great American strategy" are analyzed by Spykman (1942).

7. Charles F. Brower IV, "Sophisticated Strategist: General George A. Lincoln and the Defeat of Japan, 1944–45," *Diplomatic History* 15, no. 3 (Summer 1991): 317–337.

8. The attack on Pearl Harbor has been the subject of an enormous literature and endless controversy. See Slackman (1990); Lord (1970); Prange (1981); Delmas (1990); Conroy and Wray (1990); Herde (1980); Trefousse (1982); Wohlstetter (1962); and, on the broader context, Borg and Okamoto (1973).

9. The air of superiority assumed by the Americans was close to arrogance: they believed themselves to be physically fitter, faster to react in an unexpected situation, and more experienced. In a book published before Pearl Harbor, Captain W. D. Puleston (1941) showed a sublime self-confidence when he wrote: "The American Fleet would cross the Pacific at about the speed of translation of a cyclone—between 10 and 15 knots. It would resemble the cyclone in a more important phase, leveling everything in its path except a stronger fleet—and a

stronger fleet does not at present exist. Moreover, the American Fleet belongs to a continental nation that can risk it in battle without jeopardizing the national security. The United States is the only continental nation since Imperial Rome and Spain under Charles V to have a preponderant navy" (p. 242).

10. MacArthur had time to send out his B-17 bombers against the Japanese airfields on Formosa. As for Admiral Kimmel, at Pearl Harbor, what could he have done even if he had had foreknowledge of the enemy's intentions? Send his battleships off to California? Order his B-17s to bomb the Japanese aircraft carriers? But he had too few available, since a good number of the bomber squadrons had recently been transferred to the Philippines. Use his two available aircraft carriers? They were much too weak to confront Vice Admiral Nagumo's six carriers.

11. Between January and June 1942 the allies lost 2,783,000 tons (506 ships) in the Atlantic, 172,000 tons in the Indian Ocean (37 ships), and only 85,000 tons in the Pacific (21 ships). Cf. *The West Point Military History*, p. 117.

12. Ibid., p. 248. Tankers, on the other hand, were neglected until 1944.

13. Francis (1977).

14. Thorne (1978, 1985).

15. It has been argued that the taking of Peleliu served no useful purpose.

16. These figures are cited by Dower (1986), pp. 297–298, 300. Those killed in the Pacific account for 37 percent of all American deaths in World War II (24 percent for the Army and the Air Force, 81 percent for the Navy, 98 percent for the Marines). Out of almost 300,000 killed, 60 percent belonged to the Army and Air Force in the European theater.

17. On the aircraft carrier revolution, Clark G. Reynolds (1968); Humble (1984); Brown (1974); Roskill (1954–1961).

18. The final blow to the battleship occurred on April 7, 1945, off Okinawa, when the *Yamato*, a vessel of monstrous size, considered invincible before 1941, was sunk by Mitscher's pilots, without ever having used its enormous 460 mm guns in a naval battle.

19. For a while, in April 1943, the Americans were obliged to borrow a British aircraft carrier. But by the end of August, they had eleven carriers, against six for the Japanese.

20. Selden's article (1991) provides a good synthesis of the extensive literature; Sigal (1988).

21. Dower (1986).

22. Tojo's Japan was, however, less totalitarian than Hitler's Germany.

23. Benedict (1946). This study, commissioned during the war, was not published until after the end of hostilities. It has been the target of severe criticism for the inadequacy of its sources and for its oversimplifications.

24. Daniels (1962, 1981, 1986, 1988); Irons (1983); Takaki (1989), pp. 379–405; Collins (1985); Christgau (1985).

25. *The Japan Yearbook, 1943–1944* (Tokyo, 1944), p. 259.

26. Iriye (1981). If he is right to restore some balance, it still seems to me that Akira Iriye pushes his argument too far by dwelling on the *form* of the actions and deliberations rather than on the substance, where the antagonisms were substantial.

27. Takaki (1989), p. 378: Between 1944 and 1953, there was an average of 59 Chinese immigrants per annum. Between 1944 and 1952, only 1,428 Chinese became United States citizens.

28. Moore (1981); Potts and Potts (1985).

29. Michener (1946).

30. Oscar Hammerstein, II, and Joshua Logan, *South Pacific. A Musical Play* (New York: Random House, 1949). Part of the text is in French. See the article by Philip D. Beidler, "South Pacific and American Remembering"; or "Josh, We're Going to Buy This Son of a Bitch," *Journal of American Studies*, vol. 27, part 2 (August 1993): 207–222.

31. White and Lindstrom (1990); Lindstrom and White (1990); Weeks (1987).

32. Worsley (1968).

TEN. STORMY WATERS

1. Nagai and Iriye (1977).

2. During World War II, the Japanese, as part of their attack on Midway in June 1942, carried out a diversionary maneuver against the Aleutians. In fact, this was one of the reasons for the disaster which befell their aircraft carriers, for they deprived themselves of the services of their Fifth Fleet, which had been dispatched to take the islands of Attu and Kiska, which had no strategic importance. Between May and July 1943, the Americans retook the territories they had lost (the only ones, apart from Wake Island, to have been occupied by the enemy).

3. Dingman (1981). It was only after 1947 that the Americans set their sights on Yokosuka, for as the Cold War developed Japan became the most important strategic location. However, it was not until 1950 that the idea of making Japan a major military base was accepted. See Schaller (1985).

4. The lack of any explicit undertaking to defend Japan was a consequence of Prime Minister Yoshida's refusal to rearm his country and join in a regional defense pact. This was the consequence of an American diplomatic failure, see Yoshitsu (1983).

5. Cited in Joyaux (1985), p. 360. On the treaty negotiations, see Welfield (1988), chap. 2; Schonberger (1989), chap. 8, pp. 236–278; Dower (1979).

6. Dean Acheson, Truman's secretary of state, wrote (1969, p. 696) on the subject of the Treaty of San Francisco, signed by 49 nations (but not by the USSR, nor by China, India, or Burma), that such a good peace treaty had never been so disliked by so many of its signatories.

7. Responding to the fury provoked by his policy, Yoshida claimed that he was granting only limited recognition, which would not prevent relations with the Peking regime from being established later, when circumstances might change.

8. Cited in Welfield (1988), pp. 141–142. On the revision, see Tadashi Aruga, "The Security Treaty Revision of 1960," in Iriye and Cohen (1989), pp. 61–79; Welfield (1988), chap. 6.

9. Packard (1966).

10. Welfield (1988), chap. 9; Watanabe (1970).

11. Cited by Welfield (1988), p. 248.

12. Chang (1990) points out that a change in thinking took place between January 5, 1950, when Truman declared he would not give any military aid to the troops which had withdrawn to Taiwan, and the month of May, when Washington was examining ways to prevent the communists from taking the island.

13. Quemoy (Jinmen), nine miles long, lies at the entrance to the port of Xiamen (Amoy) in Fujian province. Matsu (Mazu), a smaller island, lies off Fuzhou, in the same province, nine to twelve miles from the coast. For an interpretation by an orthodox Chinese communist historian, see Jiang (1988), pp. 146–152.

14. Brands (1992); Karnow (1989).

15. With the Laurel-Langley agreement (1955), the Filipinos were granted parity with the United States—a purely symbolic concession, but gratifying to their nationalist feelings.

16. Shalom (1990); Pringle (1980), chap. 3; Brands (1992), pp. 221–223, 231–233, 270, 275, 286. The 1947 agreement was replaced in 1966 by an executive agreement valid for 25 years, expiring on September 16, 1991.

17. This conclusion is established in a document of the National Security Council, NSC, $^{84}/2$, dated November 9, 1950.

18. The origins of the trusteeship system are well analyzed by Chand (1991) and Sbrega (1986). Among the studies on the application of the system: Nufer (1978); Goodman and Moos (1981), chap. 3; D. F. McHenry (1975). The trusteeship agreement is reproduced, with commentary by Ambassador Warren R. Austin in Richard (1957), vol. 3, pp. 1121–1129.

19. On the place of Micronesia in American strategy, see John W. Dower, "Occupied Japan and the American Lake, 1945–1950," in Friedman and Selden (1971), pp. 146–206.

20. In 1957, Australia ended the harmonization of its armaments with the United Kingdom, henceforth conforming to the United States. On origins of this break, see Day (1989), Esthus (1964).

21. Three ground installations played a capital role. Two ground stations were linked to satellites—Pine Gap, near Alice Springs (1966), monitored the way arms controls were being put into practice, while Nurrungar (1969), in South

Australia, was to give the alarm in case of a rocket attack. A naval telecommunications station at North West Cape, in Western Australia (1963), communicated with nuclear submarines. On relations between the United States, Australia, and New Zealand, most of the studies have been done by scholars from the antipodes: Camilleri (1980, 1987); Churchward (1979); Ravenhill (1989); Barclay (1985); Barclay and Siracusa (1975); Harper (1968, 1971, 1987); Reese (1969); Livingston and Louis (1979); Albinski (1982, 1987).

22. In 1949, Dean Acheson declared:"Whether Ho Chi Minh is as much nationalist as Commie is irrelevant. . . . All Stalinists in colonial areas are nationalists" (cited in Olson and Roberts [1991], p. 32).

23. McMahon (1981); Homan (1984); Mrazek (1978).

24. McMahon (1981), pp. 313–315; Gould (1969); Louis (1977).

25. This was the opinion of the National Security Council as of February 1950.

26. Artaud and Kaplan (1989); Marks (1990).

27. Joyaux (1979).

28. On the Americans and the first Indochina war, see Olson and Roberts (1991), chap. 2; Karnow (1983), chaps. 3 and 4; Herring (1986), chap. 1; M. Young (1991).

29. Whitfield (1991).

30. On the origins of the Cold War in Asia: Nagai and Iriye (1977), Iriye (1974), Schaller (1985), Gallicchio (1988), Rose (1976). For recent bibliographical surveys: Munro-Leighton (1992), McMahon (1988). On the historiography of the Cold War: Melanson (1983).

31. Newman (1991); John N. Thomas (1974); Service (1971).

32. Tsou (1963); Rose (1976); Schaller (1979); Buhite (1973); Chern (1980); McLean (1986); Shaw (1992); Stueck (1984). American military and economic aid to Nationalist China came to over $3 billion in 1945–1949.

33. The "China White Paper" of August 1949 (*United States Relations with China, With Special Reference to the Period, 1944–1949*, vol 41), explains the American failure in China.

34. Tucker (1983); Stueck (1981); Borg and Heinrichs (1980). Chang (1990) sees greater ambiguity in Acheson's position.

35. Koen (1960). A dozen influential senators acted as a mouthpiece for the lobby: eight Republicans (including Knowland, Hickenlooper, McCarthy, Brewster, Bridges) and two Democrats (Eastland and McCarran). In the House of Representatives, the most vocal was a former missionary to China, Walter Judd, from Minnesota.

36. Shuguang Zhang (1992); Sheng (1993). In an interview with the American journalist Anna Louise Strong in August 1946, Mao predicted that future conflicts would take place between the United States and countries of the "intermediate zone" (i.e., the rest of the world, except for the USSR and its satellites).

37. On relations between China and the United States, see W. I. Cohen (1990); Schaller (1990).

38. Mayers (1986); Gaddis (1987), pp. 147–194.

39. Chang (1990); Harding and Ming (1989); Maga (1990).

40. On origins of the Korean War, see Cumings (1981, 1983, 1990); Blum (1982); Stueck (1973, 1981); Lowe (1986); McGlothen (1989). On those of the Vietnam War: Rotter (1987), Gardner (1988).

41. In 1949, Americans had evacuated South Korea, just as the Russians had left North Korea. Some revisionist historians go as far as to accuse Dean Acheson of having provoked the attack by North Korea (so as to crush it more easily?) by not including South Korea within the United States' defensive perimeter in the Pacific. In fact, in a speech to the National Press Club (January 12, 1950), the secretary of state contrasted the countries for which the United States had direct responsibility—in the North Pacific, Japan and, "to a lesser extent," Korea—with countries in the South Pacific, where the peoples themselves were expected to put up some resistance before receiving support from the UN. See *Department of State Bulletin*, vol. 22, no. 551, January 23, 1950, pp. 111–118.

42. *Statistical Abstract of the United States, 1992*, pp. 344–345. These losses were lower than were suffered during World Wars I (53,000 dead) and II (292,000 dead). They were much lower than for the indigenous population. In Vietnam, for example, 3 million people (including one million soldiers) are said to have been killed, with 4.4 million wounded, and 300,000 considered missing in action (*Newsweek*, November 22, 1993, p. 30).

43. *Statistical Abstract of the United States, 1992*, p. 341. In constant dollars, the Vietnam War cost more than double the Korean War, but between the two the American national revenue had also almost doubled.

44. On MacArthur: Schaller (1989).

45. The Korean War was not such an "unknown" (Cumings and Halliday, 1988) or "forgotten" (Blair, 1987) war as certain historians would have us think. To get an idea of the ample bibliography dealing with it, see McFarland (1986). Also worth consulting are the studies by Appleman (1960, 1987, 1989, 1990), Foot (1985, 1990), Goulden (1982), Hastings (1987), Kaufman (1986), Lee (1991), Matray (1985), Paige (1968), Rees (1964), Spanier (1959), Stueck (1986), Toland (1991), and Whelan (1990).

46. Summers (1981).

47. The Vietnam War has been the subject of an abundant historiography, an idea of which can be obtained from Burns and Leitenberg (1984), Divine (1988), Peake (1986), Sugnet, Hickey, and Crispino (1983). In French, there is an excellent study by Portes (1993). Among the immense amount of material in English, see Berman (1982, 1989), Braestrup (1984), Dobbs (1981), Fitzgerald (1972), Errington and McKercher (1991), Gettleman, Franklin, Young, and Franklin (1985), Gibbons (1986, 1989), Haley (1982), Herring (1986), Hess (1990, 1994), Karnow

(1984), Kattenburg (1980), Kolko (1985) Lewy (1978), McMahon (1990), Olson (1988), Olson and Roberts (1991), Palmer (1984), Powers (1984), Rust (1985), Smith (1983, 1985, 1991), Spector (1983), Sullivan (1985), Summers (1981, 1985), Thies (1980), Turley (1986), Young (1991).

48. Coutau-Bégarie (1983; 1987, chap. 2); Segal (1983; 1990); Hagan (1984), chap. 17; Da Cunha (1990); Drysdale and O'Hare (1991); Langdon and Ross (1990); Edward A. Kolodziej, "The Multilateralization of Regional Security in Southeast and Northeast Asia: The Role of Soviet Union," *Pacific Focus* 6, no. 1 (Spring 1991): 5–37.

49. The full text of the Vladivostok speech is given in Mikhail S. Gorbachev, *Towards a Better World* (New York: Richardson & Stierman, 1987), pp. 329–359, 355–356.

50. On the new United States policy toward China, the memoirs of the two principal protagonists, Nixon (1978) and Kissinger (1979), are essential. To situate it in its long-term context: Warren I. Cohen (1990), Kitts (1991), Schaller (1985), Iriye (1992).

51. The text (in French translation) is given in Joyaux (1985), pp. 425–428. "Acknowledge" is translated by "réaliser."

52. Iriye (1992).

53. Harding (1992); Choi (1987); Hsiao and Witunski (1983); Kusnitz (1984); Macchiarola and Oxnam (1991); Mosher (1990, 1992); Tan (1992).

54. Lasater (1993); K. H. M. Carter (1989); Myers (1989); Gary Klintworth, "Taiwan's New Role in the Asia-Pacific Regions," *Pacific Focus* 7, no. 2 (Fall 1992): 67–90.

55. Navy Secretary John Lehman declared in 1987: "China is a friend, not an ally. Our developing defense relationship with China is and will be based on common security interests, not on a formal military alliance" (Lehman, 1987, p. 23).

56. Hagan (1984), p. 336. In 1974, the distribution of the fleets (in out-of-area ship-days) in the different oceans was:

	Soviet Fleet	American Fleet	Ratio USA/USSR
Pacific	7,400	34,800	470%
Atlantic	13,800	8,500	62%
Mediterranean	20,200	15,400	76%
Indian	10,500	2,600	25%
Caribbean	1,200	—	—
Total	53,100	61,300	115%

57. The text of the treaty is given in Coutau-Bégarie (1987), pp. 333–343.

58. Guam and Eastern Samoa kept their "unincorporated" status, sending a nonvoting delegate to Congress. Their inhabitants became citizens (Guam) or "nationals" (Samoa); cf. Fallon (1991).

59. The Federated States of Micronesia includes four "states": Kosrae, Pohnpei, Chuuk (formerly known as Truk), and Yap. See Nufer (1978); Goodman and Moos (1981), chap. III.

60. Chand (1992).

61. Welfield (1988), pp. 359, 363. In 1977–1978, the United States had 48,600 troops stationed in Japan, 39,900 in South Korea, 42,700 in Hawaii, 23,900 in other Pacific bases, 25,600 on board the Seventh Fleet, and a small number in Thailand. The American Air Force in the Far East had 550 planes, 250 of them attached to the Navy.

62. American military aid, properly speaking, amounted to $108.6 billion between 1962 and 1990. The Pacific region received 31.6%, of this, of which 17% went to Indochina, 6.5% to South Korea, 2.2% to Pacific Latin America, 1.7% to Thailand and Taiwan, 1.4% to the Philippines, 0.6% to Indonesia and 0.2% to Japan. *Statistical Abstract of the United States, 1985*, p. 81; *1988*, p. 76; *1992*, p. 794.

63. On all these problems, political scientists and specialists on strategic problems and on international relations have written numerous books and articles. Many topics of interest will be found in the proceedings of the *Pacific Symposium* organized since 1979 by the National Defense University and edited by Alves. See also: Barnett (1977), Bergere et al. (1989), R. Buckley (1992), Buss (1985), Carter (1989), Steve Chan (1993), Choi (1987), Da Cunha (1990), Denoon (1993), Dreyer (1990), Harding (1992), Hayes (1987, 1991), Lee and Sato (1982), Lee, McLaurin, and Moon (1988), McIntosh (1987), McLaurin and Moon (1989), Morley (1986), Nester (1993), Ravenhill (1989), Scalapino (1972, 1987), Scalapino and Lee (1988), Scalapino and Qimao (1986), Schrader and Winnefeld (1992), Seabury (1981), Segal (1983, 1990), Sigur and Kim (1982, 1992), Thambipillai and Matuszewski (1989), Tow (1991), Tow and Feeney (1982), Walker and Sutherland (1988), West and Alting von Geusau (1987), Wilborn (1993), Winnefeld (1992).

64. Mack (1992).

65. On September 30, 1992, the Navy handed Subic Bay over to the Filipino authorities, but as Admiral David E. Jeremiah had recalled on September 27, 1991, "The bottom line is this: the United States is itself a Pacific nation, and we intend to remain so. Our interests in the Pacific are permanent, and are not subject to change because of events in Moscow or decisions in the Philippine Senate" (cited in Stevenson [1992], p. 105).

66. When South Korean President Park Chung Hee was assassinated on October 26, 1979, the United States put their forces on the alert and dispatched warships, in case North Korea should take advantage of the situation to launch an attack.

67. Simon (1993).

68. The negotiations for a new treaty lasted, intermittently, from 1964 to 1977. On September 7, 1977, everything seemed to have been settled with the signing of the Carter-Torrijos agreement. On the one hand, a general treaty did away with the 1903 treaty, providing for a progressive transfer of responsibilities to

Panama leading up to December 31, 1999, and increasing its share of the revenues. On the other hand, a neutrality pact granted the United States, in vague terms, the unilateral right to defend the canal. Accepted by the Panamanians in a referendum (held on October 23, 1977), these treaties met with strong opposition in the United States Senate. An appended declaration by the two presidents (October 14), specified that the Americans would not intervene in the country's internal affairs, but that their warships would have preferential right to pass through the canal. This last concession failed to satisfy Senator DeConcini, who felt that the United States should have the right to reopen the canal by force if necessary, should it ever be shut down because of internal unrest in Panama. In the end, the Senate incorporated this rider when it ratified the treaty (April 18, 1978), while expressly renouncing any right to intervene in the internal affairs of Panama. LaFeber (1978); Hogan (1986); Liss (1967).

69. See Stephen W. Bosworth, "The U.S. and Asia in 1992: A New Balance," *Asian Survey* 33, no. 1 (January 1993): 103–113; Yuji Miyamoto, "Towards a New Northeast Asia," *Pacific Review* 6, no. 1 (1993): 1–7; Lewis Fretz, "The United States and Asia-Pacific Security," *New Zealand International Review* 18, no. 2 (March–April 1993): 2–6; Boene and Martin (1987).

ELEVEN. THE GREAT ECONOMIC TRANSFORMATION

1. In the period 1946–1949, the export price index stood at 24.6 (base = 100 in 1982), and the import price index at 20.5. In 1990–1991, they stood at 115.6 and 111.3, respectively, i.e., they had grown by factors of 4.7 and 5.4.

2. Between 1821 and 1867, the volume of imports increased by 4.1% per annum, and of exports by 4.4%. In the period 1868–1941 the figures were 5.4% and 6.3%.

3. The other years were 1959 and 1966.

4. These conclusions are valid for trade in current dollars. By volume, on the contrary (1982 = 100), the balance of trade with the Pacific was almost always in a deficit, except in 1945–1949, 1954, 1957, 1961, the three last years being in periods of recession (cf. Appendices, Graph 4). The difference is accounted for by the evolution of the terms of trade in a direction unfavorable to the United States, taking 1982 as the base, shortly after the second oil crisis of 1979–1980.

5. With the exception of two years, 1959 and 1972. In addition, between 1957 and 1982 (with the exception of 1959), the United States also built up surpluses with the Pacific countries of Latin America.

6. Whether the American or Japanese figures are used, Graph 6 in the Appendices shows that the trade ratio had the same profile between 1962 and 1991. If the Japanese data show a higher level, this merely results from the accounting method used, imports being calculated as CIF and exports as FOB.

7. One major difference from the prewar period was that silk now played only a marginal role (comprising 1.9% of Japanese exports to the United States in 1961–1963).

8. Schlossstein (1991).

9. In 1990, the United States accounted for 34.7% of Japanese exports of machinery and capital goods (34% of electric machinery; 50% of office equipment; 38% of radios, but 13% of television sets; 28% of cathode ray tubes; 45% of automobiles and trucks, but 16% of motorbikes; 36% of scientific and optical equipment; 32% of tape recorders and players). In the opposite direction, Japan bought from the United States 33% of imported foodstuffs, 52% of cotton, 39% of wood, 32% of chemicals, and 44% of machinery and capital goods.

10. In 1986, of exports from Taiwan to the United States, clothing represented only 24% and mass market electronics 4%. In South Korea, clothing retained a high percentage (32%), well ahead of electronics (8%) and toys (5%). The "machinery" sector had become diversified, with transistors (5%) being joined by office equipment (4%) and other electrical appliances (7%). But between 1970 and 1986, South Korean exports rose from $370 million to $13,500 million, and Taiwanese exports from $550 million to $21,250 million.

11. The figures published in the *Survey of Current Business* are based on the historical cost, which in 1992 was 27% below the replacement value and 37% lower than the market value. Since American investments abroad are on average of longer standing than foreign investments in the United States, the net position at this date may seem to have fallen more than would be the case if it were based on the replacement value.

12. Portfolio investments will not be dealt with. On this topic, see Stallings (1987). Financial markets in the Pacific countries remained closed for many years. In 1984, a yen-dollar accord liberalized the control of capital in Japan.

13. The European share rose significantly. It was 20% in 1960, 45% in 1980, and 50% in 1990.

14. The December 1952 *Survey of Current Business* already reads: "Apparently the relatively low cost of unskilled labor in undeveloped countries is a less important factor in attracting American investment than a large market for finished products and a supply of skilled or semiskilled labor." In 1980, as in 1970, the Pacific accounted for only 11% of American industrial investment, 17% in 1992 (Europe = 50%). In 1992, relative to its average of 19.5%, the Pacific was underrepresented, apart from industry, in nonbanking financial industries, insurance and real estate (14%), services (12%). It is overrepresented in pumping and refining oil (30%), wholesale trade (25%), banking (22%), and others (30%). In Japan, in 1992, investments were shared in the following proportions: petroleum (18%), industry (45%), wholesale trade (21%), banking (1%), finance, insurance, and real estate (10%), services (2%), others (3%).

15. *Survey of Current Business*, August 1982, July 1993.

16. Borden (1984); Schaller (1985); Theodore Cohen (1987); Schonberger (1989); Kawai (1960).

17. Jéquier (1970). From 1958 on, exports to the United States surpassed those to Southeast Asia. On United States–Japanese trade between 1955 and 1982, see the article by Kanemitsu, in Iriye and Cohen (1989).

18. Destler (1976, 1979). For an analysis of Japanese growth between 1953 and 1971, see Denison and Chung (1976), pp. 42–43: the annualized rate of growth came to 8.81% per annum (4% in the United States between 1948 and 1969), due to an increase of 1.8% in the labor component (1.30%), 2.10% in capital (0.79%), and 4.86% in total factor productivity (1.91%), including economies of scale: 1.94% (0.42%).

19. Margolin (1989); Vogel (1991); Vandermeersch (1986).

20. Rostow (1986).

21. Kaoru Sugihara, "Patterns of Asia's Integration into the World Economy, 1880–1913," in W. Fischer et al. (1986), pp. 709–728.

22. Downen and Dickson (1984); Dirlik (1993); Hooper (1982); Kolde (1976); Vicuña (1979); Boyd (1989).

23. In 1967, ASEAN had five members: Indonesia, Malaysia, the Philippines, Singapore, and Thailand. The Sultanate of Brunei became a member in 1984. In 1990, the Swede Bert Edström rejected the idea that the extremely disparate countries of the Pacific could be seen as constituting any community, or that they could do so any time in the near future. The idea of a Pacific Basin community is no more than a Japanese dream, he suggests.

24. In 1990, Malaysia promoted the idea of a trading bloc excluding the four white countries—a proposal reduced in 1993 to a simple consultative forum, the East Asian Economic Caucus (EAEC).

25. The Andean Pact included Bolivia, Chile (which dropped out in 1976), Colombia, Ecuador, and Peru, with the addition of Venezuela in 1973.

26. Higgott et al. (1990).

27. *Time*, November 22 and 29, 1993; *Newsweek*, November 22 and 29, 1993. In Seattle, APEC decided to admit Mexico and Papua New Guinea in 1993 and Chile in 1994. The membership then came to 18, including a South American country for the first time. It rose to 21 in 1998, with the addition of Peru, Russia, and Vietnam.

28. Lucchini and Voelckel (1978); McEvoy (1986).

29. Doulman (1987).

30. There is an immense bibliography dealing with trade disagreements between the United States and Japan. Among recent studies: Borden (1984), Christopher (1986), Clapp and Halperin (1974), Stephen Cohen (1985, 1991), Cronin (1992), Cusumano (1985, 1991), Dolan and Worden (1992), Emmott (1989, 199?), Encarnation (1993), Fallows (1989,1994), Gourevitch (1989), Halberstam (1986), Hamada (1991), K. Hayashi (1989), Holstein (1990), Hulten

(1990), Ishihara (1991), Johnston, d'Andrea Tyson, and Zysman (1989), Komiya, Okuno and Suzumura (1988), Krugman (1991), Lincoln (1990, 1993), Mason (1992), Nester (1990, 1991, 1993), Nye, Pharr and Vogel (1993), Ohmae (1985), Passin (1975), *Political Economy of Japan* (1987, 1988, 1992), Prestowitz (1988), Saucier (1987), Sautter (1987), Schlosssstein (1984), Sigur and Kim (1982), Vernon (1983), Vogel (1979), Wolf (1983), van Wolferen (1989). See also the article by Van Ark and Pilat (1993).

31. Chalmers Johnson (1982). Nevertheless, the following appears in a recent publication: "Thus, despite the high living standards of many workers in larger firms, Japan in 1990 remained in general a low-wage country whose economic growth was fueled by highly skilled and educated workers who accepted poor salaries, often unsafe working conditions, and poor living standards" (Dolan and Worden, 1992, p. 203).

32. Between 1981 and 1985, the fall of the yen against the dollar (-7,5%) was much less than the decline of the French franc, (-39,4%), the German mark (-23%) or the pound sterling (-36%).

33. Lincoln (1990) shows the unique nature of Japan in the trade of manufactured goods: the intersectoral rate of exchange was much lower than anywhere else.

34. According to Japanese figures, Japan's balance of trade with the entire world (calculated in yen) was always positive after 1968, except during the two oil crises (1973–1975 and 1979–1980). Beginning in 1984, the ratio of exports to imports was exceptionally high (as much as 164% in 1986). See the article by Krasner and Okimoro, in Iriye and Cohen (1989).

35. Heffer and Serman (1992), p. 57.

36. See Lasater (1993), Mosher (1992), Myers (1989), Wheeler and Wood (1987) on the subject of relations with Taiwan.

37. Denoon (1993).

38. Between 1980 and 1993, 29 trade agreements were signed with the objective of reducing the American trade deficit, without producing any tangible results, even though Japanese imports from the United States grew considerably.

39. *Le Monde*, February 11 1994; *Time*, July 12, 1993.

40. The piracy of intellectual property is also a specialty of the Chinese of Hong Kong and Taiwan.

41. Harding (1992), Hsiao and Witunski (1983); Tan (1992); Tow (1991); Macchiarola and Oxnam (1991).

42. Wishing to increase exports, the Clinton administration placed the emphasis for the future, not on Western Europe and Japan, but on ten "Big Emerging Markets" (BEM), three of which are in the Pacific region (China, Indonesia, South Korea) and Latin America (Argentina, Brazil, Mexico), together with India, South Africa, Turkey, and Poland (*Notes on Economic Affairs*, January 26, 1994, pp. 29–40).

43. U.S. Senate (1981), Labrousse (1991), *Observatoire géopolitique des drogues* (1993).

44. For descriptions of the Pacific economy between 1985 and 1990, see Bescher (1991), Borthwick (1992), Boyd (1989), Carter (1989), Chan (1993), Clark and Chan (1992), Drysdale (1988), Institut du Pacifique (1986), Linder (1986), McCord (1991), Morley (1986, 1993), Nemerz (1987), Palar (1993), Shibusawa, Ahmad, and Bridges (1992), Sutter (1992), Van Horne (1989), CEPII (1991).

TWELVE. TOWARD CULTURAL SYMBIOSIS?

1. Talcott Parsons, "Evolutionary Universals in Society," *American Sociological Review* 29 (1964): 339–357.

2. Gillette (1985); Britsch (1989). According to Acevedo (1989), in 1986, 1.6% of Chileans were members of the Mormon Church.

3. In 1991, more than half the foreign students studying in American universities came from Pacific countries. Out of the total of 408,000, 37,000 came from Japan, 34,000 from Taiwan and 23,000 from South Korea. In 1988, 20,000 students from Communist China were studying on American campuses (*Statistical Abstract of the United States, 1992*, p. 171).

4. Whiting (1977).

5. Kawai (1960), p. 27.

6. Inoue (1991) shows that if under the 1946 Constitution the emperor's powers were scarcely changed from what they had been in the Meiji era, his *symbolic* status, on the other hand, was considerably diminished—as the Americans desired.

7. In fact, the old *zaibatsu* (Mitsui, Mitsubishi, Sumitomo) preferred a parliamentary regime until the end of the 1930s, unlike the new *zaibatsu* (Kuhara, Aikawa), which were closely linked with the military.

8. Theodore Cohen (1984), p. 457.

9. Kawai (1960), p. 248. In addition to the studies by Kawai and Inoue, there is an enormous bibliography dealing with the occupation of Japan. See, for example, Monnier (1967), Cohen (1987), Finn (1992), Schonberger (1989), Schaller (1985), Woodward (1972), Ward and Yoshikazu (1987), Williams J. (1982, 1988). The MacArthur Memorial (Old Dominion University, Norfolk, Va.) has published the proceedings of seven conferences held between 1975 and 1986. (On this topic, see Burkman, 1982.)

10. Sheila Johnson (1988); Isaacs (1958). For Latin America, see Pike (1992).

11. Warren I. Cohen, in Hunt et al. (1988), pp. 31–36; Hunt (1983); Harding (1992).

12. *The Gallup Poll: Public Opinion, 1935–1971* (New York: Random House, 1972), vol. 3: 2104–2105.

13. *The Gallup Poll: Public Opinion, 1972–1977*, vol. 1, pp. 39, 129.

14. Floris W. Wood, ed., *An American Profile: Opinion and Behavior, 1972–1989* (Detroit: Gail Research, 1990), pp. 400, 422.

15. Benedict (1946).

16. *Gallup Poll Monthly*, December 1991. The composite portrait of the American who sees Japan as a future aggressor is a gem: nonwhite, aged over 50, rural dweller, in the East, with no high school diploma, Democratic voter, and poor, with memories of World War II. Of course, these variables emerge because of their more frequent occurrence than in the national average.

17. Gourevitch (1989).

18. Rozman (1990); see also the article by Yoo Tae Gun, "Une analyse du néoconfucianisme," *Revue de Corée* 23, no. 2 (December 1991): 5–23.

19. Rozman (1960), p. 161.

20. Inoue (1992): "Had the Japanese really understood the democratic ideas that the Americans had intended, it would have been far more difficult and painful for them to accept them. Likewise, had MacArthur and his staff understood precisely how the Japanese were interpreting American democratic principles, they might have been more reluctant to approve the final version of the Constitution. The cultural and linguistic barriers in communication between the Americans and the Japanese were without a doubt a hindrance to mutual understanding. But, ironically, the same difficulties made it possible for the two sides to agree on a document without agreeing on its fundamental meaning" (p. 270).

21. Michael Kammen, "The Problem of American Exceptionalism: A Reconsideration," *American Quarterly* 45, no. 1 (March 1993): 1–43.

22. Crichton (1993), p. 74. A bibliography is included (pp. 397–399), which is unusual for a novel.

23. Alvin Z. Rubinstein and Donald E. Smith, in Thornton (1988), p. 45.

24. Kitamura (1971).

25. Nathaniel B. Thayer, in Thornton (1988): "A few Japanese say they dislike the United States. Events do not seem to change this judgment much. More Japanese say they like the United States. Some events affect this judgment. The two events that have caused the Japanese to lessen their identification with the United States have been the Vietnam War and Japan bashing, and what both have in common is the belief among Japanese that the behavior exhibited in them is atypical of the United States" (pp. 102–103).

26. Shambaugh (1991). For a longer perspective: Arkush and Lee (1989).

27. In Hunt et al. (1988), David Shambaugh describes three different images of America in Communist China: traditionalist (a mixture of goodwill and a desire for hegemony), Leninist (American imperialism), and realist (based on a study of the structures and functions of pluralism). See also Shambaugh, in Thornton (1988), pp. 142–156.

28. Nevertheless, the law of 1952 allowed Asians to become United States citizens through naturalization—something which had been impossible since 1924. This privilege was restored to the Chinese in 1943.

29. *Statistical Abstract of the United States, 1968*, p. 92: 21,665 immigrants from Latin America were recorded in the period 1941–1950 and 44,751 in 1951–1960.

30. There were 16,709 immigrants from China in 1941–1950 and 9,657 in 1951–1960. For Japan, the figures were 1,555 and 46,250, respectively. For Australia and New Zealand they were 13,805 and 11,506, respectively.

31. Daniel (1993); Takaki (1989).

32. Mexico, with 23% in 1981–1990, and the Caribbean, with 12%, were the other major sources of immigrants.

33. The largest contingent (475,000) was from Vietnam, ahead of Laos (165,000), Cambodia (122,000), Thailand (32,000), China (27,000), and Indonesia (9,000). Few refugees properly speaking came from Central America (5,629 from Nicaragua, 1,429 from El Salvador), and even fewer from Chile (955). Indochina (761,000) accounted for 43 percent of the total arriving in the United States during these three decades, mostly during the 1980s, after spending long periods in camps (*Statistical Abstract of the United States, 1992*, p. 12).

34. If the Asian Indians are excluded, the percentage falls to 2.6%.

35. Compared with the national average of 42%, California took in 52% of the Filipinos, 51% of those from Samoa and Guam, 46% of the Vietnamese, 43% of the Chinese, 37% of the Japanese, 33% of the Koreans, and 16% of the Hawaiians.

36. Kim (1986).

37. For a short analysis of these problems, see the articles by Charles S. Prebish and C. Carlyle Hazland, in Lippy and Williams (1988), 2: 669–682, 699–709. According to Prebish (p. 677), there were 500,000 Buddhists in America, divided into numerous groups.

38. *Washington Post*, October 15 1989. See also the article by Thomas Robbins and Dick Anthony, in Lippy and Williams (1988), 2: 741–754.

39. Vogel (1979); Fallows (1989, 1994).

40. Winchester (1992).

Epilogue

1. Friedman and Lebard (1991), p. 403.

2. Asian-Americans made up 3% of the population in 1994. They should amount to 5% in 2010 and 10% in 2050. For the same years, the percentages are: Whites, 76%, 68%, and 53%; Afro-Americans, 12%, 13%, and 14%; Hispanics, 9%, 14%, and 23%. Asian-Americans as a group will continue to show the strongest growth.

BIBLIOGRAPHY

An enormous bibliography has been devoted to a subject such as ours: tens of thousands of books and articles have been written about the different problems we have considered. Below we give the references for the books and other works quoted in the notes (except when a full reference is given there). A number of articles which were consulted have been omitted, so as to shorten the list. At the end, a list of the principal specialized journals is included.

[Translator's note: Where the author consulted translations into French of works originally published in English, we have given the original English-language source, listing the French translation used in square brackets.]

Acevedo, Rodolfo Antonio. 1989. *Los Mormones en Chile: 30 Años de la Iglesia de Jesuchristo de los Santos de los Ultimos Dias (1956–1986)*. Santiago: Pontificía Universitad Católica de Chile.

Acheson, Dean. 1969. *Present at the Creation: My Years in the State Department*. New York: Norton.

Adler, Jacob. 1966. *Claus Spreckels, the Sugar King in Hawaii*. Honolulu: University of Hawaii Press.

Adler, Jacob, and Kamings, Robert M. 1986. *The Fantastic Life of Walter Murray Gibson: Hawaii's Minister of Everything*. Honolulu: University of Hawaii Press.

Agoncillo, T. A., and Guerrero, Milagros C. 1974. *History of the Filipino People*. 5th ed. Quezon City: R. P. Garcia.

Agoncillo, T. A. 1974. *Filipino Nationalism, 1872–1970*. Quezon City: R. P.: Garcia.

Ahvenainen, Jorma. 1981. *The Far Eastern Telegraphs: The History of Telegraphic Communications between the Far East, Europe and America before the First World War*. Helsinki: Suomalainen Tiedeakatemia.

Albinski, Henry S. 1982. *The Australian American Security Relationship: A Regional and International Perspective*. St. Lucia: University of Queensland Press.

———. 1987. *ANZUS, the United States and Pacific Security*. Lanham, Md.: University Press of America.

———. 1989. "South Pacific Trends and U.S. Security Implications, an Introductory Overview." In Albinski, Henry S., et al., *The South Pacific: Political, Economic, and Military Trends*. Washington, D.C.: Brassey's.

Albion, Robert G. 1954. "Distant Stations." *United States Naval Institute Proceedings* 80, no. 3 (March): 265–273.

Allen, George C., and Donnithorne, Audrey G., 1957. *Western Enterprise in Indonesia and Malaya: A Study in Economic Development*. London: George Allen & Unwin.

Allen, Oliver E. 1980. *The Pacific Navigators*. Alexandria, Va.: Time-Life Books.

Alves, Dora. 1985. *Anti-nuclear Attitudes in New Zealand and Australia*. Washington, D.C.: National Defense University Press (NDU Press).

Alves, Dora, ed. 1988. *Pacific Symposium (NDU), 6th, 1985: Pacific Regional Security*. Washington, D.C.: NDU Press.

———. 1988b. *Pacific Symposium (NDU), 8th, 1987: Pacific Security toward the Year 2000*. Washington, D.C.: NDU Press.

———. 1990a. *Pacific Symposium (NDU), 9th, 1988: Cooperative Security in the Pacific Basin*. Washington, D.C.: NDU Press.

———. 1990b. *Pacific Symposium (NDU), 10th, 1989: Evolving Pacific Basin Strategies*. Washington, D.C.: NDU Press.

———. 1990c. *Pacific Symposium (NDU), 11th, 1990: Change, Interdependence and Security in the Pacific Basin*. Washington, D.C.: NDU Press.

Anderson, David L. 1985. *American Diplomats in China, 1861–1898*. Bloomington, Ind.: Indiana University Press.

———, ed. 1993. *Shadow on the White House: Presidents and Vietnam War, 1945–1975*. Lawrence, Kans.: University Press of Kansas.

Anderson, Irvine H., Jr. 1975a. *The Standard-Vacuum Oil Company and United States East Asian Policy, 1931–1941*. Princeton, N.J.: Princeton University Press.

———. 1975 b. "The 1941 *de Facto* Embargo on Oil to Japan: A Bureaucratic Reflex." *Pacific Historical Review* 44, no. 2 (May): 201–231.

Andrew, John A., III. 1976. *Rebuilding the Christian Commonwealth: New England Congregationalists and Foreign Missions, 1800–1830*. Lexington: University Press of Kentucky.

Anschel, Eugene. 1984. *Homer Lea, Sun Yat-sen and the Chinese Revolution*. New York: Praeger.

Appleman, Roy E. 1960. *South to the Naktong, North to the Yalu (June–November 1950)*. Washington, D.C.: Office of the Chief of Military History.

———. 1987. *East of Chosin: Entrapment and Breakout in Korea, 1950*. College Station: Texas A&M University Press.

―――. 1989. *Disaster in Korea: The Chinese Confront MacArthur.* College Station: Texas A&M University Press.

―――. 1990a. *Escaping the Trap: The U.S. Army X Corps in Northeast Korea, 1950.* College Station: Texas A&M University Press.

―――. 1990b. *Ridgway Duels for Korea.* College Station: Texas A&M University Press.

Arkush, R. David, and Lee, Leo D., eds. 1989. *Land without Ghosts: Chinese Impressions of America from the Mid-Nineteenth Century to the Present.* Berkeley: University of California Press.

Artaud, Denise, and Kaplan, Lawrence, eds. 1990. *Dien Bien Phu and the Crisis of Franco-American Relations, 1954–1955.* Wilmington, Del.: Scholarly Resources.

Ball, Desmond, and Downes, Cathy, eds. 1990. *Security and Defence: Pacific and Global Perspectives.* Sydney: Allen & Unwin.

Bancroft, Hubert Howe. 1900. *The New Pacific.* New York: Bancroft (rev. ed., 1912).

Barclay, Glen St. J. 1985. *Friends in High Places: Australian-American Diplomatic Relations since 1945.* New York: Oxford University Press.

Barclay, Glen St. J., and Siracusa, Joseph, eds. 1975. *Australian-American Relations since 1945: A Documentary History.* New York: Holt, Rinehart and Winston.

Barnett, Doak A. 1960. *Communist China and Asia: Challenge to American Policy.* New York: Harper.

―――. 1971. *A New U.S. Policy toward China.* Washington, D.C.: Brookings Institution.

―――. 1977. *China and the Major Powers in East Asia.* Washington, D.C.: Brookings Institution.

―――. 1981. *The FX Decision: "Another Crucial Moment in US-China-Taiwan Relations."* Washington, D.C.: Brookings Institution.

Barnett, Suzanne Wilson, and Fairbank, John K., eds. 1985. *Christianity in China: Early Protestant Missionary Writings.* Cambridge: Harvard University Press.

Barnhart, Michael A. 1987. *Japan Prepares for Total War: The Search for Economic Security, 1919–1941.* Ithaca, N.Y.: Cornell University Press.

Bartlett, Norman. 1976. *Australia and America through 200 Years.* Sydney: The Fine Arts Press.

Beaglehole, J. C. 1966. *The Exploration of the Pacific.* 3rd ed. Stanford: Stanford University Press.

Beauchamp, Edward R., and Iriye, Akira, eds. 1990. *Foreign Employees in Nineteenth-Century Japan.* Boulder, Colo.: Westview Press.

Beers, Burton F. 1962. *Vain Endeavor: Robert Lansing's Attempts to End the American-Japanese Rivalry.* Durham, N.C.: Duke University Press.

Beisner, Robert L. 1992 (1968). *Twelve against Empire: The Anti-Imperialists, 1898–1900.* Chicago: Imprint Publs.

Bell, Roger. 1983. "Testing the Open Door Thesis in Australia, 1941–1946." *Pacific Historical Review* 51, no. 3 (August): 283–311.

Benedict, Ruth. 1946. *The Chrysanthemum and the Sword: Patterns of Japanese Culture.* Boston: Houghton Mifflin.

Bergère, Marie-Claire; Cabestan, Jean-Pierre; Étévenard, Christian; Godement, François; and Joyaux, François. 1989. *La Chine et le Pacifique.* Paris: Fondation pour les études de défense nationale.

Berman, Larry. 1982. *Planning a Tragedy: The Americanization of the War in Vietnam.* New York: Norton.

———. 1989. *Lyndon Johnson's War: The Road to Stalemate in Vietnam.* New York: Norton.

Bescher, Alexander, ed. 1991. *The Pacific Rim Almanac.* New York: Harper Collins.

Billingsley, Edward Baxter. 1967. *In Defense of Neutral Rights: The United States Navy and the Wars of Independence in Chile and Peru.* Chapel Hill: University of North Carolina Press.

Blair, Clay. 1987. *The Forgotten War: America in Korea, 1950–1953.* New York: Times Books.

Blum, Robert M. 1982. *Drawing the Line: The Origins of the American Containment Policy in East Asia.* New York: Norton.

Bockstoce, John R. 1986. *Whales, Ice, and Men: The History of Whaling in the Western Arctic.* Seattle: University of Washington Press.

Boene, Bernard, and Martin, Michel Louis. 1987. *L'Amérique entre Atlantique et Pacifique: Essai de prospective stratégique.* Paris: Fondation pour les études de défense nationale.

Bolkhovitinov, Nikolay N. 1990. "The Crimean War and the Emergence of Proposals for the Sale of Russian America, 1853–1861." *Pacific Historical Review* 59, no. 1 (February).

Bolton, Geoffrey, ed. 1990. *The Oxford History of Australia,* 5 vols., 1986–1990. Melbourne: Oxford University Press. Vol. *5: 1942–1988. The Middle Way,* by G. Bolton.

Borden, William S. 1984. *The Pacific Alliance: United States Foreign Economic Policy and Japanese Trade Recovery, 1947–1955.* Madison: University of Wisconsin Press.

Borg, Dorothy. 1947. *American Policy and the Chinese Revolution, 1925–1928.* New York: Macmillan.

———. 1964. *The United States and the Far Eastern Crisis of 1933–1938: From the Manchurian Incident through the Initial Stage of the Sino-Japanese War.* Cambridge: Harvard University Press.

Borg, Dorothy, and Okamoto, Shumpei, eds. 1973. *Pearl Harbor as History: Japanese-American Relations, 1931–1941.* New York: Columbia University Press.

Borg, Dorothy, and Heinrichs, Waldo, eds. 1980. *Uncertain Years: Chinese-American Relations, 1947–1950.* New York: Columbia University Press.

Borthwick, Mark, ed. 1992. *Pacific Century: The Emergence of Modern Pacific Asia.* Boulder, Colo.: Westview Press.

Boyd, Gavin. 1989. *Pacific Trade: Investment and Politics.* New York: St. Martin's Press.

Braestrup, Peter, ed. 1984. *Vietnam as History. Ten Years after the Paris Peace Accords.* Washington, D.C.: University Press of America.

Bradley, Harold Whitman. 1942. *The American Frontier in Hawaii: The Pioneers, 1789–1843.* Stanford, Calif.: Stanford University Press.

Braisted, William R. 1971. *The United States Navy in the Pacific, 1897–1908,* 1958; *The United States Navy in the Pacific, 1909–1922.* Austin: University of Texas Press.

Brands, Jr., Harry W. 1987. "The Dwight D. Eisenhower Administration, Syngman Rhee and the Gneeva Conference of 1954." *Pacific Historical Review* 56, no. 1 (February): 59–85.

———. 1992. *Bound to Empire: The United States and the Philippines.* New York: Oxford University Press.

Breslin, Thomas A. 1980. *China, American Catholicism and the Missionary.* University Park: Pennsylvania State University Press.

Britsch, R. Lanier, 1989. *Moramona: The Mormons in Hawaii.* Laie: Institute for Polynesian Studies.

Brooke, George M. 1962. "The Vest Pocket War of Commodore Jones." *Pacific Historical Review* 31, no. 3 (August): 217–233.

Brookes, Jean Ingram. 1941. *International Rivalry in the Pacific Islands, 1800–1875.* Berkeley: University of California Press.

Brown, David. 1974. *Carrier Operations in World War II,* vol. II: *The Pacific Navies, Dec. 1941–Feb. 1943.* London: Ian Allan.

Buck, Pearl. 1931. *The Good Earth.* New York: The John Day Company.

Buckley, Roger. 1992. *U.S.-Japan Alliance Diplomacy, 1945–1990.* Cambridge: Cambridge University Press.

Buckley, Thomas H. 1970. *The United States and the Washington Conference, 1921–1922.* Knoxville: University of Tennessee Press.

Buhite, Russell D. 1973. *Patrick J. Hurley and American Foreign Policy.* Ithaca, N.Y.: Cornell University Press.

Burkman, Thomas W., ed. 1982. *The Occupation of Japan: Educational and Social Reform.* Norfolk, Va.: MacArthur Memorial Foundation.

———. 1984. *The Occupation of Japan: The International Context.* Norfolk, Va.: MacArthur Memorial Foundation.

Burks, Ardath W., ed. 1985. *The Modernizers: Overseas Students, Foreign Employees, and Meiji Japan.* Boulder, Colo.: Westview Press.

Burns, Richard Dean, ed. 1983. *Guide to American Foreign Relations since 1700.* Santa Barbara: ABC-Clio.

Burns, Richard Dean, and Leitenberg, Milton. 1984. *The Wars in Vietnam, Cambodia and Laos, 1945–1982: A Bibliographic Guide.* Santa Barbara: ABC-Clio.

Busch, Briton Cooper. 1985. *The War against the Seal: A History of the North American Seal Fishery.* Kingston & Montreal: McGill-Queen's University Press.

Buss, Claude A., ed. 1985. *National Security Interests in the Pacific Basin.* Stanford, Calif.: Hoover Institution Press.

Butel, Paul. 1989. *Histoire du thé*. Paris: Desjonquières.

Butow, Robert J. C., 1960. "The Hull-Nomura Conversations: A Fundamental Misconception." *American Historical Review,* July, pp. 822–836.

———. 1961. *Tojo and the Coming of the War.* Stanford, Calif.: Stanford University Press.

Bywater, Hector C. 1921. *Sea Power in the Pacific: A Study of the American-Japanese Naval Problem.* London: Constable. 2nd ed., 1934.

———. 1925. *The Great Pacific War: A History of the American-Japanese Campaign, 1931–1933.* London: Constable.

Cambridge History of Japan. 1989. Vol. V: *The Nineteenth Century,* Marius B. Jansch, ed. Vol. VI: *The Twentieth Century,* Peter Duus, ed. Cambridge: Cambridge University Press.

Camilleri, Joseph A. 1980. *Australian-American Relations: The Web of Dependence.* Melbourne: Macmillan.

———. 1987. *The Australia, New Zealand, U.S. Alliance: Regional Security in the Nuclear Age.* Boulder, Colo.: Westview Press.

Campbell, Charles S. 1951. *Special Business Interests and the Open Door Policy.* New Haven: Yale University Press.

———. 1976. *The Transformation of American Foreign Relations, 1865–1900.* New York: Harper Colophon.

Campbell, I. C. 1989. *A History of the Pacific Islands.* Berkeley: University of California Press.

Carlson, Ellsworth C. 1974. *The Foochow Missionaries, 1847–1880.* Cambridge: Harvard University Press.

Carpenter, Joel A., and Shenk, Wilbert R., eds. 1990. *Earthen Vessels: American Evangelicals and Foreign Missions, 1880–1980.* Grand Rapids, Mich.: Eerdmans.

Carter, Holly Maze K. 1989. *The Asian Dilemma in U.S. Foreign Policy: National Interest versus Strategic Planning.* Armonk, N.Y.: M. E. Sharpe.

Caruthers, Wade J. 1973. *American-Pacific Ocean Trade: Its Impact on Foreign Policy and Continental Expansion, 1784–1860.* New York: Exposition Press.

Centre d'études prospectives et d'informations internationals (CEPII). 1991. *Pacifique, le recentrage asiatique.* Paris: Economica.

Centre Européen de la dotation Carnegie. 1939. *Questions du Pacifique.* Paris.

Chan, Steve. 1993. *East Asia Dynamism: Growth, Order, and Security in the Pacific Region.* Boulder, Colo.: Westview Press, 2nd ed.

Chand, Ganeshwar. 1991. "The United States and the Origins of the Trusteeship System." *Review* (Fernand Braudel Center) 14, no. 2.

———. 1992. "The United States and South Pacific Regionalism: Participation or Subversion?" *Bulletin of Concerned Asian Scholars* 24, no. 3 (July–Sept.): 26–42.

Chang, Gordon H. 1990. *Friends and Enemies: The United States, China, and the Soviet Union, 1948–1972.* Stanford, Calif.: Stanford University Press.

Chaunu, Pierre. 1960. *Les Philippines et le Pacifique des Ibériques (XVIe, XVIIe, XVIIIe siècle): Introduction méthodologique et indices d'activité.* Paris: SEVPEN.

Cheng, Lucie, and Bonacich, Edna, eds. 1984. *Labor Immigration under Capitalism: Asian Workers in the United States before World War II.* Berkeley: University of California Press.

Chern, Kenneth S. 1980. *Dilemma in China: America's Policy Debate, 1945.* Hamden: Archon Books.

Chesneaux, Jean. 1987. *Transpacifiques: observations et considérations diverses sur les terres et les archipels du grand océn.* Paris: La Découverte.

Chisholm, Lawrence W. 1963. *Fenollosa: The Far East and American Culture.* New Haven: Yale University Press.

Choi, Young Jin. 1987. *L'Asie de l'Est et le rapprochement sino-américain.* Paris: Berger-Levrault.

Chong, Key Ray. 1984. *Americans and Chinese Reforms and Revolution, 1898–1922: The Role of Private Citizens in Diplomacy.* Lanham, Md.: University Press of America.

Christgau, John. 1985. "Collins versus the World: The Fight to Restore Citizenship to Japanese American Renunciants of World War II." *Pacific Historical Review,* 54, no. 1 (February): 1–31.

Christopher, Robert C. 1986. *Second to None: American Companies in Japan.* New York: Crown Publishers.

Churchward, Lloyd G. 1979. *Australia and America, 1788–1972: An Alternative History.* Sydney: APCOL.

Clapp, Priscilla, and Halpern, Norton H., eds. 1974. *United States Japanese Relations: The 1970s.* Cambridge: Harvard University Press.

Clark, Cal, and Chan, Steve, eds. 1992. *The Evolving Pacific Basin in the Global Political Economy: Domestic and International Linkages.* Boulder, Colo.: Lynne Rienner.

Clark, Victor. 1949. *History of Manufactures in the United States.* New York: Peter Smith, 3 vols.

Clayton, Lawrence A. 1986. *Grace: W. R. Grace & Co.: The Formative Years, 1850–1930.* Ottawa, Ill.: Jameson.

Cleveland, Richard J. 1842. *A Narrative of Voyages and Commercial Enterprises.* Cambridge: John Owen, 2 vols.

Clymer, Kenton J. 1986. *Protestant Missionaries in the Philippines, 1898–1916: An Inquiry into the American Colonial Mentality.* Urbana: University of Illinois Press.

Coastworth, John H. 1994. *Central America and the United States: The Clients and the Colossus.* New York: Twayne.

Cochran, Sherman. 1980. *Big Business in China: Sino-Foreign Rivalry in the Cigarette Industry, 1898–1930.* Cambridge: Harvard University Press.

Cochran, Thomas C., and Ginger, Ray. 1954. "The American-Hawaiian Steamship Company, 1899–1919." *Business History Review* 28, no. 4 (December): 343–363.

Cohen, Paul A. 1963. *China and Christianity, 1860–1870: The Missionary Movement and the Growth of Chinese Anti-Foreignism.* Cambridge: Harvard University Press.

——. 1984. *Discovering History in China: American Historical Writing on the Recent Chinese Past.* New York: Columbia University Press.

Cohen, Paul A., and Goldman, Merle, comp. 1992. *Fairbank Remembered.* Cambridge: Harvard University Press.

Cohen, Stephen D. 1985. *Uneasy Partnership: Competition and Conflict in U.S. Japanese Trade Relations.* Cambridge: Ballinger, 1985.

——. 1991. *Cowboys and Samurai: Why the United States Is Losing the Battle with the Japanese and Why It Matters.* New York: Harper Business.

Cohen, Theodore. 1987. *Remaking Japan: The American Occupation as New Deal.* New York: The Free Press.

Cohen, Warren I. 1990. *America's Response to China: A History of Sino-American Relations.* New York: Columbia University Press.

——. 1992. *East Asian Art and American Culture: A Study in International Relations.* New York: Columbia University Press.

Cohen, Warren I., ed. 1983. *New Frontiers in American-East Asian Relations: Essays Presented to Dorothy Borg.* New York: Columbia University Press.

——. 1993. *The Cambridge History of American Foreign Relations.* Cambridge: Cambridge University Press. 4 vols.: Vol. 1: Perkins, Bradford. *The Creation of a Republican Empire, 1776–1865.* Vol. 2: LaFeber, William. *The American Search for Opportunity. 1865–1913.* Vol. 3: Iriye, Akira. *The Globalizing of America, 1913–1945.* Vol. 4: Cohen, Warren I. *America in the Age of Soviet Power, 1945–1991.*

Cohen, Warren I., and Iriye, Akira, eds. 1990. *The Great Powers in East Asia, 1953–1960.* New York: Columbia University Press.

Cole, Allan B., ed. 1947. *Yankee Surveyors in the Shogun's Seas: Records of the United States Surveying Expedition to the North Pacific Ocean, 1853–1856.* Princeton, N.J.: Princeton University Press.

Cole, Bernard D. 1983. *Gunboats and Marines: The United States Navy in China, 1925–1928.* Branbury: University of Delaware Press.

Collins, Donald E. 1985. *Native American Aliens: Disloyalty and the Renunciation of Citizenship by Japanese Americans during World War II.* Westport, Conn.: Greenwood Press.

Conroy, Hilary, and Wray, Harry, eds. 1990. *Pearl Harbor Reexamined: Prologue to the Pacific War.* Honolulu: University of Hawaii Press.

Coox, Alvin D. 1985. *Nomonhan. Japan against Russia, 1939.* Stanford, Calif.: Stanford University Press.

Costa, Eduardo Ferrero, ed. 1987. *Relaciones del Peru con los Estados Unidos.* Lima: Centro Peruano de Estudios Internacionales.

Costello, John. 1981. *The Pacific War.* New York: Rawson, Wade.

Couteau-Bégarie, Hervé. 1987. *Geostratégie du Pacifique.* Paris: IFRI-Economica.

——. 1983. *La Puissance maritime soviétique.* Paris: IFRI-Economica.

Cox, Thomas R. 1973. "Harbingers of Change: American Merchants and the Formosa Annexation Scheme." *Pacific Historical Review* 42, no. 2, (May): 163–184.

Crane, Daniel M., and Breslin, Thomas A. 1986. *An Ordinary Relationship: American Opposition to Republican Revolution in China.* Miami: Florida International University Press.

Craven, Wesley Franck, and Cate, James Lea, eds. 1948–1958. *The Army Air Forces in World War II.* Chicago: University of Chicago Press. 7 vols.

Crichton, Michael. 1993. *Rising Sun.* New York: Ballantine.

Cronin, Richard P. 1992. *Japan, the United States, and Prospects for the Asia-Pacific Century: Three Scenarios for the Future.* New York: St. Martin's Press.

Crowley, James B. 1966. *Japan's Quest for Autonomy: National Security and Foreign Policy, 1930–1938.* Princeton, N.J.: Princeton University Press.

Cumings, Bruce C., ed. 1983. *Child of Conflict: The Korean-American Relationship, 1943–1953.* Seattle: University of Washington Press.

———. 1990. *The Origins of the Korean War,* 2 vols. Vol. 1: *Liberation and the Emergence of Separate Regimes, 1945–1947, 1981.* Vol. 2, *The Roaring of the Cataract, 1947–1950.* Princeton, N.J.: Princeton University Press.

Cumings, Bruce, and Halliday, Jon. 1988. *Korea: The Unknown War.* New York: Pantheon Books.

Curtis, Stephen, Jr. 1844. *Brief Extracts from the Journal of a Voyage Performed by the Whaleship M of New Bedford, Mass., Commencing May 25, 1841, and Terminating August 1, 1844.* Boston: S. N. Dickinson.

Cutler, Carl C. 1984. *Greyhounds of the Sea: The Story of the American Clipper Ship.* 3rd ed. Annapolis, Md.: Naval Institute Press.

Cusumano, Michael A. 1985. *The Japanese Automobile Industry: Technology and Management at Nissan and Toyota.* Cambridge: Harvard University Press.

———. 1991. *Japan's Software Factories: A Challenge to U.S. Management.* Oxford: Oxford University Press.

Da Cunha, Derek. 1990. *Soviet Naval Power in the Pacific.* Boulder, Colo.: Lynne Riennen.

Dana, Richard Henry. 1969 (1840). *Two Years before the Mast.* New York: Dutton (Everyman's Library).

Dangerfield, George. 1965. *The Awakening of American Nationalism, 1815–1828.* New York: Harper & Row.

Daniel, Dominique. 1993. *La Réunification familiale aux États-Unis, 1965–1990: politiques et pratiques migratoires.* Thesis, University of Paris VII, 2 vols.

Daniels, Roger. 1962. *The Politics of Prejudice: The Anti-Japanese Movement in California and the Struggle for Japanese Exclusion.* Berkeley: University of California Press.

———. 1974. "American Historians and East Asian Immigrants." *Pacific Historical Review* 43, no. 4 (November), pp. 449–472.

———. 1975. *The Decision to Relocate the Japanese Americans.* Philadelphia: Lippincott.

———. 1981. *Concentration Camps, North America: Japanese in the United States and Canada During World War II.* Malabar: Krieger.

————. 1988. *Asian America: Chinese and Japanese in the United States since 1850.* Seattle: University of Washington Press.

————. 1990. *Coming to America: A History of Immigration and Ethnicity in America.* New York: Harper Collins.

Davis, Lance E.; Gallman, Robert E.; and Hutchins, Theresa D. 1987. "The Structure of the Capital Stock in Economic Growth and Decline: The New Bedford Whaling Fleet in the Nineteenth Century." In Peter Kilby, ed., *Quantity and Quiddity: Essays in U.S. Economic History*, pp. 336–398. Middletown: Wesleyan University Press.

————. 1988. "The Decline of U.S. Whaling: Was the Stock of Whales Running Out?" *Business History Review* 62, no. 4 (Winter): 569–595.

————. 1989. "Productivity in American Whaling: The New Bedford Fleet in the Nineteenth Century." In David W. Galenson, ed., *Markets in History: Economic Studies of the Past*, pp. 97–147. Cambridge: Cambridge University Press.

————. 1991. "Call Me Ishmael—Not Domingo Floresta: The Rise and Fall of the American Whaling Industry." *Research in Economic History*, Suppl. 6, pp. 191–233.

Davis, William Heath. 1989. *Sixty Years in California: A History of Events and Life in California: Personal, Political and Military.* San Francisco: A. J. Leary.

Day, David. 1989. *The Great Betrayal, Britain, Australia, and the Onset of the Pacific War, 1939–42.* New York: Norton.

Debenedetti, Charles. 1990. *An American Ordeal: The Antiwar Movement of the Vietnam Era.* Syracuse, N.Y.: Syracuse University Press.

Delano, Amasa. 1970 (1817). *A Narrative of Voyages and Travels in the Northern and Southern Hemispheres.* Upper Saddle River, N.J.: Gregg Press.

De La Pedraja, René. 1992. *The Rise and Decline of U.S. Merchant Shipping in the Twentieth Century.* New York: Twayne.

Delgado, James P. 1990. *To California by Sea: A Maritime History of the California Gold Rush.* Columbia: University of South Carolina Press.

Delmas, Claude. 1990. *1941, Pearl Harbor. La guerre devient mondiale.* Brussels: Éd. Complexe.

Denison, Edward F., and Chung, William K. 1976. *How Japan's Economy Grew So Fast: The Sources of Postwar Expansion.* Washington, D.C.: Brookings Institution.

Dennett, Tyler. 1922. *Americans in Eastern Asia: A Critical Study of United States' Policy in the Far East in the Nineteenth Century.* New York: Macmillan.

Denoon, David B. H. 1993. *Real Reciprocity: Balancing U.S. Economic and Security Policies in the Pacific Basin.* New York: Council of Foreign Relations Press.

Dermigny, Louis. 1964. *La Chine et l'Occident. Le commerce à Canton au XVIIIe siècle, 1719–1833.* Paris: SEVPEN. 3 vols.

————. 1964. *Les Mémoires de Charles de Constant sur le commerce à la Chine.* Paris: SEVPEN.

Destler, I. M., et al. 1976. *Managing an Alliance: The Politics of U.S. Japanese Relations.* Washington, D.C.: Brookings Institution.

Destler, I. M.; Fukui, Haruhiro; and Sato, Hideo. 1979. *The Textile Wrangle: Conflict in Japanese-American Relations, 1969–1971.* Ithaca, N.Y.: Cornell University Press.

Detwiler, Donald S., and Burdick, Charles B., eds. 1980. *War in Asia and the Pacific, 1937–1949.* New York: Garland. 15 vols.

Dingman, Roger. 1976. *Power in the Pacific: The Origins of Naval Arms Limitation, 1914–1922.* Chicago: University of Chicago Press.

———. 1981. "The U.S. Navy and the Cold War: The Japan Case." In Symonds, Craig L., ed. *New Aspects of Naval History.*

Dirlik, Arif, ed. 1993. *What Is in a Rim? Critical Perspectives on the Pacific Region Idea.* Boulder, Colo.: Westview Press.

Divine, Robert A. 1988. "Vietnam Reconsidered." *Diplomatic History,* vol. 12, no. 1 (Winter): 79–93.

Dixon, Joe C., ed. 1980. *The American Military and the Far East,* Washington, D.C.: U.S. Air Force Academy.

Dobbs, Charles M. 1981. *The Unwanted Symbol: American Foreign Policy: the Cold War and Korea, 1945–1950.* Kent, Ohio: Kent State University Press.

Dodge, Bertha S., ed. 1986. *Marooned, Being a Narrative of the Sufferings and Adventures of Capt. Charles H. Barnard in a Voyage Round the World during the Years 1812, 1813, 1814, 1815 & 1816. . . .* Syracuse, N.Y.: Syracuse University Press.

Dodge, Ernest S. 1965. *New England and the South Seas.* Cambridge: Harvard University Press.

Doenecke, Justus D. 1984. *When the Wicked Rise: American Opinion-Makers and the Manchurian Crisis of 1931–1933.* Cranbury: Bucknell University Press.

Doi, Takeo. 1973. *The Anatomy of Dependence (Amae no kozo).* Tokyo: Kodansha International.

Dolan, Ronald E., and Worden, Robert L., eds. 1992. *Japan: A Country Study.* Washington, D.C.: Federal Research Division, Library of Congress, 1992.

Dorrance, John C. 1992. *The United States and the Pacific Islands.* Westport, Conn.: Praeger.

Doulman, David J. 1987. *Options for U.S. Fisheries Investment in the Pacific Islands Region.* Research Report Series no. 8. Honolulu: East-West Center.

Doulman, David J., ed. 1987. *The Development of the Tuna Industry in the Pacific Islands Region: An Analysis of Options.* Honolulu: East-West Center.

Dower, John W. 1979. *Empire and Aftermath: Yoshida Shigeru and the Japanese Experience, 1878–1954.* Cambridge: Harvard University Press.

———. 1986. *War without Mercy: Race and Power in the Pacific War.* New York: Pantheon Books.

———. 1993. *Japan in War and Peace: Selected Essays.* New York: New Press.

Downen, Robert L., and Dickson, Bruce J., eds. 1984. *The Emerging Pacific Community: A Regional Perspective.* Boulder, Colo.: Westview Press.

Downs, James M. 1968. "American Merchants and the China Opium Trade, 1800–1840." *Business History Review* 42, no. 4: 418–442.

Drake, Frederick C. 1984. *The Empire of the Seas: A Biography of Rear Admiral Robert Wilson Shufeldt, USN.* Honolulu: University of Hawaii Press.

Drake, Paul W. 1989. *The Money Doctor in the Andes: The Kemmerer Missions, 1923–1933.* Durham, N.C.: Duke University Press.

Drea, Edward J. 1991. *MacArthur's ULTRA: Codebreaking and the War against Japan, 1942–1945.* Lawrence: University Press of Kansas.

Dreyer, June Teufel, ed. 1990. *Asian Pacific Regional Security.* Washington, D.C.: Washington Institute Press.

Drinnon, Richard. 1980. *Facing West: The Metaphysics of Indian-Hating and Empire Building.* Minneapolis: University of Minnesota Press.

Drysdale, Peter. 1988. *International Economic Pluralism: Economic Policy in East Asia and the Pacific.* Boston: Allen & Unwin.

Drysdale, Peter, and O'Hare, Martin, eds., 1991. *The Soviets and the Pacific Challenge.* Armonk: M. E. Sharpe.

Dudden, Arthur P. 1992. *The American Pacific: From the Early China Trade to the Present.* New York: Oxford University Press.

Dulles, Foster Rhea. 1932. *America in the Pacific.* Boston: Houghton Muffin.

———. 1946. *China and America: The Story of Their Relations since 1784.* Princeton, N.J.: Princeton University Press.

———. 1965. *Yankees and Samurai: America's Role in the Emergence of Modern Japan, 1791–1900.* New York: Harper & Row.

Durán, Clemente Ruiz. 1990. "Asia and Latin America in the Pacific Era." *Foreign Relations Journal* (Manila), 5, no. 2 (June): 88–105.

Edström, Bert. 1990. "Japan's Elusive Pacific Dream." *Stockholm Journal of East Asian Studies* 2: 103–141.

Edwards, Lee. 1990. *Missionary for Freedom: The Life and Times of Walter H. Judd.* New York: Paragon House.

Emmott, Bill. 1989. *The Sun Also Sets: Why Japan Will Not Be Number One.* New York: Simon and Schuster.

———. 1993. *Japanophobia: The Myth of the Invincible Japanese.* New York: Times Books.

Encarnation, Dennis J. 1993. *Rivals beyond Trade: America Versus Japan in Global Competition.* Ithaca, N.Y.: Cornell University Press.

Errington, Elizabeth Jane, and McKercher, B. J. C., eds. 1991. *The Vietnam War as History.* New York: Praeger.

Esthus, Raymond A. 1959. "The Changing Concept of the Open Door, 1899–1910." *Mississippi Valley Historical Review* 46 (December): 435–454.

———. 1964. *From Enmity to Alliance: U.S.-Australian Relaitons, 1931–1941.* Seattle: University of Washington Press.

———. 1966. *Theodore Roosevelt and Japan.* Seattle: University of Washington Press.

———. 1988. *Double Eagle and Rising Sun: The Russians and Japanese at Portsmouth in 1905.* Durham, N.C.: Duke University Press.

Evans, Paul M. 1988. *John King Fairbank and the American Understanding of Modern China*. Oxford: Basil Blackwell.

Fairbank, John King. 1953. *Trade and Diplomacy on the China Coast: The Opening of the Treaty Ports, 1842–1854.* Cambridge: Harvard University Press. 2 vols.

——. 1974. *China Perceived: Images and Policies in Chinese-American Relations.* New York.

——. 1982. *Chinabound: A Fifty-Year Memoir.* New York: Harper & Row.

——. 1983. *The United States and China.* 4th ed. Cambridge: Harvard University Press.

——. 1987. *China Watch.* Cambridge: Harvard University Press.

Fairbank, John K., ed. 1974. *The Missionary Enterprise in China and America.* Cambridge: Harvard University Press.

Fallon, Joseph E. 1991. "Federal Policy and U.S. Territories: The Political Restructuring of the United States of America." *Pacific Affairs* 64, no. 1 (Spring): 23–41.

Fallows, James. 1989. *More Like Us: Making America Great Again.* Boston: Houghton Muffin.

——. 1994. *Looking at the Sun: The Rise of the New East Asian Economic and Political System.* New York: Pantheon Books.

Fanning, Edmund. 1833. *Voyages to the South Seas, Indian and Pacific Oceans, China Sea, North-West Coast, Feejee Islands, South Shetlands, etc.* New York: Collins & Hannay (2nd ed., 1838).

Feis, Herbert. 1950. *The Road to Pearl Harbor: The Coming of the War between the United States and Japan.* Princeton, N.J.: Princeton University Press.

——. 1953. *The China Tangle: The American Effort in China from Pearl Harbor to the Marshall Mission.* Princeton, N.J.: Princeton University Press.

Field, Jr., James A. 1978. "American Imperialism: The Worst Chapter in Almost Any Book." *American Historical Review* 83, no. 3: 644–683.

Fifield, Russell H. 1952. *Woodrow Wilson and the Far East: The Diplomacy of the Shantung Question.* New York: Thomas Y. Crowell.

——. 1973. *Americans in Southeast Asia: The Roots of Commitment.* New York: Crowell.

Findling, John E. 1987. *Close Neighbors, Distant Friends: United States-Central American Relations.* Westport, Conn.: Greenwood Press.

Finn, Richard B. 1992. *Winners in Peace: MacArthur, Yoshida, and Postwar Japan.* Berkeley: University of California Press.

Fischer, Georges. 1960. *Un cas de décolonisation: Les États-Unis et les Philippines.* Paris: Librairie générale de droit et de jurisprudence.

——. 1979. *Les États-Unis et le canal de Panama.* Paris: L'Harmattan.

Fischer, Lewis R., et al., eds. 1985. *Change and Adaptation in Maritime History: The National Fleets of the North Atlantic.* St. John's: Memorial University of Newfoundland.

Fischer, Lewis R., and Nordvik, Helge W., eds. 1990. *Shipping and Trade, 1750–1950: Essays in International Maritime History.* Pontefrack: Lofthouse Publications.

Fischer, Wolfram; McInnis, Marvin R.; and Schneider, Jürgen, eds. 1986. *The Emergence of a World Economy, 1500–1914.* 2 vols. Wiesbaden: Franz Steiner.

Fistié, Pierre. 1972. *La Rentrée en scène du Japon. La politique japonaise face aux États-Unis depuis 1945.* Paris: A. Colin.

Fitzgerald, Frances. 1972. *Fire in the Lake: The Vietnamese and the Americans in Vietnam.* Boston: Little, Brown.

Folkman, Jr., David I. 1972. *The Nicaragua Route.* Salt Lake City: University of Utah Press.

Foot, Rosemary. 1985. *The Wrong War: American Policy and the Dimensions of the Korean Conflict, 1950–1953.* Ithaca, N.Y.: Cornell University Press.

———. 1990. *A Substitute for Victory: The Politics of Peacemaking at the Korean Armistice Talks.* Ithaca, N.Y.: Cornell University Press.

Forsythe, Sidney A. 1971. *An American Missionary Community in China, 1895–1905.* Cambridge: Harvard University Press.

Foucrier, Annick. 1991. "La France et la Californie avant la ruée vers l'or (1788–1848)." Doctoral thesis: École de Hautes Études en Sciences Sociales.

———. 1992. "La Californie, noeud gordien du Pacifique Nord (1769–1848)," *Marins et océans,* vol. III, pp. 123–149.

Fox, Frank. 1912. *Problems of the Pacific.* London: Williams & Norgate.

———. 1928. *The Mastery of the Pacific: Can the British Empire and the United States Agree?* London: John Lane, The Bodley Head Ltd.

Francis, Michael J. 1977. *The Limits of Hegemony: U.S. Relations with Argentina and Chile during World War II.* Notre Dame, Ind.: University of Notre Dame Press.

Frank, Richard B. 1990. *Guadalcanal.* New York: Random House.

Freeman, Derek. 1983. *Margaret Mead and Samoa: The Making and Unmaking of an Anthropological Myth.* Cambridge: Harvard University Press.

Freeman, Otis Willard. 1951. *Geography of the Pacific.* New York: Wiley.

Friedman, Edward, and Selden, Mark, eds. 1971. *America's Asia: Dissenting Essays on Asian-American Relations.* New York: Vintage Books.

Friedman, George, and Lebard, Meredith. 1991. *The Coming War with Japan.* New York: St. Martin's Press.

Friedman, Hal M. 1993. "The Beast in Paradise: The United States Navy in Micronesia, 1943–1947. *Pacific Historical Review* 63, no. 2 (May): 173–195.

Friis, Herman R., ed. 1967. *The Pacific Basin: A History of Its Geographical Exploration.* New York: American Geographical Society.

Fritzsche, Bruno. 1968. "On Liberal Terms: The Boston Hide Merchants in California." *Business History Review* 42 no. 4 (Winter): 467–481.

Fry, Gerald W., and Rufino, Mauricio, eds. 1987. *Pacific Basin and Oceania.* Oxford and Santa Barbara, Calif.: Clio.

Gaddis, John L. 1982. *Strategies of Containment: A Critical Appraisal of Postwar National Security Policy.* New York: Oxford University Press.

————. 1987. *The Long Peace: Inquiries into the History of the Cold War.* New York: Oxford University Press.

Gallichio, Marc S. 1988. *The Cold War Begins in Asia: American East Asian Policy and the Fall of the Japanese Empire.* New York: Columbia University Press..

Gardner, Lloyd C. 1988. *Approaching Vietnam: From World War II through Dienbienphu, 1941–1954.* New York: Norton.

Gates, John Morgan. 1973. *Schoolbooks and Krags: The United States Army in the Philippines, 1898–1902.* Westport, Conn.: Greenwood.

————. 1984. "War Related Deaths in the Philippines, 1898–1902." *Pacific Historical Review* 53, no. 3 (August): 367–378.

Gettleman, Marvin E.; Franklin, Jane; Young, Marilyn; and Franklin, Bruce H. 1985. *Vietnam and America: A Documented History.* New York: Grove Press.

Gibb, George S., and Knowlton, Evelyn H. 1956. *The Resurgent Years: History of Standard Oil Company (New Jersey), 1911–1927,* vol. II. New York: Harper Brothers.

Gibbons, William Conrad. 1986–1989. *The U.S. Government and the Vietnam War: Executive and Legislative Roles and Relationships.* 3 vols. Princeton, N.J.: Princeton University Press. Vol. I: *1945–1960,* 1986; vol. II, *1961–1964,* 1986; vol. III, *January–July 1965,* 1989.

Gibney, Frank. 1992. *The Pacific Century. America and Asia in a Changing World.* New York: Scribner's-Macmillan.

Gibson, Arrell M. 1993. *Yankees in Paradise: The Paradise Frontier.* Albuquerque: University of New Mexico Press.

Gibson, James R. 1992. *Otter Skins, Boston Ships and China Goods: The Maritime Fur Trade of the Northwest Coast, 1785–1841.* Seattle: University of Washington Press.

Gillette, Alain. 1985. *Les Mormons.* Paris: Desclée de Brouwer.

Gleeck, Lewis E. 1976. *American Institutions in the Philippines, 1898–1941.* Manila: Historical Conservation Society.

Goldberg, Joyce S. 1986. *The Baltimore Affair.* Lincoln: University of Nebraska Press.

Goldstein, Jonathan. 1978. *Philadelphia and the China Trade, 1682–1846: Commercial, Cultural, and Attitudinal Effects.* University Park: Pennsylvania State University Press.

Goldstein, Jonathan; Israel, Jerry; and Conroy, Hilary, eds. 1991. *America Views China-American Images of China Then and Now.* Bethlehem, Pa.: Lehigh University Press.

Goodman, Grant K., and Moos, Felix, eds. 1981. *The United States and Japan in the Western Pacific: Micronesia and Papua New Guinea.* Boulder, Colo.: Westview Press.

Gough, Barry M. 1980. *Distant Dominion, Britain and the Northwest Coast of North America, 1579–1809.* Vancouver: University of British Columbia Press.

Gould, James W. 1969. *The United States and Malaysia.* Cambridge: Harvard University Press.

Goulden, Joseph C. 1982. *Korea: The Untold Story of the War.* New York: Times Books.

Gourevitch, Peter A., ed. 1989. "The Pacific region: Challenges to policy and theory." *The Annals of the American Academy of Political and Social Science* (September).

Graebner, Norman A., ed. 1961. *An Uncertain Tradition: American Secretaries of State in the Twentieth Century.* New York: McGraw-Hill.

———. 1983. *Empire on the Pacific: A Study in American Continental Expansion.* Santa Barbara, Calif.: ABC-Clio.

Graham, Terence. 1983. *The "Interests of Civilization?": Reaction in the United States against the Seizure of the Panama Canal Zone, 1903–1904.* Lund: Esselte Studium.

Grant, Zalin. 1991. *Facing the Phoenix: The CIA and the Political Defeat of the United States in Vietnam.* New York: Norton.

Grattan, C. Hartley. 1961. *The United States and the Southwest Pacific.* Cambridge: Harvard University Press.

———. 1963. *The Southwest Pacific to 1900. A Modern History: Australia, New Zealand, The Islands, Antarctica,* and *The Southwest Pacific since 1900.* Ann Arbor: University of Michigan Press.

Gray, James Thomas. 1987. *American Fur Seal Diplomacy: The Alaskan Fur Seal Controversy.* New York: Peter Lang.

Greenberg, Michael. 1951. *British Trade and the Opening of China, 1800–1842.* Cambridge: Cambridge University Press.

Griffin, Charles Carroll. 1937. *The United States and the Disruption of the Spanish Empire, 1810–1822. A Study of the Relations of the United States with Spain and with the Rebel Spanish Colonies.* New York: Columbia University Press.

Griffin, Eldon. 1938. *Clippers and Consuls.* Ann Arbor, Mich.: Edwards Brothers.

Grimmett, Richard F. 1973. "Who Were the Senate Isolationists?" *Pacific Historical Review,* 42, no. 4 (November): 479–498.

Griswold, Whitney A. 1938. *The Far Eastern Policy of the United States.* New York: Harcourt, Brace.

Guiness, Ralph B. 1933. "The Purpose of the Lewis and Clark Expedition," *Mississippi Valley Historical Review,* 20, no. 1 (June): 90–100.

Gulick, Edward V. 1973. *Peter Parker and the Opening of China.* Cambridge: Harvard University Press.

Haeger, John Denis. 1991. *John Jacob Astor: Business and Finance in the Early Republic.* Detroit: Wayne State University Press.

Hagan, Kenneth J. 1973. *American Gunboat Diplomacy and the Old Navy, 1877–1889.* Westport, Conn.: Greenwood.

Hagan, Kenneth J., ed. 1984. *In Peace and War: Interpretations of American Naval History, 1775–1984.* 2nd ed. Westport, Conn.: Greenwood, 1984.

Hague, Harlan, and Langum, David J. 1990. *Thomas O. Larkin: A Life of Patriotism and Profit in Old California.* Norman: University of Oklahoma Press.

Halberstam, David. 1986. *The Reckoning.* New York: Avon Books.

Haley, P. Edward. 1982. *Congress and the Fall of South Vietnam and Cambodia.* East Brunswick, N.J.: Associated Presses.

Hamada, Tomoko. 1991. *American Enterprise in Japan*. Albany: State University of New York Press.

Hanks, Robert J. 1981. *The Pacific Far East: Endangered American Strategic Position*. Washington, D.C.: Brassey's.

Hanlon, David. 1988. *Upon a Stone Altar: A History of the Island of Pohnpei to 1890*. Honolulu: University of Hawaii Press.

Hanson, Kermit O., and Roehl, Thomas W., eds. 1980. *The United States and the Pacific Economy in the 1980s*. Indianapolis: Bobbs-Merrill.

Harding, Harry. 1992. *A Fragile Relationship: The United States and China Since 1972*. Washington, D.C.: Brookings Institution.

Harding, Harry, and Ming, Yuan, eds. 1989. *Sino-American Relations, 1945–1955: A Joint Reassessment of a Critical Decade*. Wilmington, Del.: Scholarly Resources.

Harlow, Neal. 1982. *California Conquered: War and Peace on the Pacific, 1846–1850*. Berkeley: University of California Press.

Harper, Norman. 1987. *A Great and Powerful Friend: A Study of Australian-American Relations between 1900 and 1975*. St. Lucia: University of Queensland Press.

Harper, Norman, ed. 1968. *Pacific Orbit: Australian-American Relations Since 1942*. Melbourne: F.W. Cheshire.

———. 1971. *Australia and the United States. Documents and Readings in Australian History*. Melbourne: Thomas Nelson.

Harris, Paul W. 1991. "Cultural Imperialism and American Protestant Missionaries: Collaboration and Dependency in Mid-Nineteenth Century China." *Pacific Historical Review* 40, no. 3 (August): 309–338.

Harris, Ruth R. 1981. "The 'Magic' Leak of 1941 and Japanese-American Relations, 1941," *Pacific Historical Review* 50, no. 1 (February): 76–95.

Hart, Robert A. 1965. *The Great White Fleet; Its Voyage around the World, 1907–1909*. Boston: Little, Brown.

Hastings, Max. 1987. *The Korean War*. New York: Simon and Schuster.

Haushofer, Karl. 1924. *Geopolitik des Pazifischen Ozeans: Studien über die Wechselbeziehungen zwischen Geographie und Geschichte*. Berlin: Vowinckel.

Haviland, Edward Kenneth. 1956–1958. "American Steam Navigation in China, 1845–1878." *American Neptune*, vols. 16–18, no. 1 (July 1956): 157–179; no. 2 (October 1956): 243–269; no. 3 (January 1957): 38–64; no. 4 (April 1957): 134–151; no. 5 (July 1957): 212–230; no. 6 (October 1957): 298–314; no. 7 (January 1958): 59–85.

Hayashi, Kichiro, ed. 1989. *The U.S. Japanese-Economic Relationship: Can It Be Improved?* New York: New York University Press.

Hayashi, Saburo, and Coox, Alvin D. 1959. *Kogun: The Japanese Army in the Pacific War*. Quantico, Va.: The Marine Corps Association. First Japanese edition, 1951.

Hayes, Edmund, ed. 1981. *Log of the Union: John Boit's Remarkable Voyage to the Northwest Coast and around the World, 1794–1796*. Portland: Oregon Historical Society.

Hayes, Peter, et al. 1987. *American Lake: Nuclear Peril in the Pacific.* Harmondsworth: Penguin Books, 1987.

———. 1991. *Pacific Powderkeg: American Nuclear Dilemmas in Korea.* Lexington: Lexington Books.

Healy, David F. 1970. *U.S. Expansionism: The Imperialist Urge in the 1890s.* Madison: University of Wisconsin Press.

Heffer, Jean. 1986. *Le Port de New York et le commerce extérieur américain, 1860–1900.* Paris: Publications de la Sorbonne.

———. 1987. *L'Union en péril: la démocratie et l'esclavage (1829–1865).* Nancy: Presses universitaires de Nancy.

———. 1988. "Brèves remarques sur l'économie américaine et les autres." In François Crouzet, ed., *L'Extrême-Occident* XVI. Colloque de l'Institut de recherches sur les civilizations de l'Occident moderne, *Civilisations,* no. 16, pp. 134–145. Paris: Presses de l'Université de Paris-Sorbonne.

———. 1990. "Sumatra, 1832: Premier engagement américain en Asie du Sud-Est." *Archipel* 40: 49–64.

———. 1992. *Les États-Unis de Truman à Bush.* Paris: A. Colin.

———. 1993. "Le Pacifique, dernière frontière?" In Pierre Lagayette, ed., *Les Mythes de l'Ouest americain: Visions et révisions, Westways* 1: 83–99. Nanterre: Université de Paris X-Nanterre.

Heffer, Jean, and Serman, William. 1992. *Le XIXe siècle, 1815–1914.* Paris: Hachette.

Heffer, Jean, and Launay, Michel. 1992. *L'Ère des deux Grands, 1945–1973.* Paris: Hachette.

Heffer, Jean, and Weil, François, eds. 1994. *Chantiers d'histoire américaine.* Paris: Belin.

Heffernan, Thomas F. 1981. *Stove by a Whale: Owen Chase and the* Essex. Middletown: Wesleyan University Press.

Heinrichs, Waldo. 1966. *American Ambassador: Joseph C. Grew and the Development of the United States Diplomatic Tradition.* Oxford: Oxford University Press.

———. 1988. *Threshold of War: Franklin D. Roosevelt & American Entry into World War II.* New York: Oxford University Press.

Henson, Curtis T. 1982. *Commissioners and Commodores: The East Indian Squadron and American Diplomacy in China.* University, Ala.: University of Alabama Press.

Herbert, Thomas Walter. 1980. *Marquesan Encounters: Melville and the Meaning of Civilziation.* Cambridge: Harvard University Press.

Herde, Peter. 1980. *Pearl Harbor, 7 December 1941: Der Ausbruch des Krieges zwischen Japan und den Vereinigten Staaten und die Ausweitung des europäischen Krieges zum zweiten Weltkrieg.* Darmstadt: Wissenschaftliche Buchgesellschaft.

Herring, George C. 1986. *America's Longest War: The United States and Vietnam, 1950–1975.* 2nd ed. New York: John Wiley.

Hess, Gary R. 1987. *The United States Emergence as a Southeast Asian Power, 1940–1950.* New York: Columbia University Press.

———. 1994. "The Unending Debate: Historians and the Vietnam War." *Diplomatic History* 18, no. 2 (Spring): 239–264.

Hezel, Francis X., S.J. 1983. *The First Taint of Civilization. A History of the Caroline and Marshall Islands in Pre-Colonial Days, 1521–1885.* Honolulu: University of Hawaii Press.

Higgott, Richard A.; Cooper, Andrew F.; and Bonnor, Jenelle. 1990. "Asia-Pacific Economic Cooperation: An Evolving Case-Study in Leadership and Co-operation Building." *International Journal* 45, no. 4 (Autumn): 823–866.

Hilpert, Hans-Günther, et al. 1992. *Wirtschaftliche Integration and Kooperation im asiatisch-pazifischen Raum.* Munich: IFO, Institut für Wirtschaftsforschung.

Hing, Bill Ong. 1993. *Making and Remaking Asian America through Immigration Policy, 1850–1990.* Stanford, Calif.: Stanford University Press.

Hogan, J. Michael. 1986. *The Panama Canal in American Politics: Domestic Advocacy and the Evolution of Policy.* Carbondale: Southern Illinois University Press.

Holmes, Lowell D. 1987. *Quest for the Real Samoa: The Mead/Freeman Controversy and Beyond.* South Hadley: Bergin & Garvey.

Holstein, William J. 1990. *The Japanese Power Game: What It Means for America.* New York: Charles Scribner's Sons.

Homan, Gerlof D. 1984. "The United States and the Netherlands East Indies: The Evolution of American Anticolonialism." *Pacific Historical Review* 53, no. 4 (November): 423–446.

Hooper, Paul F., ed. 1982. *Building a Pacific Community: The Addresses and Papers of the Pacific Community Lecture Series.* Honolulu: University Press of Hawaii.

Howay, Frederick W.; Sage, W. N.; and Angus, H. F. 1942. *British Columbia and the United States: The North Pacific Slope from Fur Trade to Aviation.* New Haven: Yale University Press.

Hoyt, Edwin Palmer. 1986. *Japan's War: The Great Pacific Conflict, 1853 to 1952.* New York: McGraw Hill.

Hsiao, Gene T., and Witunski, Michael, eds. 1983. *Sino-American Normalization and Its Policy.* New York: Praeger.

Huffman, James J. 1987. "Edward Howard House: In the Service of Meiji Japan," *Pacific Historical Review* 56, no. 2 (May): 231–258.

Hull, Cordell. 1948. *The Memoirs of Cordell Hull.* 2 vols. New York: Macmillan.

Hulten, Charles R., ed. 1990. *Productivity Growth in Japan and the United States.* Chicago: University of Chicago Press and NBER.

Humble, Richard. 1984. *United States Fleet Carriers of World War II.* New York: Sterling.

Hunt, Michael H. 1973. *Frontier Defense and the Open Door: Manchuria in Chinese-American Relations 1895–1911.* New Haven: Yale University Press.

———. 1977. "Americans in the China Market: Economic Opportunities and Economic Nationalism, 1890s–1931." *Business History Review* 51, no. 3: 277–307.

————. 1983. *The Making of a Special Relationship: The United States and China to 1914.* New York: Columbia University Press.

————. 1987. *Ideology and U.S. Foreign Policy.* New Haven: Yale University Press.

Hunt, Michael H.; Shambaugh, David; Cohen, Warren I.; and Iriye, Akira. 1988. *Mutual Images in U.S.-China Relations.* Washington, D.C.: Woodrow Wilson Center. Occasional Paper, no. 32.

Hunter, Jane. 1989. *The Gospel of Gentility: American Women Missionaries in Turn-of-the-Century China.* New Haven: Yale University Press.

Hutchinson, William R. 1987. *Errand to the World: American Protestant Thought and Foreign Missions.* Chicago: University of Chicago Press.

Hyatt, Irwin T., Jr. 1976. *Our Ordered Lives Confess: Three Nineteenth-Century American Missionaries in East Shantung.* Cambridge: Harvard University Press.

Ienaga, Saburo. 1978. *The Pacific War: World War II and the Japanese, 1931–1945.* New York: Pantheon Books (translation of Japanese edition of 1968).

Ike, Nobutaka, ed. 1967. *Japan's Decision for War: Records of the 1941 Policy Conferences.* Stanford, Calif.: Stanford University Press.

Immerman, Richard H., ed. 1990. *John Foster Dulles and the Diplomacy of the Cold War.* Princeton, N.J.: Princeton University Press.

Ingraham. 1971. *Joseph Ingraham's Journal of the Brigantine Hope on a Voyage to the Northwest Coast of North America, 1790–1792, Illustrated with Charts and Drawings by the Author, Edited with Notes and an Introduction by Mark D. Kaplanoff.* Barre: Imprint Society.

Inoue, Kyoko. 1991. *MacArthur's Japanese Constitution: A Linguistic and Cultural Study of Its Making.* Chicago: University of Chicago Press.

Institut du Pacifique. 1986. *Le Pacifique "nouveau centre du monde."* Paris: Berger-Levrault.

Ion, Hamish A., and Hunt, Harry D., eds. 1988. *War and Diplomacy across the Pacific, 1919–1952.* Waterloo: Wilfrid Laurier University Press.

Iriye, Akira. 1972. *Pacific Estrangement, Japanese and American Expansion, 1897–1911.* Cambridge: Harvard University Press.

————. 1974. *The Cold War in Asia: A Historical Introduction.* Inglewood Cliffs, N.J.: Prentice-Hall.

————. 1981. *Power and Culture: The Japanese-American War, 1941–1945.* Cambridge: Harvard University Press.

————. 1984. "Contemporary History as History: American Expansion into the Pacific since 1941." *Pacific Historical Review* 53, no. 2 (May): 191–212.

————. 1987. *The Origins of the Second World War in Asia and the Pacific.* London: Longmans.

————. 1990 (1965). *After Imperialism: The Search for a New Order in the Far East, 1921–1931.* Chicago: Imprint Publications.

————. 1992. *Across the Pacific: An Inner History of American-East Asian Relations.* Rev. ed. Chicago: Imprint Publications.

Iriye, Akira, ed. 1975. *Mutual Images: Essays in American-Japanese Relations.* Cambridge: Harvard University Press.

Iriye, Akira, and Cohen, Warren I., eds. 1989. *United States and Japan in the Postwar World.* Lexington: University Press of Kentucky.

Irons, Peter. 1983. *Justice at War: The Story of the Japanese American Internment Cases.* Berkeley: University of California Press.

Isaacs, Harold R. 1958. *Scratches on Our Minds. American Images of China and India.* New York: John Day.

Ishihara, Shintaro. 1991. *The Japan That Can Say No.* New York: Simon & Schuster.

Israel, Jerry. 1971. *Progressivism and the Open Door: America and China, 1905–1921.* Pittsburgh: University of Pittsburgh Press.

Jéquier, Nicolas. 1970. *Le Defi industriel japonais.* Lausanne: Centre de recherches européennes.

Jiang, Arnold Xiangze. 1988. *The United States and China.* Chicago: University of Chicago Press.

Johnson, Chalmers A. 1982. *MITI and the Japanese Miracle: The Growth of Industrial Policy, 1925–1975.* Stanford, Calif.: Stanford University Press.

Johnson, Chalmers A.; D'Andrea Tyson, Laura; and Zysman, John, eds. 1989. *Politics and Productivity: The Real Story of Why Japan Works.* Cambridge: Ballinger.

Johnson, Robert Erwin. 1979. *Far China Station: The U.S. Navy in Asian Waters, 1800–1898.* Annapolis, Md.: Naval Institute Press.

———. 1980 (1963). *Thence Round Cape Horn: The Story of United States Naval Forces of Pacific Station, 1818–1923.* New York: Arno Press.

Johnson, Sheila K. 1988. *The Japanese through American Eyes.* Stanford, Calif.: Stanford University Press.

Johnston, Douglas M., and Valencia, Mark J. 1991. *Pacific Ocean Boundary Problems: Status and Solutions.* Norwell: Kluwer Academic Publ.

Jones, Eric; Frost, Lionel; and White, Colin. 1993. *Coming Full Circle: An Economic History of the Pacific Rim.* Boulder, Colo.: Westview.

Jones, Hazel J. 1980. *Live Machines: Hired Foreigners and Meiji Japan.* Vancouver: University of British Columbia Press.

Joyaux, François. 1979. *La Chine et le règlement de la première guerre d'Indochine, Genève, 1954.* Paris: Publications de la Sorbonne.

———. 1985, 1988. *La Nouvelle Question d'Extrême-Orient.* Paris: Payot. Vol. 1: *L'ère de la guerre froide (1945–1959);* vol. 2: *L'Ère du conflit sino-soviétique (1959–1978).*

———. 1993. *Géopolitiqué de l'Extrême-Orient.* Brussels: Éd. Complexe. 2 vols. (1. *Espaces et politiques;* 2. *Frontière et stratégies*).

———. 1993. *La politique extérieure du Japon.* Paris: Presses universitaires de France.

Kahn, Helen Dodson. 1974. "Willard Straight and the Great Game of Empire." In Merli, Frank J., and Wilson, Theodore A., eds., *Makers of American Diplomacy from Benjamin Franklin to Henry Kissinger,* pp. 333–358. New York: Scribner's.

Karnow, Stanley. 1984. *Vietnam, A History.* New York: Penguin Books.

———. 1989. *In Our Image: America's Empire in the Philippines.* New York: Ballantine.

Kattenburg, Paul M. 1980. *The Vietnam Trauma in American Foreign Policy, 1945–1975.* New Brunswick, N.J.: Transaction Books.

Kaufman, Burton I. 1986. *The Korean War: Challenges in Crisis, Credibility and Command.* New York: Alfred A. Knopf.

Kawai, Kuzuo. 1960. *Japan's American Interlude.* Chicago: University of Chicago Press.

Kawakami, K. K. 1924. *Le Problème du Pacifique et la politique japonaise.* Paris: Éd. Bossard.

Kay, Robin, ed. 1985. *Documents on the New Zealand External Relations,* vol. III: *The ANZUS Pact and the Treaty of Peace with Japan.* Wellington: GPO.

Keenan, Barry. 1977. *The Dewey Experiment in China: Educational Reform and Political Power in the Early Republic.* Cambridge: Harvard University Press.

Kemble, John Haskell. 1943. *The Panama Route, 1848–1869.* Berkeley: University of California Press.

———. 1950. "A Hundred Years of the Pacific Mail." *American Neptune* (April).

Kennan, George F. 1951. *American Diplomacy, 1900–1950.* Chicago: University of Chicago Press.

———. 1967–1972. *Memoirs,* 2 vols. (1: *1925–1950;* 2: *1950–1963*). Boston: Little, Brown.

Kennedy, Paul N. 1974. *The Samoan Tangle: A Study in Anglo-German-American Relations, 1878–1900.* Dublin: Irish University Press.

Kentaro, Awaya. 1991. "Emperor Showa's Accountability for War." *Japan Quarterly* 38, no. 4 (Oct.–Dec.): 386–398.

Kester, W. Carl. 1991. *Japanese Takeovers: The Global Contest for Corporate Control.* Boston: Harvard Business School Press.

Kim, Hyung-chan, ed. 1986. *Dictionary of Asian American History.* Westport, Conn.: Greenwood Press.

Kissinger, Henry. 1979. *White House Years.* Boston: Little, Brown.

———. 1982. *Years of Upheaval.* Boston: Little, Brown.

Kitamura, Hiroshi. 1971. *Psychological Dimensions of U.S.–Japanese Relations.* Cambridge: Harvard University, Center for International Affairs.

Kitano, Harry H. L. 1974. "Japanese Americans: The Development of a Middleman Minority." *Pacific Historical Review* 43, no. 4 (November): 500–519.

Kitts, Charles R. 1991. *The United States Odyssey in China, 1784–1990.* Lanham, Md.: University Press of America.

Koen, Ross Y. 1960. *The China Lobby in American Politics.* New York: Macmillan.

Kolde, Endel Jakob. 1976. *The Pacific Quest: The Concept and Scope of an Oceanic Community.* Lexington: Lexington Books.

Kolko, Gabriel. 1985. *Anatomy of a War: Vietnam, the United States and the Modern Historical Experience.* New York: Pantheon Books.

Komiya, Ryutaro; Okuno, Masahiro; and Suzumura, Kotaro, eds. 1988. *Industrial Policy of Japan*. San Diego: Academic Press.

Konvitz, Milton R. 1946. *The Alien and the Asiatic in American Law*. Ithaca, N.Y.: Cornell University Press.

Koo, Youngnok, and Suh, Dae-Sook, eds. 1984. *Korea and the United States: A Century of Cooperation*. Honolulu: University of Hawaii Press.

Krugman, Paul, ed. 1991. *Trade with Japan: Has the Door Opened Wider?* Chicago: University of Chicago Press.

Kushner, Howard I. 1975. *Conflict on the Northwest Coast: American-Russian Rivalry in the Pacific Northwest, 1790–1867*. Westport, Conn.: Greenwood Press.

Kusnitz, Leonard A. 1984. *Public Opinion and Foreign Policy: America's China Policy, 1949–1979*. Westport, Conn.: Greenwood.

Kuykendall, Ralph Simpson. 1938. 1953. 1967. *The Hawaiian Kingdom*. Honolulu: University of Hawaii Press, 3 vols. Vol. 1. *1778–1854: Foundation and Transformation;* vol. 2, *1854–1874: Twenty Critical Years*, 1953; vol. 3. *1874–1893, The Kalakahua Dynasty*.

Kwak, Tae-hwan, ed. 1982. *U.S.-Korean Relations, 1882–1982*. Seoul: Institute for Far Eastern Studies, Kyungnam University.

Labonne, Roger. 1936. *Le Tapis vert du Pacifique*. Paris: Berger-Levrault.

Labrousse, Alain. 1991. *La Drogue, l'argent et les armes*. Paris: Fayard.

Lael, Richard L. 1987. *Arrogant Diplomacy: U.S. Policy toward Columbia, 1903–1922*. Wilmington, Del.: Scholarly Resources.

LaFeber, Walter. 1963. *The New Empire: An Interpretation of American Expansion, 1860–1898*. Ithaca, N.Y.: Cornell University Press.

————. 1983. *Inevitable Revolutions: The United States in Central America*. New York: Norton.

————. 1987. *The American Age: United States Foreign Policy at Home and Abroad since 1750*. New York: W. W. Norton.

————. 1989. *The Panama Canal: The Crisis in Historical Perspective*. New York: Oxford University Press.

Lancaster, Clay. 1963. *The Japanese Influence in America*. New York: Walton H. Rawls.

Langdon, Frank C., and Ross, Douglas A., eds. 1990. *Superpower Maritime Strategy in the Pacific*. New York: Routledge.

Langer, William L., and Gleason, Everett S. 1952. *The Challenge to Isolation: The World Crisis of 1937–1941 and American Foreign Policy*. New York: Council of Foreign Relations.

————. 1953. *The Undeclared War, 1940–1941*. New York: Council of Foreign Relations.

Langley, Lester D. 1983. *The Banana Wars: An Inner History of American Empire, 1900–1934*. Lexington: University Press of Kentucky.

Lareau, William. 1991. *American Samurai: A Warrior for the Coming Dark Ages of American Business*. Clinton: New Win Publishing.

Larkin, Thomas Oliver. 1951–55 (1822–46). *The Larkin Papers: Personal, Business, and Official Correspondence of Thomas Oliver Larkin, Merchant and United States Consul in California*. 5 vols. Ed. by George P. Hammond. Berkeley: University of California Press.

Lasater, Martin L. 1993. *U.S. Interests in New Taiwan*. Boulder, Colo.: Westview Press.

Latourette, Kenneth Scott. 1917. "The History of Early Relations between the United States and China, 1784–1844." *Transactions of the Connecticut Academy of Arts and Sciences,* vol. 22 (August): 1–209.

———. 1929. *A History of Christian Missions in China*. New York: Macmillan.

Lea, Homer. 1909. *The Valor of Ignorance*. New York: Harper.

Lee, Chae-Jin, ed. 1991. *The Korean War: 40-Year Perspectives*. Claremont: The Keck Center for International and Strategic Studies.

Lee, Chae-Jin, and Sato, Hideo, eds. 1982. *U.S. Policy Toward Japan and Korea: A Changing Influence Relationship*. New York: Praeger.

Lee, Manwoo; McLaurin, Ronald D.; and Moon, Chung-in. 1988. *Alliance under Tension: The Evolution of South Korean-U.S. Relations*. Boulder, Colo.: Westview Press.

Lee, Yur-Bok, and Patterson, Wayne, eds. 1986. *One Hundred Years of Korean-American Relations, 1882–1982*. Tuscaloosa: University of Alabama Press.

Legarda, Benito, Jr. 1957. "The American Entrepreneurs in the Nineteenth Century Philippines." *Explorations in Entrepreneurial History*, 60: 142–159.

Lehman, John. 1987. "Successful Naval Strategy in the Pacific: How We Are Achieving It; How We Can Afford It?" *Naval War College Review* (Winter): 20–27.

Leighton, Richard M., and Coakley, Robert W. 1954. 1968. *Global Logistics and Strategy*. 2 vols. Vol. 1: *1940–1943*; vol. 2: *1943–1945*. Washington, D.C.: GPO.

Lewin, Ronald. 1982. *The American Magic: Codes, Ciphers and the Defeat of Japan*. New York: Farrar, Straus, Giroux.

Lewis, Cleona. 1938. *America's Stake in International Investments*. Washington, D.C.: Brookings Institution.

Lewy, Guenter. 1978. *America in Vietnam*. New York: Oxford University Press.

Light, Ivan. 1974. "From Vice District to Tourist Attraction: The Moral Career of American Chinatowns, 1880–1940." *Pacific Historical Review* 43, no. 3 (August): 367–394.

Lincoln, Edward J. 1990. *Japan's Unequal Trade*. Washington, D.C.: Brookings Institution.

———. 1993. *Japan's New Global Role*. Washington, D.C.: Brookings Institution.

Linder, Staffan Burenstam. 1986. *The Pacific Century: Economic and Political Consequences of Asian-Pacific Dynamism*. Stanford, Calif.: Stanford University Press.

Lindstrom, Lamont, and White Geoffrey M. 1990. *Island Encounters: Black and White Memories of the Pacific War*. Washington, D.C.: Smithsonian Institution Press.

Linn, Brian McAllister. 1989. *The U.S. Army and Counterinsurgency in the Philippine War, 1899–1902*. Chapel Hill: University of North Carolina Press.

Lippy, Charles H., and Williams, Peter W., eds. 1988. *Encyclopedia of the American Religious Experience: Studies of Traditions and Movements*. 3 vols. New York: Scribner's.

Liss, Sheldon B. 1967. *The Canal: Aspects of United States Panamanian Relations*. Notre Dame, Ind.: University of Notre Dame Press.

Liu, Kwang-ching. 1962. *Anglo-American Steamship Rivalry in China, 1862–1874*. Cambridge: Harvard University Press.

———. 1963. *Americans and Chinese: A Historical Essay and A Bibliography*. Cambridge: Harvard University Press.

Liu, Kwang-ching., ed. 1970. *American Missionaries in China: Papers from Harvard Seminars*. Cambridge: Harvard University Press.

Livermore, Seward W. 1943. "American Strategy Diplomacy in the South Pacific, 1890–1914." *Pacific Historical Review* 12, no. 1 (March): 33–51.

———. 1944. "American Naval-Base Policy in the Far East, 1850–1914." *Pacific Historical Review* 13 no. 2 (June): 113–135.

Livingston, William S., and Louis, William Roger, eds. 1979. *Australia, New Zealand and the Pacific Islands since the First World War*. Austin: University of Texas Press.

Lockwood, Stephen Chapman. 1971. *Augustine Heard and Company, 1858–1862: American Merchants in China*. Cambridge: Harvard University Press.

Logan, Frenise A. 1962. "Activities of the *Alabama* in Asian Waters." *Pacific Historical Review* 31, no. 2 (May): 143–150.

Lord, Walter. 1957. *Day of Infamy*. New York: Holt.

Louis, William Roger. 1977. *Imperialism at Bay, 1941–1945: The United States and the Decolonization of the British Empire*. Oxford: Oxford University Press.

Lowe, Peter. 1986. *The Origins of the Korean War*. London: Longman.

Lucchini, Laurent, and Voelckel, Michel. 1978. *Les États et la mer: le nationalisme maritime*. Paris: La Documentation française, Notes et Études documentaires, no. 4451-2, 10 (January).

Lutz, Jessie G. 1971. *China and the Christian Colleges, 1850–1950*. Ithaca, N.Y.: Cornell University Press.

McClellan, Robert. 1971. *The Heathen Chinese: A Study of American Attitudes toward China, 1890–1905*. Columbus: Ohio State University Press.

McCord, William Maxwell. 1991. *The Dawn of the Pacific Century: Implications for Three Worlds of Development*. New Brunswick, N.J.: Transaction Publishers.

McCormick, Thomas J. 1970. *China Market: America's Quest for Informal Empire, 1893–1901*. Chicago: Quadrangle Books.

McCullough, David. 1977. *The Path between the Seas: The Creation of the Panama Canal, 1870–1914*. New York: Simon & Schuster.

McDougall, Walter A. 1993. *Let the Sea Make a Noise: A History of the North Pacific from Magellan to MacArthur*. New York: Basic Books.

McEvoy, Arthur F. 1986. *The Fisherman's Problem: Ecology and Law in the California Fisheries, 1850–1980.* Cambridge: Cambridge University Press.

McFarland, Keith D. 1986. *The Korean War: An Annotated Bibliography.* New York: Garland Publ.

McGlothen, Ronald. 1989. "Acheson, Economics, and the American Commitment in Korea, 1947–1950." *Pacific Historical Review* 58, no. 1 (February): 23–54.

McHenry, D. F. 1977. *Micronesia: Trust Betrayed.* New York: Carnegie Foundation for International Peace.

McIntosh, Malcolm. 1987. *Arms across the Pacific: Security and Trade Issues across the Pacific.* London: Frances Pinter.

McKee, Delber L. 1986. "The Chinese Boycott of 1905–1906 Reconsidered: The Role of Chinese Americans." *Pacific Historical Review* 55, no. 2 (May): 165–191.

McLaurin, Ronald D., and Moon, Chung-in. 1989. *The United States and the Defense of the Pacific.* Boulder, Colo.: Westview Press.

McLean, David. 1986. "American Nationalism, the China Myth, and the Truman Administration: The Question of Accommodation with Peking, 1949–50." *Diplomatic History* 10, no. 1 (Winter): 25–42.

McMahon, Robert J. 1981. *Colonialism and Cold War: The United States and the Struggle for Indonesian Independence, 1945–1949.* Ithaca, N.Y.: Cornell University Press.

———. 1988. "The Cold War in Asia: Toward a New Synthesis?" *Diplomatic History* 12, no. 3 (Summer): 307–327.

McMahon Robert J., ed. 1990. *Major Problems in the History of the Vietnam War: Documents and Essays.* Lexington: D. C. Heath.

Mabon, David W. 1988. "Elusive Agreements: The Pacific Pact Proposals of 1949–1951." *Pacific Historical Review* 57, no. 2 (May): 147–177.

Macchiarola, Frank K., and Oxnam, Robert B., eds. 1991. *The China Challenge: American Policies in East Asia.* New York: Academy of Political Science.

MacIntyre, Donald G. F. W. 1972. *Sea Power in the Pacific: A History from the Sixteenth Century to the Present Day.* London: A. Barker.

———. 1975. *The Battle for the Pacific.* London: Seven House.

Mack, Andrew. 1992. "Security Cooperation in Northeast Asia: Problems and Prospects." *Journal of Northeast Asian Studies* 11, no. 2 (Summer): 21–34.

Maga, Timothy P. 1990. *John F. Kennedy and the New Pacific Community, 1961–63.* New York: St. Martin's.

Malamson, Scott C. 1990. *Tuturami: A Political Journey in the Pacific Islands.* New York: Poseidon.

Margolin, Jean-Louis. 1989. *Singapour, 1959–1987, genèes d'un nouveau pays industriel.* Paris: L.'Harmattan.

Marichal, Carlos. 1989. *A Century of the Debt Crises in Latin America: From Independence to the Great Depression, 1820–1930.* Princeton, N.J.: Princeton University Press.

———. 1990. *Les Marines de Guerre du Dreadnought an nucléaire, Actes du Colloque international; Paris, ex-École polytechnique, 23–25 novembre 1988.* Paris: Service historique de la Marine.

Marks, Frederic W., III. 1984. "The Origins of FDR's Promise to Support Britain Militarily in the Far East—A New Look." *Pacific Historical Review* 53, no. 4 (November): 447–462.

———. 1990. "The Real Hawk of Dienbienphu: Dulles or Eisenhower?" *Pacific Historical Review* 49, no. 3 (August): 297–321.

Masland, John W. 1941. "Missionary Influence upon American Far Eastern Policy." *Pacific Historical Review* 10, no. 3 (September): 279–296.

Mason, Mark. 1992. *American Multinationals and Japan: The Political Economy of Japanese Capital Controls, 1899–1980.* Cambridge: Harvard University Press.

Matray, James Irving. 1985. *The Reluctant Crusade: American Foreign Policy in Korea, 1941–1950.* Honolulu: University of Hawaii Press.

Matsuo, Kinoaki. 1942. *How Japan Plans to Win.* Boston: Little, Brown.

Matzloff, Maurice. 1959. *Strategic Planning for Coalition Warfare, 1943–1944.* Washington, D.C.: GPO.

Matzloff, Maurice, and Snell, Edwin M. 1953. *Strategic Planning for Coalition Warfare, 1941–1942.* Washington, D.C.: GPO.

May, Ernest R. 1961. *Imperial Democracy: The Emergence of America as a Great Power.* New York: Harcourt, Brace, and World.

———. 1968. *American Imperialism: A Speculative Essay.* New York: Atheneum.

———. 1975. *The Making of the Monroe Doctrine.* Cambridge: Harvard University Press.

May, Ernest R., and Thomson, James C., Jr., eds. 1972. *American-East Asian Relations: A Survey.* Cambridge: Harvard University Press.

May, Ernest R., and Fairbank, John K., eds. 1986. *America's China Trade in Historical Perspective: The Chinese and American Performance.* Cambridge: Harvard University Press.

May, Glenn Anthony. 1980. *Social Engineering in the Philippines: The Aims, Execution and Impact of American Colonial Policy, 1900–1913.* Westport, Conn.: Greenwood.

Mayers, David Allan. 1986. *Cracking the Monolith: U.S. Policy against the Sino-Soviet Alliance, 1949–1955.* Baton Rouge: Louisiana State University Press.

Mazuzan, George T. 1974. "The American International Corporation in China." *Pacific Historical Review* 43, no. 2 (May): 212–232.

Mead, Margaret. 1936. *Coming of Age in Samoa.* New York: Blue Ribbon Books (with a foreword by Franz Boas).

———. 1937. *Cooperation and Competition among Primitive Peoples.* New York: McGraw Hill.

Meech, Julia, and Weisberg, Gabriel P. 1990. *Japonisme Comes to America: The Japanese Impact on the Graphic Arts, 1876–1925.* New York: Abrams.

Meinig, Donald W. 1993. *The Shaping of America: A Geographical Perspective on 500 Years of History;* vol. 2: *Continental America, 1800–1867.* New Haven: Yale University Press.

Melanson, Richard A. 1983. *Writing History and Making Policy: The Cold War, Vietnam, and Revisionism.* Lanham, Md.: University Press of America.

Meneses, Ciuffardi Emilio. 1989. *El factor naval en las relaciones entre Chile y los Estados Unidos, 1881–1951.* Santiago: Ediciones Pedagogicas Chilenas.

Merk, Frederick. 1967. *The Oregon Question: Essays in Anglo-American Diplomacy and Politics.* Cambridge: Harvard University Press.

Metallo, Michael V. 1978. "American Missionaries, Sun Yat-sen, and the Chinese Revolution." *Pacific Historical Review* 47, no. 2 (May): 261–282.

Michener, James. 1952 (1946). *Tales of the South Pacific.* New York: Macmillan.

———. 1950. *Return to Paradise.* New York: Random House.

Miller, Char, ed. 1985. *Missions and Missionaries in the Pacific.* Lewiston: Edwin Mellen Press.   ✓

Miller, Edward S. 1991. *War Plan Orange. The U.S. Strategy to Defeat Japan, 1897–1945.* Annapolis, Md.: Naval Institute Press.

Miller, Stuart Creighton. 1969. *The Unwelcome Immigrant: The American Image of the Chinese, 1785–1882.* Berkeley: University of California Press.

———. 1982. *"Benevolent Assimilation": The American Conquest of the Philippines, 1899–1903.* New Haven: Yale University Press.

Monnier, Claude. 1967. *Les Américains et Sa Majesté l'Empereur. Étude du conflit culturel d'où naquit la constitution japonaise de 1946.* Neuchatel: Éd. de La Baconnière.

Moore, John Hammond, ed. 1977. *Australians in America, 1876–1976.* St. Lucia: University of Queensland Press.

———. 1981. *Over-Sexed, Over-Paid and Over Here. Americans in Australia, 1941–1945.* St. Lucia: University of Queensland Press.

Morison, Samuel Eliot. 1921. *The Maritime History of Massachusetts, 1783–1860.* Boston: Houghton Muffin.

———. 1947–1960. *History of United States Naval Operations in World War II.* Boston: Little, Brown. 15 vols.

———. 1963. *The Two-Ocean War: A Short History of the United States Navy in the Second World War.* Boston: Little, Brown.

———. 1967. *"Old Bruin": Commodore Matthew Galbraith Perry, 1794–1858.* Boston: Little, Brown.

Morley, James William. 1957. *The Japanese Thrust into Siberia, 1918.* New York: Columbia University Press.

Morley, James William, ed. 1971. *Dilemmas of Growth in Prewar Japan.* Princeton, N.J.: Princeton University Press.

———. 1974. *Japan's Foreign Policy, 1868–1941: A Research Guide.* New York: Columbia University.

———. 1976. *Deterrent Diplomacy: Japan, Germany, and the USSR, 1935–1940.* New York: Columbia University Press.

———. 1980. *The Fateful Choice: Japan's Advance into Southeast Asia, 1939–1941.* New York: Columbia University Press.

———. 1983. *The China Quagmire: Japan's Expansion on the Asian Continent, 1933–1941.* New York: Columbia University Press.

———. 1984. *Japan Erupts: The London Naval Conference and the Manchurian Incident, 1928–1932.* New York: Columbia University Press.

———. 1986a. *The Pacific Basin: New Challenges for the United States.* New York: Columbia University Press.

———, ed. 1986b. *Security Interdependence in the Asia Pacific Region.* Lexington, Mass.: Lexington Books.

———. 1994. "Selected translations from *Taiheiyo senso e no michi, kaisen gaiko shi.*" New York: Columbia University Press.

———. 1993. *Driven by Growth: Political Change in the Asia-Pacific Region.* Armonk, N.Y.: M. E. Sharpe.

Morrell, Abby Jane. 1970. *Narrative of a Voyage to the Ethiopic and South Atlantic, Indian Ocean, Chinese Sea, North and South Pacific Ocean in the Years 1829, 1830, 1831.* Upper Saddle River, N.J.: Gregg Press.

Morrell, Benjamin. 1832. *A Narrative of Four Voyages to the South Sea, North and South Pacific Ocean, Chinese Sea, Ethiopic and Southern Atlantic Ocean, Indian and Antarctic Ocean, from the Years 1822 to 1831.* New York: J. & J. Harper.

Morse, Hosea Ballou. 1926–1929. *The Chronicles of the East India Company Trading to China, 1635–1834.* 5 vols. Oxford: Oxford University Press.

———. 1910–1918. *The International Relations of the Chinese Empire.* 3 vols. London: Longmans, Green, 1910–1918.

Morse, Ronald A.; Oksenberg, Michel; Gordon, Bernard K.; and Borthwick, Mark. 1986. *Pacific Basin: Concept and Challenge.* Washington, D.C.: Center for National Policy (July).

Morton, Louis. 1962. *Strategy and Command: The First Two Years (United States Army in World War II. The War in the Pacific).* Washington, D.C.: Office of the Chief of Military History, Department of the Army.

Mosher, Steven W. 1990. *China Misperceived: American Illusions and Chinese Realities.* New York: Basic Books.

———. 1992. *The United States and the Republic of China: Democratic Friends, Strategic Allies, and Economic Partners.* New Brunswick, N.J.: Transaction Publ.

Mrazek, Rudolf. 1978. *The United States and the Indonesian Military, 1945–1965: A Study of an Intervention.* 2 vols. Prague: Czechoslovak Academy of Sciences, Oriental Institute.

Munoz, Heraldo, and Portales, Carlos. 1991. *Elusive Friendship: A Survey of U.S. Chilean Relations.* Boulder, Colo.: Lynne Rienner.

Munro-Leighton, Judith. 1992. "A Postrevisionist Scrutiny of America's Role in the Cold War in Asia, 1945–1950." *The Journal of American-East Asian Relations* 1 (Spring): 73–98.

Myers, Ramón H., ed. 1989. *A Unique Relationship: The United States and the Republic of China under the Taiwan Relations Act.* Stanford, Calif.: Hoover Institution Press.

Nagai, Yonosuke, and Iriye, Akira, eds. 1977. *The Origins of the Cold War in Asia.* New York: Columbia University Press.

Nemetz, Peter N., ed. 1987. *The Pacific Rim: Investment, Development, and Trade.* Vancouver: University of British Columbia Press.

Nester, William R. 1990. *Japan's Growing Power over Asia and the World Economy: Ends and Means.* London: Macmillan.

———. 1991. *Japanese Industrial Targeting: The Neomercantilist Path to Economic Superpower.* Houndmills: Macmillan.

———. 1993. *American Power: The New World Order and the Japanese Challenge.* Houndmills: Macmillan.

Neu, Charles E. 1967. *An Uncertain Friendship: Theodore Roosevelt and Japan, 1906–1909.* Cambridge: Harvard University Press.

Neumann, William L. 1963. *American Encounters Japan: From Perry to MacArthur.* Baltimore: Johns Hopkins Press.

Newman, Robert P. 1992. *Owen Lattimore and the "Loss" of China.* Berkeley: University of California Press.

Ninkovich, Frank. 1980. "Cultural Relations and American China Policy, 1942–1945." *Pacific Historical Review* 49, no. 3 (August): 471–498.

———. 1984. "The Rockefeller Foundation, China, and Cultural Change." *Journal of American History* 70, no. 4 (March): 799–820.

Nixon, Richard Milhous. 1978. *RN, the Memoirs of Richard Nixon.* New York: Grosset & Dunlap.

Nufer, Harold F. 1978. *Micronesia under American Rule: An Evaluation of the Strategic Trusteeship (1949–1977).* Hicksville, N.Y.: Exposition Press.

Nye, Jr., Joseph S.; Pharr, Susan J.; and Vogel, Ezra. 1993. *Contentious Issues and Policy Choices in U.S.–Japan Relations: Proceedings of Harvard Faculty Study Group on U.S.–Japan Relations, 1990–1992.* Cambridge: Center for International Affairs, Harvard University Press.

Observatoire géopolitique des drogues. 1993. *La Drogue, nouveau désordre mondial: rapport 1992–1993.* Paris: Hachette.

O'Connell, Robert L. 1991. *Sacred Vessels: The Cult of the Battleship and the Rise of the U.S. Navy.* New York: Oxford University Press.

O'Connor, Raymond G. 1962. *Perilous Equilibrium: The United States and the London Naval Conference of 1930.* Lawrence: University Press of Kansas.

O'Donnell, Dan. 1993. "The Pacific Guano Islands: The Stirring of American Empire in the Pacific Ocean." *Pacific Studies* 16, no. 1 (March): 43–66.

Ogden, Adele. 1941. *The California Sea-Otter Trade, 1784–1848.* Berkeley: University of California Press.

———. 1929. "Boston Hide Droghers along California Shores." *California Historical Society Quarterly* 8 (December): 289–305.

Ohmae, Kenichi. 1985. *Triad Power: The Coming Shape of Global Competition.* New York: The Free Press.

Oksenberg, Michel, and Oxnam, Robert B., eds. 1978. *Dragon and Eagle: United States-China Relations, Past and Present.* New York: Basic Books.

Olson, James S. 1988. *Dictionary of the Vietnam War.* Westport, Conn.: Greenwood Press.

Olson, James S., and Roberts, Randy. 1991. *Where the Domino Fell: America and Vietnam, 1945 to 1990.* New York: St. Martin's Press.

Owen, Norman G., ed. 1971. *Compadre Colonialism. Studies on the Philippines under American Rule.* Ann Arbor, University of Michigan Center for South and Southeast Asian Studies.

————. 1983. *The Philippine Economy and the United States: Studies in Past and Present Interactions.* Ann Arbor: University of Michigan Center for South and Southeast Asian Studies.

————. 1984. *Prosperity without Progress: Manila Hemp and Material Life in the Colonial Philippines.* Berkeley: University of California Press.

Packard III, George R. 1966. *Protest in Tokyo: The Security Treaty Crisis of 1960.* Princeton, N.J.: Princeton University Press.

Paige, Glenn D. 1968. *The Korean Decision (June 24–30, 1950).* New York: The Free Press.

Paik L., George. 1970 (1929). *The History of Protestant Missions in Korea, 1832–1910.* Seoul: Yonsei University Press.

Palat, Ravi Arvind, ed. 1993. *Pacific-Asia and the Future of the World-System.* Westport, Conn.: Greenwood Press.

Palmer, Bruce Jr. 1984. *The 25-Years War: America's Military Role in Vietnam.* Lexington: University Press of Kentucky.

Palmer, Spencer J. 1962. "American Gold Mining in Korea's Unsan District." *Pacific Historical Review* 31, no. 4 (November): 379–391.

Paolino, Ernest N. 1973. *The Foundations of the American Empire: William Henry Seward and U.S. Foreign Policy.* Ithaca, N.Y.: Cornell University Press.

Passin, Herbert, ed. 1975. *The United States and Japan.* 2nd ed. Washington, D.C.: Columbia Books.

Paterson, D. G. 1977. "The North Pacific Seal Hunt, 1886–1910: Rights and Regulations." *Explorations in Economic History* 14, no. 2, (April).

Paterson, D. G., and Wilen, J. 1977. "Depletion and Diplomacy: The North Pacific Seal Hunt, 1886–1910." *Research in Economic History,* vol. 2: 81–139.

Paterson, Thomas G. 1990. *Major Problems in American Foreign Policy.* 2 vols. Lexington: Heath.

Patterson, Samuel. 1967. *Narrative of the Adventures and Sufferings of Samuel Patterson, Who Made Three Voyages to the Northwest Coast of America and Who Sailed to the Sandwich Islands, and to many Other Parts of This World before Being Shipwrecked on the Feejee Islands.* Fairfield, Conn.: Ye Gallon Press.

Paullin, Charles Oscar. 1971. *American Voyages to the Orient, 1690–1865: An Account of Merchant and Naval Activities in China, Japan, and the Various Pacific Islands.* Annapolis, Md.: United States Naval Institute.

Peake, Louis A. 1986. *The United States in the Vietnam War, 1945–1975: A Selected Annotated Bibliography.* New York: Garland Publications.

Pelz, Stephen E. 1974. *Race to Pearl Harbor: The Failure of the Second London Naval Conference and the Onset of World War II*. Cambridge: Harvard University Press.

Penlington, Norman. 1972. *The Alaska Boundary Dispute: A Critical Appraisal*. Scarborough: McGraw Hill Ryerson Ltd.

Pereira Salas, Eugenio. 1971. *Los primeros contactos entre Chile y los Estados Unidos, 1778–1809*. Santiago de Chili: Editorial Andres Bello.

Perez, Louis A., Jr. 1982. "Intervention, Hegemony, and Dependency: The United States in the Circum-Caribbean, 1898–1980." *Pacific Historical Review* 51, no. 2 (May): 165–194.

Pethick, Derek. 1981. *The Nootka Connection: Europe and the Northwest Coast, 1790–1796*. Seattle: University of Washington Press.

Phillips, Clifton Jackson. 1969. *Protestant America and the Pagan World: The First Half Century of the American Board of Commissioners for Foreign Missions, 1810–1860*. Cambridge: Harvard University Press.

Pike, David Wingeate, ed. 1991. *The Opening of the Second World War*. New York: Peter Lang.

Pike, Fredrick B. 1963. *Chile and the United States, 1880–1962: The Emergence of Chile's Social Crisis and the Challenge to United States Diplomacy*. Notre Dame, Ind.: University of Notre Dame Press.

———. 1977. *The United States and the Andean Republics: Peru, Bolivia, and Ecuador*. Cambridge: Harvard University Press.

———. 1992. *The United States and Latin America: Myths and Stereotypes of Civilization and Nature*. Austin: University of Texas Press.

Pletcher, David M. 1973. *The Diplomacy of Annexation: Texas, Oregon and the Mexican War*. Columbia: University of Missouri Press.

Political Economy of Japan. 1987. 1988. 1992. 3 vols. 1. Yamamura, Kozo, and Yasuba, Yasukichi, eds., *The Domestic Transformation*, 1987; 2. Inoguchi, Takashu, and Okimoto, Daniel I., eds., *The Changing International Context*, 1988; 3. Kumon, Shumpei, and Rosovsky, Henry, eds., *Cultural and Social Dynamics*, 1992. Stanford, Calif.: Stanford University Press.

Pons, Xavier. 1988. *Le Géant du Pacifique*. Paris: Economica.

Porter, David. 1986. *Journal of a Cruise*. Annapolis, Md.: Naval Institute Press.

Portes, Jacques. 1993. *Les Américains et la guerre du Vietnam*. Brussels: Éditions Complexe.

Potts, E. Daniel, and Potts, Annette. 1985. *Yanks Down Under, 1941–1945: The American Impact on Australia*. New York: Oxford University Press, 1985.

Powers, Thomas. 1984. *Vietnam: The War at Home. Vietnam and the American People, 1964–1968*. Boston: G. K. Hall.

Prange, Gordon W.; Goldstein, David M.; and Dillon, Katherine V. 1981. *At Dawn We Slept: The Untold Story of Pearl Harbor*. New York: McGraw-Hill.

Pratt, Julius W. 1932. "The Hawaiian Revolution: A Re-Interpretation." *Pacific Historical Review* 1, no. 4 (September): 273–295.

————. 1936. *Expanionists of 1898: The Acquisition of Hawaii and the Spanish Islands.* Baltimore: Johns Hopkins University Press.

Prestowitz, Clyde V., Jr. 1988. *Trading Places: How We Allowed Japan to Take the Lead.* New York: Basic Books.

Pringle, Robert. 1980. *Indonesia and the Philippines: American Interests in Island Southeast Asia.* New York: Columbia University Press.

Pugach, Noel H. 1969. "Making the Open Door Work: Paul S. Reinsch in China, 1913–1919." *Pacific Historical Review* 38, no. 2 (May): 157–175.

————. 1973. "Embarrassed Monarchist: Frank J. Goodnow and Constitutional Development in China, 1913–1915." *Pacific Historical Review* 42, no. 4 (November): 499–517.

————. 1987. "Second Career: James A. Thomas and the Chinese American Bank of Commerce." *Pacific Historical Review* 56, no. 2 (May): 195–229.

Puleston, Capt. W. D., U.S.N. 1941. *The Armed Forces of the Pacific: A Comparison of the Military and Naval Power of the United States and Japan.* New Haven: Yale University Press.

Rabe, Valentin H. 1978. *The Home Base of American China Missions, 1880–1920.* Cambridge: Harvard University Press.

Radius, Walter A. 1944. *United States Shipping in Transpacific Trade, 1922–1938.* Palo Alto, Calif.: Stanford University Press.

Randall, Stephen J. 1977. *The Diplomacy of Modernization: Colombian-American Relations, 1920–1940.* Toronto: University of Toronto Press.

————. 1992. *Colombia and the United States: Hegemony and Interdependence.* Athens: University of Georgia Press.

Rappaport, Armin. 1963. *Henry L. Stimson and Japan, 1931–1933.* Chicago: University of Chicago Press.

Ravenhill, John, ed. 1989. *No Longer an American Lake? Alliance Problems in the South Pacific.* Berkeley: Institute of International Studies, University of California.

Rea, Kenneth W. 1977. *Early American Relations, 1841–1912: The Collected Articles of Earl Swisher.* Boulder, Colo.: Westview Press.

Reboul, Lt. Col. 1922. *Le Conflit du Pacifique et notre Marine de guerre.* Paris: Berger-Levrault.

Reckner, James R. 1988. *Teddy Roosevelt's Great White Fleet.* Annapolis, Md.: Naval Institute Press.

Reed, James. 1983. *The Missionary Mind and American East Asia Policy, 1911–1915.* Cambridge: Harvard University Press.

Reed, Peter Mellish. 1958. "Standard Oil in Indonesia, 1898–1928." *Business History Review* 32, (Autumn): 329–337.

Rees, David. 1964. *Korea: The Limited War.* New York: St. Martin's Press.

Reese, Trevor R. 1969. *Australia, New Zealand and the United States: A Survey of International Relations 1941–1968.* London: Oxford University Press.

Reischauer, Edwin Oldfather. 1965. *The United States and Japan*. 3rd ed. Cambridge: Harvard University Press, (1950).

———. 1986. *My Life between Japan and America*. New York: Harper & Row.

Reischauer, Haru Matsukata. 1986. *Samurai and Silk: A Japanese and American Heritage*. Cambridge: Belknap Press of Harvard University Press.

Reynolds, Clark G. 1968. *The Fast Carriers: The Forging of an Air Navy*. New York: McGraw Hill.

———. 1982. *The Carrier War*. Alexandria, Va.: Time-Life Books.

Reynolds, David. 1981. *The Creation of the Anglo-American Alliance, 1937–1941: A Study in Competitive Cooperation*. Chapel Hill: University of North Carolina Press.

Reynolds, Stephen W. 1970. *The Voyage of the New Hazard to the Northwest Coast, Hawaii and China, 1810–1813*. Ed. by Judge F. W. Howay. Fairfield, Conn.: Ye Gallon Press.

Richard, Dorothy E. 1957. *U.S. Naval Administration of the Trust Territory of the Pacific Islands*. 3 vols. Washington, D.C.: GPO.

Roberts, Priscilla, ed. 1991. *Sino-American Relations since 1900*. Hong Kong: Centre of Asian Studies, University of Hong Kong.

Robinson, Alfred. 1970. *Life in California during a Residence of Several Years in that Territory, including Narrative of Events which have transpired since that period when California was an Independent Government*. Santa Barbara, Calif.: Peregrine Publishers.

Roden, Donald. 1986. "In Search of the Real Horace Capron: An Historiographical Perspective in Japanese-American Relations." *Pacific Historical Review* 55, no. 4 (November): 549–575.

Ronda, James P. 1990. *Astoria and Empire*. Lincoln: University of Nebraska Press.

Rose, Lisle A. 1976. *Roots of Tragedy: The United States and the Struggle for Asia, 1945–1953*. Westport, Conn.: Greenwood.

Rosenberg, Emily S. 1982. *Spreading the American Dream. American Economic and Cultural Expansion, 1890–1945*. New York: Hill & Wang.

Rosenstone, Robert A. 1988. *Mirror in the Shrine: American Encounters with Meiji Japan*. Cambridge: Harvard University Press.

Roskill, Stephen W. 1954–1961. *The War at Sea, 1939–1945*. 4 vols. London: H. M. Stationery Office.

Rostow, Walt Whitman. 1986. *The United States and the Regional Organization of Asia and the Pacific, 1965–1985*. Austin: University of Texas Press.

Roth, Rodris. 1961. *Tea Drinking in Eighteenth Century America: Its Etiquette and Equipage*. United States National Museum Bulletin 225. Washington, D.C.: Smithsonian Institution.

Roth, Russell. 1981. *Muddy Glory: America's "Indian Wars" in the Philippines, 1899–1935*. West Hanover: Christopher Publishing House.

Rotter, Andrew J. 1987. *The Path to Vietnam: Origins of the American Commitment to Southeast Asia*. Ithaca, N.Y.: Cornell University Press.

Rozman, Gilbert, ed. 1990. *The East Asian Region: Confucian Heritage and Its Modern Adaptation.* Princeton, N.J.: Princeton University Press.

Russ, William Adam, Jr. 1943. "The Role of Sugar in Hawaiian Annexation." *Pacific Historical Review* 12, no. 4, pp. 339–350.

———. 1959. *The Hawaiian Revolution, 1893–94.* Selingsgrove: Susquehannah University Press.

———. 1961. *The Hawaiian Republic, 1894–1898 and Its Struggle to Win Annexation.* Selingsgrove: Susquehannah University Press.

Rust, William J. 1985. *Kennedy in Vietnam: American Vietnam Policy, 1960–1963.* New York: Charles Scribner's Sons.

Rydell, Robert. 1983. "Visions of Empire: International Expositions in Portland and Seattle, 1905–1909." *Pacific Historical Review* 52, no. 1 (February): 37–65.

Ryden, George Herbert. 1933. *The Foreign Policy of the United States in Relation to Samoa.* New Haven: Yale University Press.

Salamanca, Bonifacio S. 1968. *The Filipino Reaction to American Rule, 1901–1913.* Hamden, Conn.: Shoe String Press.

Sater, William F. 1990. *Chile and the United States: Empires in Conflict.* Athens: University of Georgia Press.

Saucier, Philippe. 1987. *Spécialisation internationale et compétitivité de l'économie japonaise.* Paris: Economica.

Sautter, Christian. 1987. *Les Dents du géant: le Japon à la conquête du monde.* Paris: Orban.

Sbrega, John J. 1986. "Determination versus Drift: The Anglo-American Debate over the Trusteeship Issue, 1941–1945." *Pacific Historical Review* 55, no. 2 (May): 256–280.

———. 1989. *The War against Japan, 1941–1945: An Annotated Bibliography.* New York: Garland.

Scalapino, Robert A. 1987a. *American-Japanese Relations in a Changing Era.* New York: Library Press.

———. 1987b. *Major Power Relations in Northeast Asia.* Lanham, Md.: University Press of America.

Scalapino, Robert A., and Chufrin, Gonnady I., eds. 1991. *Asia in the 1990s: American and Soviet Perspectives.* Berkeley, Calif.: Institute of East Asian Studies.

Scalapino, Robert A., and Lee, Hongkoo, eds. 1988. *Korea-U.S. Relations: The Politics of Trade and Security.* Berkeley, Calif.: Institute of East Asian Studies.

Scalapino, Robert A., and Chen, Qimao, eds. 1986. *Pacific-Asian Issues: American and Chinese Views.* Berkeley, Calif.: Institute of Asian Studies.

Schaller, Michael. 1979. *The U.S. Crusade in China, 1938–1945.* New York: Columbia University Press.

———. 1985. *The American Occupation of Japan: The Origins of the Cold War in Asia.* New York: Oxford University Press.

———. 1989. *Douglas MacArthur: The Far Eastern General.* New York: Oxford University Press.

———. 1990. *The United States and China in the Twentieth Century.* 2nd ed. New York: Oxford University Press.

Schlight, John, ed. 1986. *The Second Indochina War.* Washington, D.C.: Center of Military History, U.S. Army.

Schlossstein, Steven. 1984. *Trade War: Greed, Power, and Industrial Policy on Opposite Sides of the Pacific.* New York: Congdon & Weed.

———. 1989. *The End of the American Century.* Chicago: Congdon & Weed.

———. 1991. *Asia's New Little Dragons: The Dynamic Emergence of Indonesia, Thailand, and Malaysia.* Chicago: Contemporary Books, 1991.

Schonberger, Howard B. 1989. *Aftermath of War: Americans and the Remaking of Japan, 1945–1952.* Kent, Ohio: Kent State University Press.

Schoonover, Thomas. 1989. "A United States Dilemma: Economic Opportunity and Anti-Americanism in El Salvador, 1901–1911." *Pacific Historical Review* 58, no. 4 (November): 403–428.

Schrader, John V., and Winnefeld, James A. 1992. *Understanding the Evolving U.S. Role in Pacific Rim Security: A Scenario-Based Analysis.* Santa Monica, Calif.: Rand.

Schroeder, John H. 1985. *Shaping a Maritime Empire: The Commercial and Diplomatic Role of the American Navy, 1829–1861.* Westport, Conn.: Greenwood.

Schurmann, Franz. 1974. *The Logic of World Power: An Inquiry into the Origins, Currents, and Contradictions of World Politics.* New York: Pantheon Books.

Schwantes, Robert S. 1955. *Japanese and Americans: A Century of Cultural Relations.* New York: Harper & Brothers.

Seaburg, Carl, and Paterson, Stanley. 1971. *Merchant Prince of Boston: Colonel T. H Perkins, 1764–1854.* Cambridge: Harvard University Press.

Seabury, Paul. 1981. *America's Stake in the Pacific.* Washington, D.C.: Ethics and Public Policy Century.

Seager, Robert II. 1977. *Alfred Thayer Mahan: The Man and His Letters.* Annapolis, Md.: Naval Institute Press.

Segal, Gerald. 1990a. *The Soviet Union and the Pacific.* Boston: Unwin Hyman.

———. 1990b. *Rethinking the Pacific.* Oxford: Oxford University Press.

Segal, Gerald, ed. 1983. *The Soviet Union in East Asia: Predicaments of Power.* London: Heinemann.

———. 1989. *Political and Economic Encyclopaedia of the Pacific.* Harlow: Longman.

Seidel, Robert N. 1972. "American Reformers Abroad: The Kemmerer Missions in South America, 1923–1931." *Journal of Economic History* 32, no. 2 (June): 520–545.

Selden, Mark. 1974. *Remaking Asia: Essays on the American Uses of Power.* New York: Pantheon Books.

———. 1991. "The United States, Japan, and the Atomic Bomb." *Bulletin of the Concerned Asian Scholars* 23, no. 1 (January–March): 3–12.

Service, John S. 1971. *The Amerasia Papers: Some Problems in the History of U.S.-China Relations.* Berkeley, Calif.: Center for Chinese Studies.

Shalom, Stephen R. 1990. "Securing the U.S. Philippine Military Bases Agreement of 1947." *Bulletin of the Concerned Asian Scholars* 22 (October–December): 3–12.

Shambaugh, David. 1991. *Beautiful Imperialist: China Perceives America, 1972–1990.* Princeton, N.J.: Princeton University Press.

Sharp, Daniel A., ed. 1972. *U.S. Foreign Policy and Peru.* Austin: University of Texas Press.

Shavit, David. 1990. *The United States in Asia. A Historical Dictionary.* Westport, Conn.: Greenwood.

Shaw, Yu-Ming. 1992. *An American Missionary in China: John Leighton Stuart and Chinese-American Relations.* Cambridge: Harvard University Press.

Sheehan, Neil. 1988. *A Bright Shining Lie: John Paul Vann and America in Vietnam.* New York: Random House.

Sheng, Michael M. 1993. "America's Lost Chance in China? A Reappraisal of Chinese Communist Policy toward the United States before 1945." *Australian Journal of Chinese Affairs* 29 (January): 135–157.

Shibusawa, Masahide; Ahmad, Zakaria Haji; and Bridges, Brian. 1992. *Pacific Asia in the 1990s.* London: Routledge.

Shineberg, Dorothy. 1967. *They Came for Sandalwood.* Melbourne: Melbourne University Press; London, New York: Cambridge University Press.

Shuguang, Zhang. 1992. "'Preparedness Eliminates Mishaps': the CCP's Security Concerns in 1949–1950 and the Origins of Sino-American Confrontation." *Journal of American-East Asian Relations* 1 (Spring): 42–72.

Siegfried, André. 1940. *Suez et Panama et les routes maritimes mondiales.* Paris: A. Colin.

Sigal, Leon V. 1988. *Fighting to a Finish: The Politics of War Termination in the United States and Japan, 1945.* Ithaca, N.Y.: Cornell University Press.

Sigur, Gaston, and Kim, Young C., eds. 1982. *Japanese and U.S. Policy in Asia.* New York: Praeger.

———. 1992. *Asia and the Decline of Communism.* New Brunswick, N.J.: Transaction Publishers.

Simon, Sheldon W. 1993. "U.S. Strategy and Southeast Asian Security: Issues of Compatibility." *Contemporary Southeast Asia* 14, no. 4 (March): 301–313.

Slackman, Michael. 1990. *Target: Pearl Harbor.* Honolulu: University of Hawaii Press.

Smith, Hugh, and Bergin, Anthony, eds. 1993. *Naval Power in the Pacific: Toward the Year 2000.* Boulder, Colo.: Lynne Rienner.

Smith, Philip C. F. 1984. *The Empress of China.* Philadelphia: Philadelphia Maritime Museum.

Smith, Ralph Bernard. 1983–1991. *An International History of the Vietnam War.* 3 vols. New York: St. Martin's Press. Vol. 1: *Revolution versus Containment, 1955–61* (1983). Vol. 2: *The Struggle for South-East Asia, 1961–65* (1985). Vol. 3: *The Making of a Limited War, 1965–66* (1991).

Smythe, S. J., Donald. 1962. "Pershing and the Disarmament of the Moros." *Pacific Historical Review* 31, no. 3 (August): 241–256.

Spanier, John W. 1959. *The Truman-MacArthur Controversy and the Korean War.* Cambridge: Harvard University Press.

Spate, O. H. K. 1979, 1983, 1988. *The Pacific since Magellan.* 3 vols. Vol. 1: *The Spanish Lake.* Vol. 2: *Monopolists and Freebooters.* London: Croom Helm. Vol. 3: *Paradise Found and Lost.* Minneapolis: University of Minnesota Press.

Spector, Ronald H. 1983. *Advice and Support: The Early Years of the U.S. Army in Vietnam, 1941–1960.* Washington, D.C.: Center of Military History, U.S. Army.

———. 1985. *Eagle against the Sun.* New York: Vintage Books.

Spoehr, Luther W. 1973. "Sambo and the Heathen Chinee: Californians' Racial Stereotypes in the Late 1870s." *Pacific Historical Review,* 42, no. 2 (May): 185–204.

Spykman, Nicholas John. 1942. *America's Strategy in World Politics: The United States and the Balance of Power.* New York: Harcourt, Brace.

Stackpole, Edouard A. 1953. *The Sea-Hunters: The New England Whalemen during Two Centuries, 1635–1835.* Philadelphia: Lippincott.

Stallings, Barbara. 1987. *Banker to the Third World: U.S. Portfolio Investment in Latin America, 1900–1986.* Berkeley: University of California Press.

Stallings, Barbara, and Szekely, Gabriel, eds. 1993. *Japan, the United States and Latin America: Toward a Trilateral Relationship in the Western Hemisphere.* Baltimore: Johns Hopkins University Press.

Stanley, Peter W. 1974. *A Nation in the Making: The Philippines and the United States, 1899–1921.* Cambridge: Harvard University Press.

———, ed. 1984. *Reappraising an Empire: New Perspectives on Philippine-American History.* Cambridge: Harvard University Press.

Stanton, William. 1975. *The Great United States Expedition of 1838–1842.* Berkeley: University of California Press.

Starbuck, Alexander. 1878. *Senate Miscellaneous Documents,* 44th Congress, 1st Session, no. 107, U.S. Commission of Fish and Fisheries, Part IV, Report of the Commissioner for 1875–1876, Appendix A: *The Sea Fisheries. A History of the American Whale-Fishery from Its Earliest Inception to the Year 1877.* Washington, D.C.: GPO.

Stelle, Charles C. 1940. "American Trade in Opium to China, Prior to 1820." *Pacific Historical Review* 9, no. 4 (December): 425–444.

———. 1941. "American Trade in Opium to China, 1821–1839." *Pacific Historical Review* 10, no. 1 (March): 57–74.

Stephan, John J. 1983. *Hawaii under the Rising Sun: Japan's Plans for Conquest after Pearl Harbor.* Honolulu: University of Hawaii Press.

Stephen, John J., and Chichkanov, V. P., eds. 1986. *Soviet-American Horizons on the Pacific.* Honolulu: University of Hawaii Press.

Stevens, Sylvester K. 1968. *American Expansion in Hawaii, 1842–1898.* New York: Russell & Russell (reprint 1945).

Stevenson, Charles H. 1992. "U.S. Foreign Policy in Southeast Asia: Implications for Current Regional Issues," *Contemporary Southeast Asia* 14, no. 4 (September): 87–111.

Stewart, Walt. 1946. *Henry Meiggs, Yankee Pizarro.* Durham, N.C.: Duke University Press.

Strauss, W. Patrick. 1962. "Pioneer American Diplomats in Polynesia, 1820–1840." *Pacific Historical Review* 31, no. 1 (February): 21–30.

———. 1964. *Americans in Polynesia, 1783–1842.* East Lansing: Michigan State University Press.

Stueck, William Whitney, Jr. 1973. Cold War Revisionism and the Origins of the Korean Conflict: The Kolko Thesis," and replies by Joyce and Gabriel Kolko. *Pacific Historical Review* 42, no. 4 (November): 537–575.

———. 1981. *The Road to Confrontation: American Policy Toward China and Korea, 1947–1950.* Chapel Hill: University of North Carolina Press.

———. 1984. *The Wedemeyer Mission: American Politics and Foreign Policy during the Cold War.* Athens: University of Georgia Press.

———. 1986. "The Limits of Influence: British Policy and American Expansion of the War in Korea." *Pacific Historical Review* 55 no. 1 (February): 65–95.

Sugnet, Christopher L.; Hickey, John T.; and Crispino, Robert. 1983. *Vietnam War, Bibliography Selected from Cornell University's Echols Collection.* Lexington: Lexington Books.

Sullivan, Michael P. 1985. *The Vietnam War: A Study in the Making of American Policy.* Lexington: University of Kentucky Press.

Summers, Harry G., Jr. 1981. *On Strategy: The Vietnam War in Context.* Carlisle Barracks: Strategic Studies Institute.

———. 1985. *Vietnam War Almanac.* New York: Facts on File Publications.

Sutter, Robert G. 1992. *East Asia and the Pacific Challenges for U.S. Policy.* Boulder, Colo.: Westview Press.

Swisher, Earl, ed. 1953. *China's Management of the American Barbarians: A Study of Sino-American Relations, 1841–1861, with Documents.* New Haven: Yale University Press.

Tabrah, Ruth M. 1980. *Hawaii: A Bicentennial History.* New York: Norton.

Takaki, Ronald. 1990. *Strangers from a Different Shore. A History of Asian Americans.* New York: Penguin Books.

Tan, Quingshan. 1992. *The Making of U.S.-China Policy: From Normalization to the Post-Cold War Era.* Boulder, Colo.: Lynne Rienner.

Tate, E. Mowbray. 1986. *Transpacific Steam: The Story of Steam Navigation from the Pacific Coast of North America to the Far East and the Antipodes, 1867–1941.* New York: Cornwall Books.

Tate, Merze. 1967. *The United States and the Hawaiian Kingdom: A Political History.* New Haven: Yale University Press.

———. 1968. *Hawaii: Reciprocity or Annexation.* East Lansing: Michigan State University Press.

Taylor, Sandra C. 1984. "The Ineffectual Voice: Japan Missionaries and American Foreign Policy, 1870–1941." *Pacific Historical Review* 53, no. 2 (February): 20–38.

Tenhula, John. 1991. *Voices from Southeast Asia: The Refugee Experience in the United States.* New York: Holmes & Meier.

Terkel, Studs. 1984. *The "Good War": An Oral History of World War Two.* New York: Pantheon Books.

Thambipillai, Pushpa, and Matuszewski, Daniel, eds. 1989. *The Soviet Union and the Asia-Pacific Region: Views from the Region.* New York: Praeger.

Thies, Wallace J. 1980. *When Governments Collide: Coercion and Diplomacy in the Vietnam Conflict, 1964–1968.* Berkeley: University of California Press.

Thomas, John N. 1974. *The Institute of Pacific Relations: Asian Scholars and American Politics.* Seattle: University of Washington Press.

Thomas, Winburn T. 1959. *Protestant Beginnings in Japan.* Rutland: Charles C. Tuttle.

Thompson, Richard A. 1978. *The Yellow Peril, 1890–1924.* New York: Arno Press.

Thomson, James C., Jr. 1969. *While China Faces West: American Reformers in Nationalist China, 1928–1937.* Cambridge: Harvard University Press.

Thomson, James C., Jr.; Stanley, Peter W.; and Perry, John Curtis. 1981. *Sentimental Imperialists: The American Experience in East Asia.* New York: Harper & Row.

Thorne, Christopher G. 1972. *The Limits of Foreign Policy: The West, the League, and the Far Eastern Crisis of 1931–1933.* London: Hamilton.

———. 1978. *Allies of a Kind: The United States, Britain, and the War against Japan, 1941–1945.* New York: Oxford University Press.

———. 1985. *The Issue of War: State, Societies, and the Far Eastern Conflict of 1941–1945.* London: H. Hamilton.

Thornton, Thomas Perry, ed. 1988. "Anti-Americanism: Origins and Context." *Annals of the American Academy of Political and Social Sciences* (May).

Tobin, Joseph Jay. 1983. "Strange Foreigners: American Reactions to Living in Japan." Ph.D. diss., University of Chicago.

Toland, John. 1991. *In Mortal Combat: Korea, 1950–1953.* New York: William Morrow.

Tong, Te-Kong. 1964. *United States Diplomacy in China, 1844–1860.* Seattle: University of Washington Press.

Tow, William T. ed. 1989. *Building Sino-American Relations: An Analysis for the 1990s.* New York: Pergamon House.

———. 1991. *Encountering the Dominant Player: U.S. Extended Deterrence Strategy in the Asia-Pacific.* New York: Columbia University Press.

Tow, William T., and Feeney, William R., eds. 1982. *U.S. Foreign Policy and Asian-Pacific Security: A Transregional Approach.* Boulder, Colo.: Westview Press.

Trani, Eugene. 1971. "Woodrow Wilson, China, and the Missionaries, 1913–1921." *Journal of Presbyterian History* 49: 328–351.

Trask, David F. 1981. *The War with Spain in 1898.* New York: Macmillan.

Trautmann, Frederick, ed. 1990. *With Perry to Japan: A Memoir by William Heine.* Honolulu: University of Hawaii Press.

Treat, Payson Jackson. 1917. *The Early Diplomatic Relations between the United States and Japan, 1853–1865.* Baltimore: Johns Hopkins University Press.

———. 1963 (1938). *Diplomatic Relations between the United States and Japan, 1853–1895.* 3 vols. 1: *1853–1875*; 2: *1876–1895*; 3: *1895–1905.* Gloucester, Mass.: Peter Smith.

Trefousse, Hans L. 1982. *Pearl Harbor: The Continuing Controversy.* Malabar: Krieger.

Tsai, Shih-Shan. 1974. "Chinese Immigration through Communist Chinese Eyes: An Introduction to the Historiography." *Pacific Historical Review* 43, no. 3 (August): 395–408.

———. 1983. *China and the Overseas Chinese in the United States, 1868–1911.* Fayetteville: University of Arkansas Press.

Tsou, Tang. 1963. *America's Failure in China, 1941–1950.* Chicago: University of Chicago Press.

Tucker, Nancy Bernkopf. 1983. *Patterns in the Dust: Chinese-American Relations and the Recognition Controversy, 1949–1950.* New York: Columbia University Press.

Turley, William S. 1986. *The Second Indochina War: A Short Political and Military History, 1954–1975.* Boulder, Colo.: Westview Press.

Turner, Frederick Jackson. 1920. *The Frontier in American History.* New York: Henry Holt.

Tyler, David B. 1968. *The Wilkes Expedition: The First United States Exploring Expedition (1838–1842).* Philadelphia: American Philosophical Society.

United Nations. *Foreign Trade Statistics of Asia and the Pacific,* series A (New York: 1969–). Combined with *Foreign Trade Statistics of Asia and the Pacific,* series B (1975–1986) with the title *Foreign Trade Statistics of Asia and the Pacific.*

U.S. Department of State. *Foreign Relations of the United States.*

United States Military Academy, West Point. Bradley, John H. 1984. *The Second World War: Asia and the Pacific.* Wayne, N.J.: Avery Publishing.

———. 1985. *Atlas of the Second World War: Asia and the Pacific,* 52 maps.

U.S. Senate, 97th Congress, 1st Session, Committee on Governmental Affairs. Permanent Subcommittee on Investigations. 1981. *International Narcotics Trafficking: Hearings.* Washington, D.C. GPO.

Unterberger, Betty M. 1956. *America's Siberian Expedition, 1918–1920: A Study of National Policy.* Durham, N.C.: Duke University Press.

Utley, Jonathan G. 1985. *Going to War with Japan, 1937–1941.* Knoxville: University of Tennessee Press.

Van Alstyne, Richard W. 1973. *The United States and East Asia.* London: Thames and Hudson.

Van Ark, Bart, and Pilat, Dirk. 1993. "Productivity Levels in Germany, Japan, and the United States: Differences and Causes." *Brookings Papers on Economic Activity, Microeconomics* 2: 1–69.

Vandermeersch, Léon. 1986. *Le Nouveau Monde sinisé.* Paris: PUF.

Van Horne, Mike. 1989. *Pacific Rim Trade: The Definitive Guide to Exporting and Investment.* New York: Amacom.

Van Zandt, Howard F. 1981. *Pioneer American Merchants in Japan.* Tokyo: Lotus Press.

Varg, Paul A. 1952. *Open Door Diplomat: The Life of W. W. Rockhill.* Urbana: University of Illinois Press.

———. 1958. *Missionaries, Chinese, and Diplomats: The American Protestant Movement in China, 1890–1952.* Princeton, N.J.: Princeton University Press.

———. 1968. *The Making of a Myth: The United States and China.* East Lansing: Michigan State University Press.

———. 1973. *The Closing of the Door: Sino-American Relations, 1936–1946.* East Lansing: Michigan State University Press.

———. 1990. *America, from Client State to World Power: Six Major Transitions in United States Foreign Relations.* Norman: University of Oklahoma Press.

Vernon, Raymond. 1983. *Two Hungry Giants: The United States and Japan in the Quest for Oil and Ores.* Cambridge: Harvard University Press.

Vevier, Charles. 1955. "The Open Door: An Idea in Action, 1906–1913." *Pacific Historical Review* 24, no. 1 (February): 49–62.

Vicuña, Francisco Orrego, ed. 1979. *La Communidad del Pacifico en perspective.* 2 vols. Santiago de Chili: Editorial Universitaria.

Viola, Herman J., and Margolis, Carolyn, eds. 1985. *Magnificent Voyagers: The U.S. Exploring Expedition, 1838–1842.* Washington, D.C.: Smithsonian.

Vogel, Ezra F. 1986. *Japan as Number One.* Singapore: Institute of Southeast Asian Studies.

———. 1991. *The Four Little Dragons: The Spread of Industrialization in East Asia.* Cambridge: Harvard University Press.

Walker, Ranginui, and Sutherland, William, eds. 1988. *The Pacific: Peace, Security and the Nuclear Issue.* Atlantic Highlands: Zed Books.

Walker, William O., III. 1991. *Opium and Foreign Policy: The Anglo-American Search for Order in Asia, 1912–1954.* Chapel Hill: University of North Carolina Press.

Ward, Robert Edward, and Yoshikazu, Sakamoto, eds. 1987. *Democratizing Japan: The Allied Occupation.* Honolulu: University of Hawaii Press.

Ward, R. Gerard, ed. 1966. *American Activities in the Central Pacific, 1790–1860: A History, Geography and Ethnography Pertaining to American Involvement and Americans in the Pacific Taken from Contemporary Newspapers, etc.* 8 vols. Ridgewood: Gregg Press.

Watanabe, Akio. 1970. *The Okinawa Problem: A Chapter in Japan-U.S. Relations.* Carlton: Melbourne University Press.

Webb, Robert Lloyd. 1988. *On the Northwest: Commercial Whaling in the Pacific Northwest, 1790–1967.* Vancouver: University of British Columbia Press.

Weeks, Charles J., Jr. 1987. "The United States Occupation of Tonga, 1942–1945: The Social and Economic Impact." *Pacific Historical Review* 56, no. 3 (August): 399–426.

Weigle, Richard D. 1947. "Sugar and the Hawaiian Revolution." *Pacific Historical Review* 16, no. 1 (March): 41–58.

Welch, Richard E., Jr. 1974. "American Atrocities in the Philippines: The Indictment and the Response." *Pacific Historical Review* 43, no. 2 (May): 233–253.

———. 1979. *Response to Imperialism: The United States and the Philippine-American War, 1899–1902.* Chapel Hill: University of North Carolina Press.

Welfield, John. 1988. *An Empire in Eclipse: Japan in the Postwar American Alliance System. A Study in the Interaction of Domestic Politics and Foreign Policy.* London: Athlone Press.

Werner, Levi. 1948. "American Attitudes toward the Pacific Islands, 1914–1919." *Pacific Historical Review* 17, no. 1 (February): 55–64.

West, Philip. 1976. *Yenching University and Sino-Western Relations, 1916–1952.* Cambridge: Harvard University Press.

West, Philip, and Alting von Geusau, Frans A. M., eds. 1987. *The Pacific Rim and the Western World: Strategic, Economic, and Cultural Perspectives.* Boulder, Colo.: Westview Press.

Wheeler, Gerald E. 1963. *Prelude to Pearl Harbor: The United States Navy and the Far East, 1921–1931.* Columbia: University of Missouri Press.

Wheeler, Jimmy W., and Wood, Perry. 1987. *Beyond Recrimination: Perspectives on U.S.–Taiwan Trade Tensions.* Indianapolis: Hudson Institute.

Wheeler, Mary E. 1971. "Empires in Conflict and Cooperation: The 'Bostonians' and the Russian-American Company." *Pacific Historical Review* 40, no. 4 (November): 419–441.

Whelan, Richard. 1990. *Drawing the Line: The Korean War, 1950–1953.* Boston: Little, Brown.

Whipple, Addison B. C. 1979. *The Whalers.* Alexandria, Va.: Time-Life Books.

———. 1980. *The Clipper Ships.* Alexandria, Va.: Time-Life Books.

White, Geoffrey M., and Lindstrom, Lamont, eds. 1989. *The Pacific Theater: Island Representations of World War II.* Honolulu: University of Hawaii Press.

White, Philo. 1965. *Philo White's Narrative of a Cruise in the Pacific to South America and California on the U.S. Sloop of War "Dale," 1841–1843.* ed. Charles C. Camp. Denver: Old West.

Whitfield, Stephen J. 1991. *The Culture of the Cold War.* Baltimore: Johns Hopkins University Press.

Whiting, Robert. 1977. *The Chrysanthemum and the Bat: Baseball Samurai Style.* New York: Dodd, Mead.

———. 1989. *You Gotta Have Wa.* New York: Macmillan.

Wiest, Jean-Paul. 1988. *Maryknoll in China: A History (1918–1955).* Armonk: ME Sharpe.

Wilborn, Thomas L. 1993. *Stability, Security Structures and U.S. Policy for East Asia and the Pacific.* Carlisle Barracks, Pa.: Strategic Studies Institute, U.S. Army War College.

Wiley, Peter Booth (with Korogi Ichiro). 1990. *Yankees in the Land of the Gods: Commodore Perry and the Opening of Japan.* New York: Viking.

Wilkes, Charles. 1844. *Narrative of the United States Expedition During the Years 1838, 1839, 1840, 1841, 1842.* 5 vols. and an atlas. Philadelphia: C. Sherman.

Wilkins, Mira. 1970. *The Emergence of Multinational Enterprise: American Business Abroad from the Colonial Era to 1914.* Cambridge: Harvard University Press.

———. 1974. *The Maturing of Multinational Enterprise. American Business Abroad from 1914 to 1970.* Cambridge: Harvard University Press.

Williams, Frances L. 1963. *Matthew Fontaine Maury: Scientist of the Sea.* New Brunswick, N.J.: Rutgers University Press.

Williams, Gerald O. 1984. *The Bering Sea Fur Seal Dispute: A Monograph on the Maritime History of Alaska.* Eugene: Alaska Maritime Publ.

Williams, Justin, Sr. 1982. "From Charlottesville to Tokyo: Military Government Training and Democratic Reforms in Occupied Japan." *Pacific Historical Review* 51, no. 4 (November): 407–422.

———. 1988. "American Democratization Policy of Occupied Japan: Correcting the Revisionist Version." *Pacific Historical Review* 57, no. 2 (May): 179–202, with 2 "Rejoinders" by John W. Dower and Howard Schonberger: 202–218.

Williams, Mary W. 1965. *Anglo-American Isthmian Diplomacy, 1815–1915.* Gloucester, Mass.: Peter Smith.

Williams, William Appleman. 1962. *The Tragedy of American Diplomacy.* New York: Delta Books.

Williams, William Appleman; McCormick, Thomas; Gardner, Lloyd; and LaFeber, Walter, eds. 1989. *America in Vietnam: A Documentary History.* New York: Norton.

Williamson, Harold F., and Daum, Arnold R. 1959. *The American Petroleum Industry: The Age of Illumination, 1859–1899.* Evanston, Ill.: Northwestern University Press.

Williamson, Harold F.; Andreano, Ralph L.; Daum, Arnold R.; and Klose, Gilbert C. 1963. *The American Petroleum Industry: The Age of Energy, 1899–1959.* Evanston, Ill.: Northwestern University Press.

Willmott, H. P. 1982. *Empires in the Balance: Japanese and Allied Pacific Strategies to April 1942.* Annapolis, Md.: Naval Institute Press.

———. 1983. *The Barrier and the Javelin: Japanese and Allied Pacific Strategies February to June 1942.* Annapolis, Md.: Naval Institute Press.

Wilson, Joan Hoff. 1971. *American Business and Foreign Policy, 1920–1933.* Lexington: University of Kentucky Press.

Wilson, Joe F. 1979. *The United States, Chile, and Peru in the Tacna and Arica Plebiscite.* Washington, D.C.: University Press of America.

Wilz, John Edward. 1985. "Did the United States Betray Korea in 1905?" *Pacific Historical Review* 54, no. 3 (August) 243–270.

Winchester, Simon. 1992. *Pacific Rising: The Emergence of a New World Culture.* New York: Simon and Schuster.

Winnefield, James, et al. 1992. *A New Strategy, and Fewer Forces: The Pacific Dimension*. 2 vols. Santa Monica, Calif.: Rand.

Wittman, Sandra M. 1989. *Writing about Vietnam: A Bibliography of the Literature of the Vietnam Conflict*. Boston: G. K. Hall.

Wohlstetter, Roberta. 1962. *Pearl Harbor: Warning and Decision*. Palo Alto, Calif.: Stanford University Press.

Wolf, Marvin J. 1983. *The Japanese Conspiracy: The Plot to Dominate Industry Worldwide—and How to Deal with It*. New York: Empire Books.

Wolferen, Karel van. 1989. *The Enigma of Japanese Power: People and Politics in a Stateless Nation*. New York: A. A. Knopf.

Woodward, William P. 1972. *The Allied Occupation of Japan, 1945–1952, and Japanese Religions*. Leiden: E.J. Brill.

Worsley, Peter. 1968. *The Trumpet Shall Sound: A Study of "Cargo" Cults in Melanesia*. 2nd ed. New York: Schocken Books.

Wright, Philip G. 1935. *Trade and Trade Barriers in the Pacific*. London: P. S. King & Son.

Wu, William F. 1982. *The Yellow Peril: Chinese Americans in American Fiction, 1850–1940*. Hamden: Archon Books.

Yasuba, Yasukichi. 1993. "Natural Resources in Japanese Economic History, 1800–1940." *The Newsletter of the Cliometric Society* 8, no. 3 (October): 14–16.

Yavenditti, Michael J. 1974. "John Hersey and the American Conscience: The Reception of *Hiroshima*." *Pacific Historical Review* 43, no. 1 (February): 24–49.

Yoshitsu, Michael M. 1983. *Japan and the San Francisco Peace Settlement*. New York: Columbia University Press.

Young, Marilyn Blatt. 1968a. *The Rhetoric of Empire: American China Policy, 1895–1901*. Cambridge: Harvard University Press.

———. 1968b. "American Expansion, 1870–1900. The Far East," In Barton, Bernstein, ed. *Towards a New Past*, pp. 176–201.

———. 1991. *The Vietnam Wars, 1945–1990*. New York: Harper Perennial.

Yui, Tsumehiko, and Nakagawa, Keiichiro, eds. 1985. *Business History of Shipping: Strategy and Structure*. Tokyo: University of Tokyo Press.

Zhang Yongjin. 1991. *China in the International System, 1918–1920: The Middle Kingdom at the Periphery*. New York: St. Martin's Press.

SPECIALIST JOURNALS

American Neptune (1941–)

Asian Affairs (1970–)

Asian Development Review. Studies of Asian and Pacific Economic Issues (Manila, 1982–)

Asian Survey (monthly) (1961–)

Australian Journal of Chinese Affairs (1975–)

Beijing Review (1958–)

Bulletin of Concerned Asian Scholars (1968–)

China Briefing (1980–)

China Quarterly (1960–)

Contemporary Pacific. A Journal of Island Affairs (1989–)

Contemporary Southeast Asia. A Quarterly Journal of International and Strategic Affairs (Singapore, 1980–)

Diplomatic History (1977–)

Far Eastern Economic Review (1946–)

Japan Quarterly (1954–)

Japan Review of International Affairs (1987–)

Journal of American History (1964–), formerly *Mississippi Valley Historical Review.*

Journal of American-East Asian Relations (1992–)

Journal of Asian Studies (1942–)

Journal of Economic Development and Social Change in Asia and Pacific (Singapore, 1976–)

Journal of Northeast Asian Studies (1982–)

Journal of Pacific History (Canberra, 1966–)

Modern Asian Studies (1967–)

Modern China (1975–)

Pacific Affairs. An International Review of Asia and the Pacific (1928–)

Pacific Economic Bulletin (1986–)

Pacific Focus. Inha Journal of International Studies (Inchon, 1986–)

Pacific Historical Review (1932–)

Pacific Islands Monthly (1930–)

Pacific Islands Yearbook (1974–)

Pacific Review (1988–)

Pacific Studies. A Journal Devoted to the Study of the Pacific, Its Islands and Adjacent Countries (Brigham Young University, 1978–)

Revue du Pacifique (1922–1939)

INDEX

THE UNITED STATES AND THE PACIFIC

was composed in 11.5/13 Bembo
on a Macintosh G4 using QuarkXPress 4.1
at Coghill Composition Company;
printed by sheet-fed offset
on 60# EB Natural stock
(an acid-free paper) using soy-based ink,
smyth sewn over binder's boards
in Holliston ICG cloth with Multicolor endsheets,
and wrapped with dust jackets printed in two colors
on 80# enamel stock finished with film lamination
by Edwards Brothers, Inc.;
designed by Wendy McMillen;
published by

THE UNIVERSITY OF NOTRE DAME PRESS

Notre Dame, Indiana 46556